I0489422

Specifications, Tolerances, and Other Technical Requirements for Weighing and Measuring Devices

as adopted by the 101ᵗʰ National Conference on Weights and Measures 2016

Editors:

Tina Butcher
Linda Crown
Rick Harshman

Georgia Harris, Acting Chief
Office of Weights and Measures
Physical Measurement Laboratory

December 2016

U. S. Department of Commerce
Penny Pritzker, Secretary

National Institute of Standards and Technology
Willie May, Under Secretary of Commerce for Standards and Technology and Director

NIST Handbook **44**

2017 Edition

Supersedes NIST Handbook 44, 2016 Edition

Certain commercial entities, equipment, or materials may be identified in this document in order to describe an experimental procedure or concept adequately. Such identification is not intended to imply recommendation or endorsement by the National Institute of Standards and Technology, nor is it intended to imply that the entities, materials, or equipment are necessarily the best available for the purpose.

National Institute of Standards and Technology Handbook 44, 2017 Edition
Natl. Inst. Stand. Technol. Handb. 44, 2017 Ed., 526 pages (December 2016)
CODEN: NIHAE2

WASHINGTON: 2016

Foreword

NIST Handbook 44 was first published in 1949, having been preceded by similar handbooks of various designations and in several forms, beginning in 1918.

NIST Handbook 44 is published in its entirety each year following the Annual Meeting of the National Conference on Weights and Measures (NCWM). The Committee on Specifications and Tolerances of the NCWM developed the 2017 edition with the assistance of the Office of Weights and Measures (OWM) of the National Institute of Standards and Technology (NIST). This handbook includes amendments endorsed by the 101th National Conference on Weights and Measures during its Annual Meeting in 2016.

NIST has a statutory responsibility for "cooperation with the states in securing uniformity of weights and measures laws and methods of inspection." In partial fulfillment of this responsibility, NIST is pleased to publish these recommendations of the NCWM.

This handbook conforms to the concept of primary use of SI (metric) measurements recommended in the Omnibus Trade and Competitiveness Act of 1988 by citing SI units before U.S. customary units where both units appear together and placing separate sections containing requirements in SI units before corresponding sections containing requirements in U.S. customary units. In some cases, however, trade practice is currently restricted to the use of U.S. customary units; therefore, some requirements in this handbook will continue to specify only U.S. customary units until the NCWM achieves a broad consensus on the permitted SI units.

In accord with NIST policy, the meter/liter spellings are used in this document. However, the metre/litre spellings are acceptable, and are preferred by the NCWM.

It should be noted that a space has been inserted instead of commas in all numerical values greater than 9999 in this document, following a growing practice, originating in tabular work, to use spaces to separate large numbers into groups of three digits. This avoids conflict with the practice in many countries to use the comma as a decimal marker.

> You are invited to provide online feedback regarding this document
> at: **https://www.surveymonkey.com/s/customer-satisfaction-pub**
> or by e-mail to: owm@nist.gov.

Committee on Specifications and Tolerances
of the 101ᵗʰ Conference

Mahesh Albuquerque, Colorado
Matthew Curran, Florida
Ivan Hankins, Iowa
Rachelle Miller, Wisconsin
Jane Zulkiewicz, Town of Barnstable, Massachusetts
Luciano Burtini, Measurement Canada, Technical Advisor
Rick Harshman, NIST, Technical Advisor
Darrell Flocken, NCWM, NTEP Specialist

Past Chairmen of the Committee

Conference	Chairman	Conference	Chairman
8-11	L.A. Fischer, NBS	73	K. Butcher, MD
12-28	F.S. Holbrook, NBS	74	R. Andersen, NY
29-38	J.P. McBride, MA	75	D. Watson, TX
39-42	R.E. Meek, IN	76	J. Truex, OH
43-44	J.E. Brenton, CA	77-78	C. Carroll, MA
45-47	C.L. Jackson, WI	79	J. Jeffries, FL
48	T.C. Harris, VA	80	R. Suiter, NE
49-50	R.E. Meek, IN	81	G. West, NM
51-52	G.L. Johnson, KY	82-83	R. Murdock, NC
53	H.D. Robinson, ME	84	D. Brown, IA
54-55	R. Rebuffo, NE	85	M. Hopper, CA
56-57	D.E. Konsoer, WI	86	G. Shefcheck, OR
58	J.C. Mays, FL	87	M. Coyne, MA
59	T.F. Brink, VT	88	R. Wotthlie, MD
60	W.S. Watson, CA	89	C. VanBuren, MI
61	K.J. Simila, OR	90	J. Kane, MT
62	W.E. Czaia, MN	91	C. Cooney, OR
63	M.L. Kinlaw, NC	92	M. Sikula, NY
64	J.A. Bird, NJ	93	C. Fulmer, SC
65	D.A. Guensler, CA	94	T. R. Lucas, OH
66	G.A. Delano, MT	95	B. Saum, San Luis Obispo Co., CA
67	F.C. Nagele, MI	96-97	S. Giguere, ME
68	L.H. DeGrange, MD	98	K. Ramsburg, MD
69	S.A. Colbrook, IL	99	B. Gurney, UT
70	D.A. Guensler, CA	100	M. Albuquerque, CO
71-72	F. Gerk, NM		

iv

Table of Contents

Table of Contents (continued)

2016 Amendments

The following table lists the codes, paragraphs, and pages in which the 101th National Conference on Weights and Measures made amendments. In the column headed "Action," changes are noted as "added," "amended," "deleted," or "renumbered." Each code, section, or paragraph that has been changed will be noted as "Added 2016" or "Amended 2016."

SECTION	CODE	S&T ITEM NO.	PARAGRAPH	ACTION	PAGE
1.10.	General Code	310-1	G-S.1. Identification.	Amended	1-4
		310-2	G-S.9. Metrologically Significant Software Updates.	Added	1-9
2.20.	Scales	320-2	S.5.4. Relationship of Load Cell Verification Interval to the Scale Division.	Amended	2-19
		320-3	N.1.3.3.2. Prescribed Test Pattern and Test Loads for Livestock Scales with More Than Two Sections and Combination Vehicle.	Amended	2-29
			N.1.3.3.3. Prescribed Test Patterns and Test Loads for Two-Section Livestock Scales.	Amended	2-29
3.30.	Liquid Measuring Devices	330-1	S.1.6.3. Return to Zero.	Amended	3-7
		330-2	S.1.6.10. Automatic Timeout – Pay-At-Pump Retail Motor-Fuel Devices.	Subsection Added	3-12
		330-3	N.4.5. Verification of Linearization Factors.	Added	3-20
			UR.4. Maintenance Requirements.	Added	3-25
			U.R.4.1. Use of Adjustments.	Added	3-25
3.31.	Vehicle-Tank Meters	331-1	S.1.1.5. Return to Zero.	Amended	3-32
		331-2	Table S.2.2.	Amended	3-35
		331-4	N.4.6. Verification of Linearization Factors.	Added	3-39
			UR.3. Maintenance Requirements.	Subsection Added	3-43
			UR.3.1. Use of Adjustments.	Added	3-43

SECTION	CODE	S&T ITEM NO.	PARAGRAPH	ACTION	PAGE
3.32.	Liquefied Petroleum Gas and Anhydrous Ammonia Liquid-Measuring Devices	332-1	S.1.4.2. Return to Zero.	Amended	3-49
		332-2	S.1.4.1. Indication of Delivery.	Amended	3-49
			S.1.5.1. For Stationary Retail Devices Only.	Amended	3-49
			S.1.5.3. Agreement Between Indications.	Added	3-51
			S.1.5.3. Recorded Representations Point of Sale System	Renumbered to S.1.5.4. and Title Amended	3-51
			S.1.5.5. Recorded Representations for Transactions Where a Post-Delivery Discount(s) is Provided.	Added	3-51
			S.1.5.6. Transaction Information, Power Loss.	Added	3-52
			S.1.5.7. Totalizers for Retail Motor-Fuel Dispensers.	Added	3-52
			S.2.5. Zero-Set-Back Interlock for Stationary Retail Motor-Fuel Devices.	Added	3-53
			S.2.5. Thermometer Well.	Renumbered to S.2.6.	3-54
			S.2.6. Automatic Temperature Compensation.	Renumbered to S.2.7.	3-54
			S.2.6.1. Provision for Deactivating.	Renumbered to S.2.7.1.	3-54
			S.2.6.2. Provision for Sealing	Renumbered to S.2.7.2.	3-54
			UR.2.7. For Stationary Retail Computing Type Systems Only, Installed After January 1, 2016.	Added	3-59
			UR.2.7.1. Unit Price and Product Identity.	Added	3-59
			UR.2.7.2. Computing Device	Added	3-60
		332-3	S.2.1. Vapor Elimination	Amended	3-52
		332-4	Table S.2.2. Categories of Device and Methods of Sealing	Amended	3-53
		332-7	UR.2.3. Vapor-Return Line	Amended	3-58

SECTION	CODE	S&T ITEM NO.	PARAGRAPH	ACTION	PAGE
3.39.	Hydrogen Gas-Measuring Devices – Tentative Code	339-2	Table T.2. Accuracy Classes and Tolerances for Hydrogen Gas-Metering Devices.	Amended	3-144
5.54.	Taximeters	354-1	A.2. Exceptions.	Amended	5-27
			S.1.2. Advancement of Indicating Elements.	Amended	5-27
		254-2	S.2. Basis of Fare Calculations.	Amended	5-30
		354-3	S.3.2. Flag.	Deleted	5-30
			S.3.3. Control for Extras Mechanism.	Renumbered to S.3.2.	5-30
5.58.	Multiple Dimension Measuring Devices	358-1	S.1.5. Value of Dimension/Volume Division Units.	Amended	5-78
		358-2	Table S.4.1.a. Marking Requirements for Multiple Dimension Measuring Systems.	Amended	5-82
			Table S.4.1.b. Multiple Dimension Measuring Systems Notes for Table S.4.1.a.	Amended	5-83
		358-3	S.2.2.1. Maximum Value of Tare for Multi-Interval (Variable Division-Value) Devices.	Added	5-81
			S.2.2.2. Net Values, Mathematical Agreement.	Added	5-81
			Table 1. Examples of Acceptable Altering of Tare to Achieve Accurate Net Indication	Added	5-81
			T.2.3. Multi-Interval (Variable Division-Value) Devices.	Amended	5-84
5.58.	Multiple Dimension Measuring Devices (cont.)	358-3	T.2.4. Mixed-Interval Devices.	Added	5-84

SECTION	CODE	S&T ITEM NO.	PARAGRAPH	ACTION	PAGE
Appendix D	Definitions	337-2	diesel gallon equivalent (DGE)	Added	D-10
			gasoline gallon equivalent (GGE)	Amended	D-13
			gasoline liter equivalent (GLE)	Deleted	D-13
		354-3	flag	Deleted	D-13
		354-4	flat rate	Added	D-13
			negotiated rate	Added	D-19
		360-4	calibration parameter	Amended	D-8

2016 Editorial Changes

SECTION	CODE	PARAGRAPH	ACTION	PAGE
Introduction		Form 15	Updated to include current version	4
Appendix C	General Tables of Units of Measurement	1 liter	Corrected 61.02~~5~~4 cubic inches	C-18
Appendix D		User Requirement	Corrected reference (Also see Introduction, Section D~~Q~~.)	D-27

Introduction

A. Source.

The specifications, tolerances and other technical requirements in this handbook comprise all of those adopted by the National Conference on Weights and Measures, Inc. (NCWM). Contact NCWM at:

1135 M Street, Suite 100   Phone: (402) 434-4880   E-mail: info@ncwm.net
Lincoln, NE 68508   Fax: (402) 434-4878   URL: **www.ncwm.net**

The NCWM is supported by the National Institute of Standards and Technology (NIST), which provides its Executive Secretary and publishes some of its documents. NIST also develops technical publications for use by weights and measures agencies; these publications may subsequently be endorsed or adopted by NCWM.

All of the specifications, tolerances, and other technical requirements given herein are recommended by NCWM for official promulgation in and use by the states in exercising their control of commercial weighing and measuring apparatus. A similar recommendation is made with respect to the local jurisdictions within a state in the absence of the promulgation of specifications, tolerances, and other technical requirements at the state level.
(Amended 2015)

B. Purpose.

The purpose of these technical requirements is to eliminate from use, weights and measures and weighing and measuring devices that give readings that are false, that are of such construction that they are faulty (that is, that are not reasonably permanent in their adjustment or will not repeat their indications correctly), or that facilitate the perpetration of fraud, without prejudice to apparatus that conforms as closely as practicable to the official standards.

C. Amendments.

Proposed amendments to NIST Handbook 44 are deliberated and developed by NCWM's Committee on Specifications and Tolerances before presentation to the general membership for a vote. In some instances, amendments that significantly affect other NIST Handbooks may be processed jointly by two or more committees.

Amendments to the handbooks are made in accordance with NCWM procedures and policies. The process begins at the regional weights and measures association meetings in the fall of each year and is culminated at the NCWM Annual Meeting in July. After passing through one or more of the regional associations the proposed amendment is placed on the agenda of the appropriate NCWM committee for consideration at NCWM's Interim Meeting in January and after final deliberation and development by the committee the amendment may be presented to the membership for a vote at the annual NCWM meeting in July. NCWM policy provides for exceptions to the process to accommodate urgent or priority items. NIST staff provides technical assistance and advice throughout the process.

The policy is available on the NCWM website at **www.ncwm.net**. For information on the regional weights and measures associations, visit **www.ncwm.net/resource/regional_associations**.
(Amended 2015)

D. System of Paragraph Designation.

In order that technical requirements of a similar nature, or those directed to a single characteristic, may be grouped together in an orderly fashion, and to facilitate the location of individual requirements, the paragraphs of each code are divided into sections. Each section is designated by a letter and a name, and each subsection is given a letter-number designation and a side title.

The letter that appears first in a paragraph designation has a specific meaning, as follows:

G.   The letter G is a prefix and indicates that the requirement is part of the General Code.

A.   **Application.** These paragraphs pertain to the application of the requirements of a code.

S.   **Specification.** These paragraphs relate to the design of equipment. Specification paragraphs are directed particularly to manufacturers of devices.

N.   **Note.** These paragraphs apply to the official testing of devices.

T.   **Tolerance.** Tolerances are performance requirements. They fix the limit of allowable error or departure from true performance or value.

     Sensitivity. The sensitivity requirements, applicable only to nonautomatic-indicating scales, are performance requirements and are lettered with a "T."

UR. **User Requirement.** These paragraphs are directed particularly to the owner and operator of a device. User requirements apply to the selection, installation, use, and maintenance of devices.

D.   **Definitions of Terms.** A definitions section appears in Appendix D to provide the definition of the terms having a special meaning.

The numerical designation after a letter follows the decimal system of paragraph identification that fixes both the relationship and the limitation of the requirements of the paragraph. For example, in the Scales Code, under Specifications, the following numerical designations occur:

S. Specifications

S.1. Design of Indicating and Recording Elements and of Recorded Representations.

    S.1.1. Zero Indication.
        S.1.1.1. Digital Indicating Elements.
        S.1.1.2. No-Load Reference Value.
    S.1.2. Value of Scale Division Units.
        S.1.2.1. Digital Indicating Scales.
    S.1.3. Graduations.
        S.1.3.1. Length.
        S.1.3.2. Width.
        S.1.3.3. Clear Space Between Graduations.

In this example, paragraphs S.1.1., S.1.2., and S.1.3. are directed and limited to paragraph S.1., which pertains to the design of indicating and recording elements and of recorded representations. Paragraphs S.1.1.1. and S.1.1.2. are directly related to each other, but they are limited to the design of zero indication. Likewise, paragraphs S.1.3.1., S.1.3.2., and S.1.3.3. are directly related to each other, but they are limited to the design of graduations.

This handbook conforms to the concept of primary use of SI (metric) measurements recommended in the Omnibus Trade and Competitiveness Act of 1988 by citing SI metric units before U.S. customary units where both units appear together and placing separate sections containing requirements for metric units before corresponding sections containing requirements for customary units. Occasionally, a paragraph or table carries the suffix "M" because the requirement in SI units is shown as a separate statement, rather than combined with the U.S. customary units. In these few instances, separate requirements were judged to be more easily understood than attempting to combine SI and U.S. customary units in a single paragraph or table. In some cases, however, trade practice is currently restricted to the use of customary units; therefore, some requirements in this handbook will continue to specify only customary units until the Conference achieves a broad consensus on the permitted metric units.

E. Classification of Requirements.

The classification of requirements into "retroactive" and "nonretroactive" status is made in order that the requirements may be put into force and effect without unnecessary hardship and without wholesale condemnation of apparatus. Retroactive requirements are enforceable with respect to all equipment and are printed in upright roman type. Nonretroactive requirements are those that, while clearly desirable, are not so vital that they should at once be enforced with respect to all apparatus. Nonretroactive requirements are printed in *italic type.*

It is not expected that, after their promulgation in a given jurisdiction, nonretroactive requirements will always remain nonretroactive. It is entirely proper that a weights and measures official, following a careful analysis of existing conditions, fix reasonable periods for the continuance of the nonretroactive application of particular requirements, after which such requirements will become retroactive. These periods should be long enough to avoid undue hardship to the owners or operators of apparatus and, in the case of some requirements, should approximate the average useful life of the apparatus in question.

In order that all interested parties may have timely and ample notice of impending changes in the status of requirements, the following procedure is suggested for the official who plans to change the classification of requirements. If sufficient data are available to make such action feasible, publish in combination with the codes themselves the date or dates at which nonretroactive requirements are to become retroactive. In other cases, give equally effective notice at the earliest practicable date.

A nonretroactive requirement, in italic type, will indicate the year from which it should be enforced and, in some cases, the date the requirement shall be changed to retroactive status. For example, *[Nonretroactive as of 1978 and to become retroactive on January 1, 1985].* As a general rule, each nonretroactive requirement is reviewed after it has been in effect for 10 years to determine the appropriateness of its nonretroactive status.

F. Using the Handbook.

Handbook 44 is designed to be a working tool for federal, state, and local weights and measures officials, the equipment manufacturers, installers, and service agencies/agents. As noted in Section 1.10. General Code paragraph G-A.1. Commercial and Law-Enforcement Equipment, applicable portions of Handbook 44 may be used by the weights and measures official to test noncommercial weighing and measuring equipment upon request. Additionally, applicable language in Handbook 44 may be cited as a standard in noncommercial applications, for example, when the handbook is referenced or cited as part of a quality system or in multiple-party contract agreements where noncommercial weighing or measuring equipment is used.

The section on Fundamental Considerations (Appendix A) should be studied until its contents are well known. The General Code, with general requirements pertaining to all devices, obviously must be well known to a user of the handbook. The makeup of the specific codes, the order of paragraph presentation, and particularly paragraph designation are worthy of careful study. It is not deemed advisable for a user to attempt to commit to memory tolerances or tolerance tables, even though these are used frequently. For the handbook to serve its purpose, it should be available when any of its requirements are to be applied. Direct reference is the only sure way to apply a requirement properly and to check whether other requirements may be applicable.

This handbook supplies criteria which enable the user to determine the suitability, accuracy, and repetitive consistency of a weighing or measuring device, both in the laboratory and in the field. However, not all code sections can be appropriately applied in both settings. Since some sections are designed to be applied specifically to tests performed under laboratory conditions, it would be impractical or unrealistic to apply them to field tests. Not all tests described in the "Notes" section of the handbook are required to be performed in the field as an official test. An inspector may officially approve or reject a device which has been tested in accordance with those sections applicable to the type of test being conducted.

(Paragraph added 1996)

National Conference on Weights and Measures / National Type Evaluation Program

Form 15: Proposal to Amend NIST Handbooks

Submit proposals at least two weeks prior to regional meetings. See meeting dates at **www.ncwm.net/meetings**. If the item is deemed by a region to have merit, that region will forward the item to NCWM for national consideration. See **http://www.ncwm.net/standards-development/idea** for more information on the Form 15 process. Submit in Microsoft Word Format to the NCWM Executive Director via email at **don.onwiler@ncwm.net** for review and dispersal to selected regions.

General Information	
1. Date:	**2. Regional Association(s):** (Not applicable for proposals to the Board of Directors or NTEP Committee) ___ Central (CWMA) ___ Northeastern (NEWMA) ___ Southern (SWMA) ___ Western (WWMA)

3. Standing Committee:

__Laws & Regulations   __ Specifications & Tolerances   __Professional Development   __Board of Directors   __NTEP Committee

4. Submitter's Name:	**Submitter's Organization:**

5. Address:

6. City:	**7. State:**	**8. Zip Code:**	**9. Country:**

10. Phone Number:	**11. Fax Number:**	**12. Email Address:**

Proposal Information

13. Purpose: Concise statement as to the intent or purpose of this proposal, such as problem being fixed. (Do not include justification here.)

14. Document to be Amended:

__ NIST Handbook 44   __ NIST Handbook 130   __ NIST Handbook 133   ___ NCWM Guidance Document
__ NCWM Bylaws   __NTEP Administrative Policy

15. Cite portion to be Amended:

Section:
Paragraph:

16. Proposal: Please use ~~strikeout~~ to show words to be deleted and <u>underline</u> to show new words.

17. Justification: Please include national importance, background on the issue, and reference to supporting data or documents.

18. Possible Opposing Argument's: Please demonstrate that you are aware and have considered possible opposition.

19. Requested Action if Considered for NCWM Agenda:

___ Voting Item   ___ Developing Item   ___ Informational Item   ___ Other (Please Describe):

20. List of Attachments:

Submit form via e-mail to don.onwiler@ncwm.net   Revised: May 2016
1135 M Street, Suite 110 / Lincoln, Nebraska 68508
P. 402.434.4880   F. 402.434.4878   E. info@ncwm.net   W. www.ncwm.net/

4

Table of Contents

Section 1.10.   General Code

G-A.   Application

G-A.1.   Commercial and Law-Enforcement Equipment. – These specifications, tolerances, and other technical requirements apply as follows:

(a)   To commercial weighing and measuring equipment; that is, to weights and measures and weighing and measuring devices commercially used or employed in establishing the size, quantity, extent, area, composition (limited to meat and poultry), constituent values (limited to grain), or measurement of quantities, things, produce, or articles for distribution or consumption, purchased, offered, or submitted for sale, hire, or award, or in computing any basic charge or payment for services rendered on the basis of weight or measure.

(Amended 2008)

(b)   To any accessory attached to or used in connection with a commercial weighing or measuring device when such accessory is so designed that its operation affects the accuracy of the device.

(c)   To weighing and measuring equipment in official use for the enforcement of law or for the collection of statistical information by government agencies.

(These requirements should be used as a guide by the weights and measures official when, upon request, courtesy examinations of noncommercial equipment are made.)

G-A.2.   Code Application. – This General Code shall apply to all classes of devices as covered in the specific codes. The specific code requirements supersede General Code requirements in all cases of conflict.

(Amended 1972)

G-A.3.   Special and Unclassified Equipment. – Insofar as they are clearly appropriate, the requirements and provisions of the General Code and of specific codes apply to equipment failing, by reason of special design or otherwise, to fall clearly within one of the particular equipment classes for which separate codes have been established. With respect to such equipment, code requirements and provisions shall be applied with due regard to the design, intended purpose, and conditions of use of the equipment.

G-A.4.   Metric Equipment. – Employment of the weights and measures of the metric system is lawful throughout the United States. These specifications, tolerances, and other requirements shall not be understood or construed as in any way prohibiting the manufacture, sale, or use of equipment designed to give results in terms of metric units. The specific provisions of these requirements and the principles upon which the requirements are based shall be applied to metric equipment insofar as appropriate and practicable. The tolerances on metric equipment, when not specified herein, shall be equivalent to those specified for similar equipment constructed or graduated in the U.S. customary system.

G-A.5.   Retroactive Requirements. – "Retroactive" requirements are enforceable with respect to all equipment. Retroactive requirements are printed herein in upright roman type.

G-A.6.   Nonretroactive Requirements. – "Nonretroactive" requirements are enforceable on or after the effective date for devices:

(a)   manufactured within a state after the effective date;

(b)   both new and used, brought into a state after the effective date;

(c)   used in noncommercial applications which are placed into commercial use after the effective date; and

(d) undergoing type evaluation, including devices that have been modified to the extent that a new NTEP Certificate of Conformance (CC) is required.

Nonretroactive requirements are not enforceable with respect to devices that are in commercial service in the state as of the effective date or to new equipment in the stock of a manufacturer or a dealer in the state as of the effective date. *[Nonretroactive requirements are printed in italic type.]*

(Amended 1989 and 2011)

G-A.7.   Effective Enforcement Dates of Code Requirements. – Unless otherwise specified, each new or amended code requirement shall not be subject to enforcement prior to January 1 of the year following the adoption by the National Conference on Weights and Measures and publication by the National Institute of Standards and Technology.

G-S.   Specifications

G-S.1.   Identification. – All equipment, except weights and separate parts necessary to the measurement process but not having any metrological effect, shall be clearly and permanently marked for the purposes of identification with the following information:

(a)   the name, initials, or trademark of the manufacturer or distributor;

(b)   a model identifier that positively identifies the pattern or design of the device;

   (1)   The model identifier shall be prefaced by the word "Model," "Type," or "Pattern." These terms may be followed by the word "Number" or an abbreviation of that word. The abbreviation for the word "Number" shall, as a minimum, begin with the letter "N" (e.g., No or No.). The abbreviation for the word "Model" shall be "Mod" or "Mod." Prefix lettering may be initial capitals, all capitals, or all lower case.
   [Nonretroactive as of January 1, 2003]

   (Added 2000) (Amended 2001)

   (c)   a nonrepetitive serial number, except for equipment with no moving or electronic component parts and software;
   [Nonretroactive as of January 1, 1968]

   (Amended 2003 and 2016)

   (1)   The serial number shall be prefaced by words, an abbreviation, or a symbol, that clearly identifies the number as the required serial number.
   [Nonretroactive as of January 1, 1986]

   (2)   Abbreviations for the word "Serial" shall, as a minimum, begin with the letter "S," and abbreviations for the word "Number" shall, as a minimum, begin with the letter "N" (e.g., S/N, SN, Ser. No., and S. No.).
   [Nonretroactive as of January 1, 2001]

(d)   the current software version or revision identifier for not-built-for-purpose, software-based devices manufactured as of January 1, 2004, and all software-based devices (or equipment) manufactured as of January 1, 2022;

   (Added 2003) (Amended 2016)

(1)   *The version or revision identifier shall be:*

    i.   *prefaced by words, an abbreviation, or a symbol, that clearly identifies the number as the required version or revision.*
    [Nonretroactive as of January 1, 2007]

    (Added 2006)

        Note: *If the equipment is capable of displaying the version or revision identifier, but is unable to meet the formatting requirements, through the NTEP type evaluation process, other options may be deemed acceptable and described in the CC.*

        (Added 2016)

    ii.   *continuously displayed or be accessible via the display. Instructions for displaying the version or revision identifier shall be described in the CC. As an alternative, permanently marking the version or revision identifier shall be acceptable providing the device does not always have an integral interface to communicate the version or revision identifier.*
    [Nonretroactive as of January 1, 2022]

    (Added 2016)

(2)   *Abbreviations for the word "Version" shall, as a minimum, begin with the letter "V" and may be followed by the word "Number." Abbreviations for the word "Revision" shall, as a minimum, begin with the letter "R" and may be followed by the word "Number." The abbreviation for the word "Number" shall, as a minimum, begin with the letter "N" (e.g., No or No.). Prefix lettering may be initial capitals, all capitals, or all lowercase.*
[Nonretroactive as of January 1, 2007]

(Added 2006) (Amended 2016)

(e)   a National Type Evaluation Program (NTEP) Certificate of Conformance (CC) number or a corresponding CC Addendum Number for devices that have a CC.

    (1)   *The CC Number or a corresponding CC Addendum Number shall be prefaced by the terms "NTEP CC," "CC," or "Approval." These terms may be followed by the word "Number" or an abbreviation of that word. The abbreviation for the word "Number" shall, as a minimum, begin with the letter "N" (e.g., No or No.).*
    [Nonretroactive as of January 1, 2003]

The required information shall be so located that it is readily observable without the necessity of the disassembly of a part requiring the use of any means separate from the device.

(Amended 1985, 1991, 1999, 2000, 2001, 2003, 2006, and 2016)

G-S.1.1.   Location of Marking Information for Not-Built-For-Purpose, Software-Based Devices. *– For not-built-for-purpose, software-based devices either:*

    (a)   *The required information in G-S.1 Identification. (a), (b), (d), and (e) shall be permanently marked or continuously displayed on the device; or*

    (b)   *The Certificate of Conformance (CC) Number shall be:*

        (1)   *permanently marked on the device;*

        (2)   *continuously displayed; or*

(3) *accessible through an easily recognized menu and, if necessary, a submenu. Examples of menu and submenu identification include, but are not limited to, "Help," "System Identification," "G-S.1. Identification," or "Weights and Measures Identification."*

Note: For (b), clear instructions for accessing the information required in G-S.1. (a), (b), and (d) shall be listed on the CC, including information necessary to identify that the software in the device is the same type that was evaluated.
[Nonretroactive as of January 1, 2004]
(Added 2003) (Amended 2006)

G-S.1.2. Devices and Main Elements Remanufactured as of January 1, 2002. – All devices and main elements remanufactured as of January 1, 2002, shall be clearly and permanently marked for the purposes of identification with the following information:

(a) the name, initials, or trademark of the last remanufacturer or distributor; and

(b) the remanufacturer's or distributor's model designation, if different than the original model designation.
(Added 2001) (Amended 2011)

Note: Definitions for "manufactured device," "repaired device," and "repaired element" are included (along with definitions for "remanufactured device" and "remanufactured element") in Appendix D, Definitions.

G-S.2. Facilitation of Fraud. – All equipment and all mechanisms, software, and devices attached to or used in conjunction therewith shall be so designed, constructed, assembled, and installed for use such that they do not facilitate the perpetration of fraud.
(Amended 2007)

G-S.3. Permanence. – All equipment shall be of such materials, design, and construction as to make it probable that, under normal service conditions:

(a) accuracy will be maintained;

(b) operating parts will continue to function as intended; and

(c) adjustments will remain reasonably permanent.

Undue stresses, deflections, or distortions of parts shall not occur to the extent that accuracy or permanence is detrimentally affected.

G-S.4. Interchange or Reversal of Parts. – Parts of a device that may readily be interchanged or reversed in the course of field assembly or of normal usage shall be:

(a) so constructed that their interchange or reversal will not affect the performance of the device; or

(b) so marked as to show their proper positions.

G-S.5. Indicating and Recording Elements.

G-S.5.1. General. – All weighing and measuring devices shall be provided with indicating or recording elements appropriate in design and adequate in amount. Primary indications and recorded representations shall be clear, definite, accurate, and easily read under any conditions of normal operation of the device.

G-S.5.2. Graduations, Indications, and Recorded Representations.

G-S.5.2.1. Analog Indication and Representation. – Graduations and a suitable indicator shall be provided in connection with indications designed to advance continuously.

G-S.5.2.2.   Digital Indication and Representation. – Digital elements shall be so designed that:

(a) All digital values of like value in a system agree with one another.

(b) A digital value coincides with its associated analog value to the nearest minimum graduation.

(c) A digital value "rounds off" to the nearest minimum unit that can be indicated or recorded.

(d) *A digital zero indication includes the display of a zero for all places that are displayed to the right of the decimal point and at least one place to the left. When no decimal values are displayed, a zero shall be displayed for each place of the displayed scale division.*
[Nonretroactive as of January 1, 1986]

(Amended 1973 and 1985)

G-S.5.2.3.   Size and Character. – In any series of graduations, indications, or recorded representations, corresponding graduations and units shall be uniform in size and character. Graduations, indications, or recorded representations that are subordinate to, or of a lesser value than others with which they are associated, shall be appropriately portrayed or designated.

[Made retroactive as of January 1, 1975]

G-S.5.2.4.   Values. – If graduations, indications, or recorded representations are intended to have specific values, these shall be adequately defined by a sufficient number of figures, words, symbols, or combinations thereof, uniformly placed with reference to the graduations, indications, or recorded representations and as close thereto as practicable, but not so positioned as to interfere with the accuracy of reading.

G-S.5.2.5.   Permanence. – Graduations, indications, or recorded representations and their defining figures, words, and symbols shall be of such character that they will not tend easily to become obliterated or illegible.

G-S.5.3.   Values of Graduated Intervals or Increments. – In any series of graduations, indications, or recorded representations, the values of the graduated intervals or increments shall be uniform throughout the series.

G-S.5.3.1.   On Devices That Indicate or Record in More Than One Unit. – On devices designed to indicate or record in more than one unit of measurement, the values indicated and recorded shall be identified with an appropriate word, symbol, or abbreviation.

(Amended 1978 and 1986)

G-S.5.4.   Repeatability of Indications. – A device shall be capable of repeating, within prescribed tolerances, its indications and recorded representations. This requirement shall be met irrespective of repeated manipulation of any element of the device in a manner approximating normal usage (including displacement of the indicating elements to the full extent allowed by the construction of the device and repeated operation of a locking or relieving mechanism) and of the repeated performance of steps or operations that are embraced in the testing procedure.

G-S.5.5.   Money Values, Mathematical Agreement. – Any recorded money value and any digital money-value indication on a computing-type weighing or measuring device used in retail trade shall be in mathematical agreement with its associated quantity representation or indication to the nearest 1 cent of money value. This does not apply to auxiliary digital indications intended for the operator's use only, when these indications are obtained from existing analog customer indications that meet this requirement.

(Amended 1973)

G-S.5.6.   Recorded Representations. – Insofar as they are appropriate, the requirements for indicating and recording elements shall also apply to recorded representations. All recorded values shall be printed digitally. In applications where recorded representations are required, the customer may be given the option of not receiving

the recorded representation. For systems equipped with the capability of issuing an electronic receipt, ticket, or other recorded representation, the customer may be given the option to receive any required information electronically (e.g., via cell phone, computer, etc.) in lieu of or in addition to a hard copy.
(Amended 1975 and 2014)

G-S.5.6.1.   Indicated and Recorded Representation of Units. – Appropriate abbreviations.

(a) For equipment manufactured on or after January 1, 2008, the appropriate defining symbols are shown in NIST Special Publication SP 811 "Guide for the Use of International System of Units (SI)" and Handbook 44 Appendix C – General Tables of Units of Measurement.

Note: SP 811 can be viewed or downloaded at http://physics.nist.gov/cuu/pdf/sp811.pdf or by going to http://www.nist.gov/pml/wmd/index.cfm and selecting Weights and Measures Publications and the link to Special Publications (SP 811), "Guide for the Use of the International System of Units (SI)."
(Added 2007)

(b) The appropriate defining symbols on equipment manufactured prior to January 1, 2008, with limited character sets are shown in Table 1. Representation of SI Units on Equipment Manufactured Prior to January 1, 2008, with Limited Character Sets.

(Added 1977) (Amended 2007)

Table 1.
Representation of SI Units on Equipment Manufactured Prior to January 1, 2008, with Limited Character Sets

Name of Unit	International Symbol (common use symbol)	Representation		
		Form I	Form II	
		(double case)	(single case lower)	(single case upper)
Base SI Units				
meter	m	m	m	M
kilogram	kg	kg	kg	KG
Derived SI Units				
newton	N	N	n	N
pascal	Pa	Pa	pa	PA
watt	W	W	w	W
volt	V	V	v	V
degree Celsius	°C	°C	°c	°C
Other Units				
liter	l or L	L	l	L
gram	g	g	g	G
metric ton	t	t	tne	TNE
bar	bar	bar	bar	BAR

(Table Amended 2007)

G-S.5.7.   Magnified Graduations and Indications. – All requirements for graduations and indications apply to a series of graduations and an indicator magnified by an optical system or as magnified and projected on a screen.

G-S.6.   Marking Operational Controls, Indications, and Features. – *All operational controls, indications, and features, including switches, lights, displays, push buttons, and other means, shall be clearly and definitely identified. The use of approved pictograms or symbols shall be acceptable.*
[Nonretroactive as of January 1, 1977]
(Amended 1978 and 1995)

G-S.7.   Lettering. – All required markings and instructions shall be distinct and easily readable and shall be of such character that they will not tend to become obliterated or illegible.

G-S.8.   Provision for Sealing Electronic Adjustable Components. – *A device shall be designed with provision(s) for applying a security seal that must be broken, or for using other approved means of providing security (e.g., data change audit trail available at the time of inspection), before any change that detrimentally affects the metrological integrity of the device can be made to any electronic mechanism.*
[Nonretroactive as of January 1, 1990]

A device may be fitted with an automatic or a semi-automatic calibration mechanism. This mechanism shall be incorporated inside the device. After sealing, neither the mechanism nor the calibration process shall facilitate fraud.
(Added 1985) (Amended 1989 and 1993)

G-S.8.1.   Multiple Weighing or Measuring Elements that Share a Common Provision for Sealing. – *A change to any metrological parameter (calibration or configuration) of any weighing or measuring element shall be individually identified.*
[Nonretroactive as of January 1, 2010]

Note: For devices that utilize an electronic form of sealing, in addition to the requirements in G-S.8.1., any appropriate audit trail requirements in an applicable specific device code also apply. Examples of identification of a change to the metrological parameters of a weighing or measuring element include, but are not limited to:

(1) a broken, missing, or replaced physical seal on an individual weighing, measuring, or indicating element or active junction box;

(2) a change in a calibration factor or configuration setting for each weighing or measuring element;

(3) a display of the date of calibration or configuration event for each weighing or measuring element; or

(4) counters indicating the number of calibration and/or configuration events for each weighing or measuring element.

(Added 2007)

G-S.9.   Metrologically Significant Software Updates. – A software update that changes the metrologically significant software shall be considered a sealable event.
(Added 2016)

G-N.   Notes

G-N.1.   Conflict of Laws and Regulations. – If any particular provisions of these specifications, tolerances, and other requirements are found to conflict with existing state laws, or with existing regulations or local ordinances relating to health, safety, or fire prevention, the enforcement of such provisions shall be suspended until conflicting requirements can be harmonized. Such suspension shall not affect the validity or enforcement of the remaining provisions of these specifications, tolerances, and other requirements.

G-N.2.   Testing With Nonassociated Equipment. – Tests to determine conditions, such as radio frequency interference (RFI) that may adversely affect the performance of a device shall be conducted with equipment and under conditions that are usual and customary with respect to the location and use of the device.
(Added 1976)

G-T. Tolerances

G-T.1. Acceptance Tolerances. – Acceptance tolerances shall apply to equipment:

(a) to be put into commercial use for the first time;

(b) that has been placed in commercial service within the preceding 30 days and is being officially tested for the first time;

(c) that has been returned to commercial service following official rejection for failure to conform to performance requirements and is being officially tested for the first time within 30 days after corrective service;

(d) that is being officially tested for the first time within 30 days after major reconditioning or overhaul; and

(e) undergoing type evaluation.
    (Amended 1989)

G-T.2. Maintenance Tolerances. – Maintenance tolerances shall apply to equipment in actual use, except as provided in G-T.1. Acceptance Tolerances.

G-T.3. Application. – Tolerances "in excess" and tolerances "in deficiency" shall apply to errors in excess and to errors in deficiency, respectively. Tolerances "on overregistration" and tolerances "on underregistration" shall apply to errors in the direction of overregistration and of underregistration, respectively. (Also see Appendix D, Definitions.)

G-T.4. For Intermediate Values. – For a capacity, indication, load, value, etc., intermediate between two capacities, indications, loads, values, etc., listed in a table of tolerances, the tolerances prescribed for the lower capacity, indication, load, value, etc., shall be applied.

G-UR. User Requirements

G-UR.1. Selection Requirements.

G-UR.1.1. Suitability of Equipment. – Commercial equipment shall be suitable for the service in which it is used with respect to elements of its design, including but not limited to its weighing capacity (for weighing devices), its computing capability (for computing devices), its rate of flow (for liquid-measuring devices), the character, number, size, and location of its indicating or recording elements, and the value of its smallest unit and unit prices.
(Amended 1974)

G-UR.1.2. Environment. – Equipment shall be suitable for the environment in which it is used including, but not limited to, the effects of wind, weather, and RFI.
(Added 1976)

G-UR.1.3. Liquid-Measuring Devices. – To be suitable for its application, the minimum delivery for liquid-measuring devices shall be no less than 100 divisions, except that the minimum delivery for retail analog devices shall be no less than 10 divisions. Maximum division values and tolerances are stated in the specific codes.
(Added 1995)

G-UR.2. Installation Requirements.

G-UR.2.1. Installation. – A device shall be installed in accordance with the manufacturer's instructions, including any instructions marked on the device. A device installed in a fixed location shall be installed so that neither its operation nor its performance will be adversely affected by any characteristic of the foundation, supports, or any other detail of the installation.

G-UR.2.1.1. Visibility of Identification. – Equipment shall be installed in such a manner that all required markings are readily observable.
(Added 1978)

G-UR.2.2. Installation of Indicating or Recording Element. – A device shall be so installed that there is no obstruction between a primary indicating or recording element and the weighing or measuring element; otherwise there shall be convenient and permanently installed means for direct communication, oral or visual, between an individual located at a primary indicating or recording element and an individual located at the weighing or measuring element. (Also see G-UR.3.3. Position of Equipment.)

G-UR.2.3. Accessibility for Inspection, Testing, and Sealing Purposes. – A device shall be located, or such facilities for normal access thereto shall be provided, to permit:

(a) inspecting and testing the device;

(b) inspecting and applying security seals to the device; and

(c) readily bringing the testing equipment of the weights and measures official to the device by customary means and in the amount and size deemed necessary by such official for the proper conduct of the test.

Otherwise, it shall be the responsibility of the device owner or operator to supply such special facilities, including such labor as may be needed to inspect, test, and seal the device, and to transport the testing equipment to and from the device, as required by the weights and measures official.
(Amended 1991)

G-UR.3. Use Requirements.

G-UR.3.1. Method of Operation. – Equipment shall be operated only in the manner that is obviously indicated by its construction or that is indicated by instructions on the equipment.

G-UR.3.2. Associated and Nonassociated Equipment. – A device shall meet all performance requirements when associated or nonassociated equipment is operated in its usual and customary manner and location.
(Added 1976)

G-UR.3.3. Position of Equipment. – A device or system equipped with a primary indicating element and used in direct sales, except for prescription scales, shall be positioned so that its indications may be accurately read and the weighing or measuring operation may be observed from some reasonable "customer" and "operator" position. The permissible distance between the equipment and a reasonable customer and operator position shall be determined in each case upon the basis of the individual circumstances, particularly the size and character of the indicating element.
(Amended 1974 and 1998)

G-UR.3.4. Responsibility, Money-Operated Devices. – Money-operated devices, other than parking meters, shall have clearly and conspicuously displayed thereon, or immediately adjacent thereto, adequate information detailing the method for the return of monies paid when the product or service cannot be obtained. This information shall include the name, address, and phone number of the local responsible party for the device. This requirement does not apply to devices at locations where employees are present and responsible for resolving any monetary discrepancies for the customer.

(Amended 1977 and 1993)

G-UR.4. Maintenance Requirements.

G-UR.4.1.   Maintenance of Equipment. – All equipment in service and all mechanisms and devices attached thereto or used in connection therewith shall be continuously maintained in proper operating condition throughout the period of such service. Equipment in service at a single place of business shall not be considered "maintained in a proper operating condition" if:

(a) predominantly, equipment of all types or applications are found to be in error in a direction favorable to the device user; or

(b) predominantly, equipment of the same type or application is found to be in error in a direction favorable to the device user.

(Amended 1973, 1991, and 2015)

G-UR.4.2.   Abnormal Performance. – Unstable indications or other abnormal equipment performance observed during operation shall be corrected and, if necessary, brought to the attention of competent service personnel.

(Added 1976)

G-UR.4.3.   Use of Adjustments. – Weighing elements and measuring elements that are adjustable shall be adjusted only to correct those conditions that such elements are designed to control, and shall not be adjusted to compensate for defective or abnormal installation or accessories or for badly worn or otherwise defective parts of the assembly. Any faulty installation conditions shall be corrected, and any defective parts shall be renewed or suitably repaired, before adjustments are undertaken. Whenever equipment is adjusted, the adjustments shall be so made as to bring performance errors as close as practicable to zero value.

G-UR.4.4.   Assistance in Testing Operations. – If the design, construction, or location of any device is such as to require a testing procedure involving special equipment or accessories or an abnormal amount of labor, such equipment, accessories, and labor shall be supplied by the owner or operator of the device as required by the weights and measures official.

G-UR.4.5.   Security Seal. – A security seal shall be appropriately affixed to any adjustment mechanism designed to be sealed.

G-UR.4.6.   Testing Devices at a Central Location.

(a) When devices in commercial service require special test facilities, or must be removed from service for testing, or are routinely transported for the purpose of use (e.g., vehicle-mounted devices and devices used in multiple locations), the official with statutory authority may require that the devices be brought to a central location for testing. The dealer or owner of these devices shall provide transportation of the devices to and from the test location.

(b) When the request for removal and delivery to a central test location involves devices used in submetering (e.g., electric, hydrocarbon vapor, or water meters), the owner or operator shall not interrupt the utility service to the customer or tenant except for the removal and replacement of the device. Provisions shall be made by the owner or operator to minimize inconvenience to the customer or tenant. All replacement or temporary meters shall be tested and sealed by a weights and measures official or bear a current, valid approval seal prior to use.

(Added 1994)

Section 2

Table of Contents

THIS PAGE INTENTIONALLY LEFT BLANK

Table of Contents

Section 2.20.   Scales

A. Application

A.1.   General. – This code applies to all types of weighing devices other than automatic bulk-weighing systems, belt-conveyor scales, and automatic weighing systems. The code comprises requirements that generally apply to all weighing devices, and specific requirements that are applicable only to certain types of weighing devices.
(Amended 1972 and 1983)

A.2.   Wheel-Load Weighers, Portable Axle-Load Weighers, and Axle-Load Scales. – The requirements for wheel-load weighers, portable axle-load weighers, and axle-load scales apply only to such scales in official use for the enforcement of traffic and highway laws or for the collection of statistical information by government agencies.

A.3.   Additional Code Requirements. – In addition to the requirements of this code, devices covered by the Scales code shall meet the requirements of Section 1.10. General Code.

S. Specifications

S.1.   Design of Indicating and Recording Elements and of Recorded Representations.

S.1.1.   Zero Indication.

(a) On a scale equipped with indicating or recording elements, provision shall be made to either indicate or record a zero-balance condition.

(b) On an automatic-indicating scale or balance indicator, provision shall be made to indicate or record an out-of-balance condition on both sides of zero.

(c) A zero-balance condition may be indicated by other than a continuous digital zero indication, provided that an effective automatic means is provided to inhibit a weighing operation or to return to a continuous digital indication when the scale is in an out-of-balance condition.
(Added 1987) (Amended 1993)
(Amended 1987)

S.1.1.1.   Digital Indicating Elements.

(a) A digital zero indication shall represent a balance condition that is within $\pm \frac{1}{2}$ the value of the scale division.

(b) *A digital indicating device shall either automatically maintain a "center-of-zero" condition to $\pm \frac{1}{4}$ scale division or less, or have an auxiliary or supplemental "center-of-zero" indicator that defines a zero-balance condition to $\pm \frac{1}{4}$ of a scale division or less. A "center-of-zero" indication may operate when zero is indicated for gross and/or net mode(s).*
[Nonretroactive as of January 1, 1993]
(Amended 1992 and 2008)

S.1.1.2.   No-Load Reference Value. – On a single draft manually operated receiving hopper scale installed below grade, used to receive grain, and utilizing a no-load reference value, provision shall be made to indicate and record the no-load reference value prior to the gross load value.
(Added 1983)

S.1.2. ***Value of Scale Division Units.*** *– Except for batching scales and weighing systems used exclusively for weighing in predetermined amounts, the value of a scale division "d" expressed in a unit of weight shall be equal to:*

(a) 1, 2, or 5; or

(b) a decimal multiple or submultiple of 1, 2, or 5; or

      Examples: scale divisions may be 10, 20, 50, 100; or 0.01, 0.02, 0.05; or 0.1, 0.2, 0.5, etc.

(c) a binary submultiple of a specific unit of weight.

      Examples: scale divisions may be ½, ¼, ⅛, ¹/₁₆, etc.
[Nonretroactive as of January 1, 1986]

    S.1.2.1. ***Digital Indicating Scales, Units.*** *– Except for postal scales, a digital-indicating scale shall indicate weight values using only a single unit of measure. Weight values shall be presented in a decimal format with the value of the scale division expressed as 1, 2, or 5, or a decimal multiple or submultiple of 1, 2, or 5.*

The requirement that the value of the scale division be expressed only as 1, 2, or 5, or a decimal multiple or submultiple of only 1, 2, or 5 does not apply to net weight indications and recorded representations that are calculated from gross and tare weight indications where the scale division of the gross weight is different from the scale division of the tare weight(s) on multi-interval or multiple range scales. For example, a multiple range or multi-interval scale may indicate and record tare weights in a lower weighing range (WR) or weighing segment (WS), gross weights in the higher weighing range or weighing segment, and net weights as follows:

55 kg	Gross Weight (WR2 d = 5 kg)	10.05 lb	Gross Weight (WS2 d = 0.05 lb)
– 4 kg	Tare Weight (WR1 d = 2 kg)	– 0.06 lb	Tare Weight (WS1 d = 0.02 lb)
= 51 kg	Net Weight (Mathematically Correct)	= 9.99 lb	Net Weight (Mathematically Correct)

[Nonretroactive as of January 1, 1989]

(Added 1987) (Amended 2008)

    S.1.2.2.   **Verification Scale Interval.**

        S.1.2.2.1.   **Class I and II Scales and Dynamic Monorail Scales.** If $e \neq d$, the verification scale interval "e" shall be determined by the expression:

$$d < e \leq 10\ d$$

If the displayed division (d) is less than the verification division (e), then the verification division shall be less than or equal to 10 times the displayed division.

The value of e must satisfy the relationship, $e = 10^k$ of the unit of measure, where k is a positive or negative whole number or zero. This requirement does not apply to a Class I device with d < 1 mg where e = 1 mg. If $e \neq d$, the value of "d" shall be a decimal submultiple of "e," and the ratio shall not be more than 10:1. If $e \neq d$, and both "e" and "d" are continuously displayed during normal operation, then "d" shall be differentiated from "e" by size, shape, color, etc. throughout the range of weights displayed as "d."

(Added 1999)

        S.1.2.2.2.   **Class III and IIII Scales.** The value of "e" is specified by the manufacturer as marked on the device. Except for dynamic monorail scales, "e" must be less than or equal to "d."

(Added 1999)

S.1.2.3. **Prescription Scale with a Counting Feature.** – A Class I or Class II prescription scale with an operational counting feature shall not calculate a piece weight or total count unless the sample used to determine the individual piece weight meets the following conditions:

(a) minimum individual piece weight is greater than or equal to 3 e; and

(b) minimum sample piece count is greater than or equal to 10 pieces.

(Added 2003)

S.1.3. Graduations.

S.1.3.1. **Length.** – Graduations shall be so varied in length that they may be conveniently read.

S.1.3.2. **Width.** – In any series of graduations, the width of a graduation shall in no case be greater than the width of the clear space between graduations. The width of main graduations shall be not more than 50 % greater than the width of subordinate graduations. Graduations shall be not less than 0.2 mm (0.008 in) wide.

S.1.3.3. **Clear Space Between Graduations.** – The clear space between graduations shall be not less than 0.5 mm (0.02 in) for graduations representing money-values, and not less than 0.75 mm (0.03 in) for other graduations. If the graduations are not parallel, the measurement shall be made:

(a) along the line of relative movement between the graduations at the end of the indicator; or

(b) if the indicator is continuous, at the point of widest separation of the graduations.

S.1.4. Indicators.

S.1.4.1. **Symmetry.** – The index of an indicator shall be of the same shape as the graduations, at least throughout that portion of its length associated with the graduations.

S.1.4.2. **Length.** – The index of an indicator shall reach to the finest graduations with which it is used, unless the indicator and the graduations are in the same plane, in which case, the distance between the end of the indicator and the ends of the graduations, measured along the line of the graduations, shall be not more than 1.0 mm (0.04 in).

S.1.4.3. **Width.** – The width of the index of an indicator in relation to the series of graduations with which it is used shall be not greater than:

(a) the width of the narrowest graduation;
   [Nonretroactive as of January 1, 2002]

(b) the width of the clear space between weight graduations; and

(c) three-fourths of the width of the clear space between money-value graduations.

When the index of an indicator extends along the entire length of a graduation, that portion of the index of the indicator that may be brought into coincidence with the graduation shall be of the same width throughout the length of the index that coincides with the graduation.

S.1.4.4. **Clearance.** – The clearance between the index of an indicator and the graduations shall in no case be more than 1.5 mm (0.06 in).

S.1.4.5. **Parallax.** – Parallax effects shall be reduced to the practicable minimum.

S.1.5. Weighbeams.

S.1.5.1.   Normal Balance Position. – The normal balance position of the weighbeam of a beam scale shall be horizontal.

S.1.5.2.   Travel. – The weighbeam of a beam scale shall have equal travel above and below the horizontal. The total travel of the weighbeam of a beam scale in a trig loop or between other limiting stops near the weighbeam tip shall be not less than the minimum travel shown in Tables 1M and 1. When such limiting stops are not provided, the total travel at the weighbeam tip shall be not less than 8 % of the distance from the weighbeam fulcrum to the weighbeam tip.

Table 1M. Minimum Travel of Weighbeam of Beam Scale Between Limiting Stops		Table 1. Minimum Travel of Weighbeam of Beam Scale Between Limiting Stops	
Distance From Weighbeam Fulcrum to Limiting Stops (centimeters)	Minimum Travel Between Limiting Stops (millimeter)	Distance From Weighbeam Fulcrum to Limiting Stops (inches)	Minimum Travel Between Limiting Stops (inch)
30 or less	10	12 or less	0.4
30+ to 50, inclusive	13	12+ to 20, inclusive	0.5
50+ to 100, inclusive	18	20+ to 40, inclusive	0.7
Over 100	23	Over 40	0.9

S.1.5.3.   Subdivision. – A subdivided weighbeam bar shall be subdivided by scale division graduations, notches, or a combination of both. Graduations on a particular bar shall be of uniform width and perpendicular to the top edge of the bar. Notches on a particular bar shall be uniform in shape and dimensions and perpendicular to the face of the bar. When a combination of graduations and notches is employed, the graduations shall be positioned in relation to the notches to indicate notch values clearly and accurately.

S.1.5.4.   Readability. – A subdivided weighbeam bar shall be so subdivided and marked, and a weighbeam poise shall be so constructed, that the weight corresponding to any normal poise position can easily and accurately be read directly from the beam, whether or not provision is made for the optional recording of representations of weight.

S.1.5.5.   Capacity. – On an automatic-indicating scale having a nominal capacity of 15 kg (30 lb) or less and used for direct sales to retail customers:

(a) the capacity of any weighbeam bar shall be a multiple of the reading-face capacity;

(b) each bar shall be subdivided throughout or shall be subdivided into notched intervals, each equal to the reading-face capacity; and

(c) the value of any turnover poise shall be equal to the reading-face capacity.

S.1.5.6.   Poise Stop. – Except on a steelyard with no zero graduation, a shoulder or stop shall be provided on each weighbeam bar to prevent a poise from traveling and remaining back of the zero graduation.

S.1.6. Poises.

S.1.6.1.   General. – No part of a poise shall be readily detachable. A locking screw shall be perpendicular to the longitudinal axis of the weighbeam and shall not be removable. Except on a steelyard with no zero graduation, the poise shall not be readily removable from a weighbeam. The knife-edge of a hanging poise shall be hard and sharp and so constructed as to allow the poise to swing freely on the bearing surfaces in the weighbeam notches.

S.1.6.2.   Adjusting Material. – The adjusting material in a poise shall be securely enclosed and firmly fixed in position; if softer than brass, it shall not be in contact with the weighbeam.

S.1.6.3.   Pawl. – A poise, other than a hanging poise, on a notched weighbeam bar shall have a pawl that will seat the poise in a definite and correct position in any notch, wherever in the notch the pawl is placed, and hold it there firmly and without appreciable movement. The dimension of the tip of the pawl that is transverse to the longitudinal axis of the weighbeam shall be at least equal to the corresponding dimension of the notches.

S.1.6.4.   Reading Edge or Indicator. – The reading edge or indicator of a poise shall be sharply defined, and a reading edge shall be parallel to the graduations on the weighbeam.

S.1.7.   Capacity Indication, Weight Ranges, and Unit Weights.

(a) **Gross Capacity.** – An indicating or recording element shall not display nor record any values when the gross load (not counting the initial dead load that has been canceled by an initial zero-setting mechanism) is in excess of 105 % of scale capacity.

(b) *Capacity Indication.* – *Electronic computing scales (excluding postal scales and weight classifiers) shall neither display nor record a gross or net weight in excess of scale capacity plus 9 d.*
[Nonretroactive as of January 1, 1993]

The total value of weight ranges and of unit weights in effect or in place at any time shall automatically be accounted for on the reading face and on any recorded representation.

This requirement does not apply to: (1) single-revolution dial scales, (2) multi-revolution dial scales not equipped with unit weights, (3) scales equipped with two or more weighbeams, nor (4) devices that indicate mathematically derived totalized values.
(Amended 1990, 1992, and 1995)

S.1.8.   Computing Scales.

S.1.8.1.   Money-Value Graduations, Metric Unit Prices. – The value of the graduated intervals representing money-values on a computing scale with analog indications shall not exceed:

(a) 1 cent at all unit prices of 55 cents per kilogram and less;

(b) 2 cents at unit prices of 56 cents per kilogram through $2.75 per kilogram (special graduations defining 5-cent intervals may be employed but not in the spaces between regular graduations);

(c) 5 cents at unit prices of $2.76 per kilogram through $7.50 per kilogram; or

(d) 10 cents at unit prices above $7.50 per kilogram.

Value figures and graduations shall not be duplicated in any column or row on the graduated chart. (Also see S.1.8.2. Money-Value Computation.)

S.1.8.2.   Money-Value Graduations, U.S. Customary Unit Prices. – The value of the graduated intervals representing money-values on a computing scale with analog indications shall not exceed:

(a) 1 cent at all unit prices of 25 cents per pound and less;

(b) 2 cents at unit prices of 26 cents per pound through $1.25 per pound (special graduations defining 5-cent intervals may be employed but not in the spaces between regular graduations);

(c) 5 cents at unit prices of $1.26 per pound through $3.40 per pound; or

(d) 10 cents at unit prices above $3.40 per pound.

Value figures and graduations shall not be duplicated in any column or row on the graduated chart. (Also see S.1.8.2. Money-Value Computation.)

S.1.8.3.   Money-Value Computation. – A computing scale with analog quantity indications used in retail trade may compute and present digital money-values to the nearest quantity graduation when the value of the minimum graduated interval is 0.005 kg (0.01 lb) or less. (Also see Sec. 1.10. General Code G-S.5.5. Money-Values, Mathematical Agreement.)

S.1.8.4.   Customer's Indications. – Weight indications shall be shown on the customer's side of computing scales when these are used for direct sales to retail customers. Computing scales equipped on the operator's side with digital indications, such as the net weight, unit price, or total price, shall be similarly equipped on the customer's side. Unit price displays visible to the customer shall be in terms of single whole units of weight and not in common or decimal fractions of the unit. Scales indicating in metric units may indicate price per 100 g.

(Amended 1985 and 1995)

> *S.1.8.4.1.   Scales that will function as either a normal round off scale or as a weight classifier shall be provided with a sealable means for selecting the mode of operation and shall have a clear indication (annunciator), adjacent to the weight display on both the operator's and customer's side whenever the scale is operating as a weight classifier.*
> *[Nonretroactive as of January 1, 2001]*
>
> (Added 1999)

S.1.8.5.   Recorded Representations, Point-of-Sale Systems. – The sales information recorded by cash registers when interfaced with a weighing element shall contain the following information for items weighed at the checkout stand:

(a) the net weight;[1]

(b) the unit price;[1]

(c) the total price; and

(d) the product class or, in a system equipped with price look-up capability, the product name or code number.

---

[1] For devices interfaced with scales indicating in metric units, the unit price may be expressed in price per 100 grams. Weight values shall be identified by kilograms, kg, grams, g, ounces, oz, pounds, or lb. *The "#" symbol is not acceptable.*
[Nonretroactive as of January 1, 2006]
(Amended 1995 and 2005)

S.1.9.   Prepackaging Scales.

S.1.9.1.   Value of the Scale Division. – On a prepackaging scale, the value of the intervals representing weight values shall be uniform throughout the entire reading face. The recorded weight values shall be identical with those on the indicator.

S.1.9.2.   Label Printer. – A prepackaging scale or a device that produces a printed ticket to be used as the label for a package shall print all values digitally and of such size, style of type, and color as to be clear and conspicuous on the label.

S.1.10.   Adjustable Components. – An adjustable component such as a pendulum, spring, or potentiometer shall be held securely in adjustment and, except for a zero-load balance mechanism, shall be located within the housing of the element.

(Added 1986)

S.1.11.   Provision for Sealing.

(a) *Except on Class I scales, provision shall be made for applying a security seal in a manner that requires the security seal to be broken before an adjustment can be made to any component affecting the performance of an electronic device.*
[Nonretroactive as of January 1, 1979]

(b) *Except on Class I scales, a device shall be designed with provision(s) for applying a security seal that must be broken, or for using other approved means of providing security (e.g., data change audit trail available at the time of inspection), before any change that detrimentally affects the metrological integrity of the device can be made to any electronic mechanism.*
[Nonretroactive as of January 1, 1990]

(c) *Except on Class I scales, audit trails shall use the format set forth in Table S.1.11. Categories of Device and Methods of Sealing.*
[Nonretroactive as of January 1, 1995]

A device may be fitted with an automatic or a semi-automatic calibration mechanism. This mechanism shall be incorporated inside the device. After sealing, neither the mechanism nor the calibration process shall facilitate fraud.

(Amended 1989, 1991, and 1993)

Table S.1.11. Categories of Device and Methods of Sealing	
Categories of Device	**Methods of Sealing**
Category 1: *No remote configuration capability.*	*Seal by physical seal or two event counters: one for calibration parameters and one for configuration parameters.*
Category 2: *Remote configuration capability, but access is controlled by physical hardware.* *The device shall clearly indicate that it is in the remote configuration mode and record such message if capable of printing in this mode.*	*The hardware enabling access for remote communication must be at the device and sealed using a physical seal or two event counters: one for calibration parameters and one for configuration parameters.*
Category 3: *Remote configuration capability access may be unlimited or controlled through a software switch (e.g., password).*	*An event logger is required in the device; it must include an event counter (000 to 999), the parameter ID, the date and time of the change, and the new value of the parameter. A printed copy of the information must be available through the device or through another on-site device. The event logger shall have a capacity to retain records equal to 10 times the number of sealable parameters in the device, but not more than 1000 records are required. (**Note:** Does not require 1000 changes to be stored for each parameter.)*

[Nonretroactive as of January 1, 1995]
(Table added 1993)

S.1.12. Manual Weight Entries. – *A device when being used for direct sale shall accept an entry of a manual gross or net weight value only when the scale gross or net* weight indication is at zero. Recorded manual weight entries, except those on labels generated for packages of standard weights, shall identify the weight value as a manual weight entry by one of the following terms: "Manual Weight," "Manual Wt," or "MAN WT." The use of a symbol to identify multiple manual weight entries on a single document is permitted, provided that the symbol is defined on the same page on which the manual weight entries appear and the definition of the symbol is automatically printed by the recording element as part of the document.*
[Nonretroactive as of January 1, 1993] [*Nonretroactive as of January 1, 2005]
(Added 1992) (Amended 2004)

S.1.13. Vehicle On-Board Weighing Systems: Vehicle in Motion. – When the vehicle is in motion, a vehicle on-board weighing system shall either:

(a) be accurate; or

(b) inhibit the weighing operation.

(Added 1993)

S.2. Design of Balance, Tare, Level, Damping, and Arresting Mechanisms.

S.2.1. Zero-Load Adjustment.

S.2.1.1. General. – A scale shall be equipped with means by which the zero-load balance may be adjusted. Any loose material used for this purpose shall be enclosed so that it cannot shift in position and alter the balance condition of the scale.

Except for an initial zero-setting mechanism, an automatic zero adjustment outside the limits specified in S.2.1.3. Scales Equipped with an Automatic Zero-Tracking Mechanism is prohibited.
(Amended 2010)

S.2.1.2.   Scales used in Direct Sales. – A manual zero-setting mechanism (except on a digital scale with an analog zero-adjustment mechanism with a range of not greater than one scale division) shall be operable or accessible only by a tool outside of and entirely separate from this mechanism, or it shall be enclosed in a cabinet. Except on Class I or II scales, a balance ball shall either meet this requirement or not itself be rotatable.

A semiautomatic zero-setting mechanism shall be operable or accessible only by a tool outside of and separate from this mechanism or it shall be enclosed in a cabinet, or it shall be operable only when the indication is stable within plus or minus:

(a)   3.0 scale divisions for scales of more than 2000 kg (5000 lb) capacity in service prior to January 1, 1981, and for all axle load, railway track, and vehicle scales; or

(b)   1.0 scale division for all other scales.

S.2.1.3.   Scales Equipped with an Automatic Zero-Tracking Mechanism.

S.2.1.3.1.   Automatic Zero-Tracking Mechanism for Scales Manufactured Between January 1, 1981, and January 1, 2007. – The maximum load that can be "rezeroed," when either placed on or removed from the platform all at once under normal operating conditions, shall be for:

(a)   bench, counter, and livestock scales: 0.6 scale division;

(b)   vehicle, axle load, and railway track scales: 3.0 scale divisions; and

(c)   all other scales: 1.0 scale division.
(Amended 2005)

S.2.1.3.2.   Automatic Zero-Tracking Mechanism for Scales Manufactured on or after January 1, 2007. – The maximum load that can be "rezeroed," when either placed on or removed from the platform all at once under normal operating conditions, shall be:

(a)   for vehicle, axle load, and railway track scales: 3.0 scale divisions; and

(b)   for all other scales: 0.5 scale division.
(Added 2005)

S.2.1.3.3.   Means to Disable Automatic Zero-Tracking Mechanism on Class III L Devices. – *Class III L devices equipped with an automatic zero-tracking mechanism shall be designed with a sealable means that would allow zero tracking to be disabled during the inspection and test of the device. [Nonretroactive as of January 1, 2001]*
(Added 1999) (Amended 2005)

S.2.1.4.   Monorail Scales. – On a static monorail scale equipped with digital indications, means shall be provided for setting the zero-load balance to within 0.02 % of scale capacity. On a dynamic monorail weighing system, means shall be provided to automatically maintain these conditions.
(Amended 1999)

S.2.1.5.   Initial Zero-Setting Mechanism. – Scales of accuracy Classes I, II, and III may be equipped with an initial zero-setting device.

(a) For weighing, load-receiving, and indicating elements in the same housing or covered on the same CC, an initial zero-setting mechanism shall not zero a load in excess of 20 % of the maximum capacity of the scale unless tests show that the scale meets all applicable tolerances for any amount of initial load compensated by this device within the specified range.

(b) *For indicating elements not permanently attached to weighing and load-receiving elements covered on a separate CC, the maximum initial zero-setting mechanism range of electronic indicators shall not exceed 20 % of the configured capacity.*
[Nonretroactive as of January 1, 2009]

(Added 2008)

(Added 1990) (Amended 2008)

S.2.1.6.   Combined Zero-Tare ("0/T") Key. – Scales not intended to be used in direct sales applications may be equipped with a combined zero and tare function key, provided that the device is clearly marked as to how the key functions. The device must also be clearly marked on or adjacent to the weight display with the statement "Not for Direct Sales."

(Added 1998)

S.2.2.   Balance Indicator. – On a balance indicator consisting of two indicating edges, lines, or points, the ends of the indicators shall be sharply defined. When the scale is in balance, the ends shall be separated by not more than 1.0 mm (0.04 in).

S.2.2.1.   Dairy-Product Test, Grain-Test, Prescription, and Class I and II Scales. – Except on digital indicating devices, a dairy-product test, grain-test, prescription, or Class I or II scale shall be equipped with a balance indicator. If an indicator and a graduated scale are not in the same plane, the clearance between the indicator and the graduations shall be not more than 1.0 mm (0.04 in).

***S.2.2.2.   Equal-Arm Scale.** – An equal-arm scale shall be equipped with a balance indicator. If the indicator and balance graduation are not in the same plane, the clearance between the indicator and the balance graduation shall be not more than 1.0 mm (0.04 in).*
[Nonretroactive as of January 1, 1989]

(Added 1988)

S.2.3.   Tare. – *On any scale (except a monorail scale equipped with digital indications and multi-interval scales or multiple range scales when the value of tare is determined in a lower weighing range or weighing segment), the value of the tare division shall be equal to the value of the scale division.* * The tare mechanism shall operate only in a backward direction (that is, in a direction of underregistration) with respect to the zero-load balance condition of the scale. *A device designed to automatically clear any tare value shall also be designed to prevent the automatic clearing of tare until a complete transaction has been indicated.* *
[*Nonretroactive as of January 1, 1983]

(Amended 1985 and 2008)

Note: *On a computing scale, this requires the input of a unit price, the display of the unit price, and a computed positive total price at a readable equilibrium. Other devices require a complete weighing operation, including tare, net, and gross weight determination**
[*Nonretroactive as of January 1, 1983]

S.2.3.1.   Monorail Scales Equipped with Digital Indications. – On a static monorail weighing system equipped with digital indications, means shall be provided for setting any tare value of less than 5 % of the scale capacity to within 0.02 % of scale capacity. On a dynamic monorail weighing system, means shall be provided to automatically maintain this condition.

(Amended 1999)

S.2.4.   Level-Indicating Means. – Except for portable wheel-load weighers and portable axle load scales, a portable scale shall be equipped with level-indicating means if its weighing performance is changed by an amount greater than the appropriate acceptance tolerance when it is tilted up to and including 5 % rise over run in any direction from a level position and rebalanced. The level-indicating means shall be readable without removing any scale parts requiring a tool.

[This requirement is nonretroactive as of January 1, 1986, for prescription, jewelers', and dairy-product test scales and scales marked Class I and II.]

Note: Portable wheel-load weighers and portable axle-load scales shall be accurate when tilted up to and including 5 % rise over run in any direction from a level position and rebalanced.

(Amended 1991 and 2008)

S.2.4.1.   Vehicle On-Board Weighing Systems. – A vehicle on-board weighing system shall operate within tolerance when the weighing system is tilted up to and including 5 % rise over run in any direction from a level position and rebalanced. If the accuracy of the system is affected by out-of-level conditions normal to the use of the device, the system shall be equipped with an out-of-level sensor that inhibits the weighing operation when the system is out of level to the extent that the accuracy limits are exceeded.

(Added 1992) (Amended 2008)

S.2.5.   Damping Means. – An automatic-indicating scale and a balance indicator shall be equipped with effective means to damp oscillations and to bring the indicating elements quickly to rest.

S.2.5.1.   Digital Indicating Elements. – Digital indicating elements equipped with recording elements shall be equipped with effective means to permit the recording of weight values only when the indication is stable within plus or minus:

(a) 3.0 scale divisions for scales of more than 2000 kg (5000 lb) capacity in service prior to January 1, 1981, hopper (other than grain hopper) scales with a capacity exceeding 22 000 kg (50 000 lb), and for all vehicle, axle load, livestock, and railway track scales; and

(b) 1.0 scale division for all other scales.

The values recorded shall be within applicable tolerances.

(Amended 1995)

S.2.5.2.   Jewelers', Prescription, and Class I, and Class II Scales. – A jewelers', prescription, Class I, or Class II scales shall be equipped with appropriate means for arresting the oscillation of the mechanism.

S.2.5.3.   Class I and Class II Prescription Scales with a Counting Feature. – A Class I or Class II prescription scale shall indicate to the operator when the piece weight computation is complete by a stable display of the quantity placed on the load-receiving element.

(Added 2003)

S.3.   Design of Load-Receiving Elements.

S.3.1.   Travel of Pans of Equal-Arm Scale. – The travel between limiting stops of the pans of a nonautomatic-indicating equal-arm scale not equipped with a balance indicator shall be not less than the minimum travel shown in Table 2M. and Table 2.

Table 2M. Minimum Travel of Pans of Nonautomatic Indicating Equal-Arm Scale without Balance Indicator	
Nominal Capacity (kilograms)	Minimum Travel of Pans (millimeters)
2 or less	9
2+ to 5, inclusive	13
5+ to 12, inclusive	19
Over 12	25

Table 2. Minimum Travel of Pans of Nonautomatic Indicating Equal-Arm Scale without Balance Indicator	
Nominal Capacity (pounds)	Minimum Travel of Pans (inch)
4 or less	0.35
4+ to 12, inclusive	0.5
12+ to 26, inclusive	0.75
Over 26	1.0

S.3.2.   Drainage. – A load-receiving element intended to receive wet commodities shall be so constructed as to drain effectively.

S.3.3.   Scoop Counterbalance. – A scoop on a scale used for direct sales to retail customers shall not be counterbalanced by a removable weight. A permanently attached scoop-counterbalance shall indicate clearly on both the operator's and customer's sides of the scale whether it is positioned for the scoop to be on or off the scale.

S.4. Design of Weighing Elements.

S.4.1.   Antifriction Means. – Frictional effects shall be reduced to a minimum by suitable antifriction elements. Opposing surfaces and points shall be properly shaped, finished, and hardened. A platform scale having a frame around the platform shall be equipped with means to prevent interference between platform and frame.

S.4.2.   Adjustable Components. – An adjustable component such as a nose-iron or potentiometer shall be held securely in adjustment. The position of a nose-iron on a scale of more than 1000 kg (2000 lb) capacity, as determined by the factory adjustment, shall be accurately, clearly, and permanently defined.
(Amended 1986)

S.4.3.   Multiple Load-Receiving Elements. – Except for mechanical bench and counter scales, a scale with a single indicating or recording element, or a combination indicating-recording element, that is coupled to two or more load-receiving elements with independent weighing systems, shall be provided with means to prohibit the activation of any load-receiving element (or elements) not in use, and shall be provided with automatic means to indicate clearly and definitely which load-receiving element (or elements) is in use.

S.5. Design of Weighing Devices, Accuracy Class.

S.5.1.   Designation of Accuracy Class. – Weighing devices are divided into accuracy classes and shall be designated as I, II, III, III L, or IIII.
[Nonretroactive as of January 1, 1986]

S.5.2.   Parameters for Accuracy Class. – The accuracy class of a weighing device is designated by the manufacturer and shall comply with parameters shown in Table 3.
[Nonretroactive as of January 1, 1986]

S.5.3.   Multi-Interval and Multiple Range Scales, Division Value. – On a multi-interval scale and multiple range scale, the value of "e" shall be equal to the value of "d."[2]

(Added 1986) (Amended 1995)

S.5.4.   Relationship of Minimum Load Cell Verification Interval Value to the Scale Division. – The relationship of the value for the minimum load cell verification scale interval, v_{min}, to the scale division, d, for a specific scale using National Type Evaluation Program (NTEP) certified load cells shall comply with the following formulae where N is the number of load cells in a single independent[1] weighing/load-receiving element (such as hopper, railroad track, or vehicle scale weighing/load-receiving elements):

(a) $v_{min} \leq \dfrac{d*}{\sqrt{N}}$ *for scales without lever systems; and*

(b) $v_{min} \leq \dfrac{d*}{\sqrt{N} \times (scale\,multiple)}$ *for scales with lever systems.*

[1]*"Independent" means with a weighing/load-receiving element not attached to adjacent elements and with its own A/D conversion circuitry and displayed weight.*

[*When the value of the scale division, d, is different from the verification scale division, e, for the scale, the value of e must be used in the formulae above.]

This requirement does not apply to complete weighing/load-receiving elements or scales, which satisfy all the following criteria:

-    *the complete weighing/load-receiving element or scale has been evaluated for compliance with T.N.8.1. Temperature under the NTEP;*

-    *the complete weighing/load-receiving element or scale has received an NTEP Certificate of Conformance; and*

-    *the complete weighing/load-receiving element or scale is equipped with an automatic zero-tracking mechanism which cannot be made inoperative in the normal weighing mode. (A test mode which permits the disabling of the automatic zero-tracking mechanism is permissible, provided the scale cannot function normally while in this mode.*

[Nonretroactive as of January 1, 1994]

(Added 1993) (Amended 1996 and 2016)

---

[2] Footnote 1 to Table 3 Parameters for Accuracy Classes.

Class	Value of the Verification Scale Division (d or e[1])	Number of Scale[4] Divisions (n)	
		Minimum	Maximum
Table 3. **Parameters for Accuracy Classes**			
SI Units			
I	equal to or greater than 1 mg	50 000	--
II	1 to 50 mg, inclusive	100	100 000
	equal to or greater than 100 mg	5 000	100 000
III[2,5]	0.1 to 2 g, inclusive	100	10 000
	equal to or greater than 5 g	500	10 000
III L[3]	equal to or greater than 2 kg	2 000	10 000
IIII	equal to or greater than 5 g	100	1 200
U.S. Customary Units			
III[5]	0.0002 lb to 0.005 lb, inclusive	100	10 000
	0.005 oz to 0.125 oz, inclusive	100	10 000
	equal to or greater than 0.01 lb	500	10 000
	equal to or greater than 0.25 oz	500	10 000
III L[3]	equal to or greater than 5 lb	2 000	10 000
IIII	greater than 0.01 lb	100	1 200
	greater than 0.25 oz	100	1 200

[1] For Class I and II devices equipped with auxiliary reading means (i.e., a rider, a vernier, or a least significant decimal differentiated by size, shape, or color), the value of the verification scale division "e" is the value of the scale division immediately preceding the auxiliary means.

[2] A Class III scale marked "For prescription weighing only" may have a verification scale division (e) not less than 0.01 g.
(Added 1986) (Amended 2003)

[3] The value of a scale division for crane and hopper (other than grain hopper) scales shall be not less than 0.2 kg (0.5 lb). The minimum number of scale divisions shall be not less than 1000.

[4] On a multiple range or multi-interval scale, the number of divisions for each range independently shall not exceed the maximum specified for the accuracy class. The number of scale divisions, n, for each weighing range is determined by dividing the scale capacity for each range by the verification scale division, e, for each range. On a scale system with multiple load-receiving elements and multiple indications, each element considered shall not independently exceed the maximum specified for the accuracy class. If the system has a summing indicator, the n_{max} for the summed indication shall not exceed the maximum specified for the accuracy class.
(Added 1997)

[5] The minimum number of scale divisions for a Class III Hopper Scale used for weighing grain shall be 2000.)

[*Nonretroactive as of January 1, 1986*]

(Amended 1986, 1987, 1997, 1998, 1999, 2003, and 2004)

S.6.   Marking Requirements. – (Also see G-S.1. Identification, G-S.4. Interchange or Reversal of Parts, G-S.6. Marking Operational Controls, Indications, and Features, G-S.7. Lettering, G-UR.2.1.1. Visibility of Identification, and UR.3.4.1. Use in Pairs.)

S.6.1.   Nominal Capacity; Vehicle and Axle-Load Scales. – *For all vehicle and axle-load scales, the marked nominal capacity shall not exceed the concentrated load capacity (CLC) times the quantity of the number of sections in the scale minus 0.5.*

As a formula, this is stated as: nominal capacity ≤ CLC × (N − 0.5)
where N = the number of sections in the scale.
[Nonretroactive as of January 1, 1989]

Note: When the device is used in a combination railway track and vehicle weighing application, the above formula shall apply only to the vehicle scale application.

(Added 1988) (Amended 1999 and 2002)

S.6.2.   Location of Marking Information. – Scales that are not permanently attached to an indicating element, and for which the load-receiving element is the only part of the weighing/load-receiving element visible after installation, may have the marking information required in Section 1.10. General Code, G-S.1. Identification and Section 2.20. Scales Code, S.6. Marking Requirements located in an area that is accessible only through the use of a tool; provided that the information is easily accessible (e.g., the information may appear on the junction box under an access plate). The identification information for these scales shall be located on the weighbridge (load-receiving element) near the point where the signal leaves the weighing element or beneath the nearest access cover.

(Added 1989)

S.6.3.   Scales, Main Elements, and Components of Scales or Weighing Systems. – Scales, main elements of scales when not contained in a single enclosure for the entire scale, load cells for which Certificates of Conformance (CC) have been issued under the National Type Evaluation Program (NTEP), and other equipment necessary to a weighing system, but having no metrological effect on the weighing system, shall be marked as specified in Table S.6.3.a. Marking Requirements and explained in the accompanying notes in Table S.6.3.b. Notes for Table S.6.3.a.

(Added 1990)

Table S.6.3.a. Marking Requirements					
	Weighing Equipment				
To Be Marked With ⇓	**Weighing, Load-Receiving, and Indicating Element in Same Housing or Covered on the Same CC[1]**	**Indicating Element not Permanently Attached to Weighing and Load-Receiving Element or Covered by a Separate CC**	**Weighing and Load-Receiving Element Not Permanently Attached to Indicating Element or Covered by a Separate CC**	**Load Cell with CC (11)**	**Other Equipment or Device (10)**
Manufacturer's ID (1)	X	X	X	X	X
Model Designation and Prefix (1)	X	X	X	X	X
Serial Number and Prefix (2)	X	X	X	X	X (16)
Certificate of Conformance Number (CC) (23)	X	X	X	X	X (23)
Accuracy Class (17)	X	X (8)	X (19)	X	
Nominal Capacity (3)(18)(20)	X	X	X		
Value of Scale Division, "d" (3)	X	X			
Value of "e" (4)	X	X			
Temperature Limits (5)	X	X	X	X	
Concentrated Load Capacity (CLC) (12)(20)(22)		X	X (9)		
Special Application (13)	X	X	X		
Maximum Number of Scale Divisions (n_{max}) (6)		X (8)	X (19)	X	
Minimum Verification Scale Division (e_{min})			X (19)		
"S" or "M" (7)				X	
Direction of Loading (15)				X	
Minimum Dead Load				X	
Maximum Capacity				X	
Safe Load Limit				X	
Load Cell Verification Interval (v_{min}) (21)				X	
Section Capacity and Prefix (14)(20)(22)(24)		X	X		

Table S.6.3.a.
Marking Requirements

Note: For applicable notes, Table S.6.3.b.
[1] Weighing/load-receiving elements and indicators which are in the same housing or which are permanently attached will generally appear on the same CC. If not in the same housing, elements shall be hard-wired together or sealed with a physical seal or an electronic link. This requirement does not apply to peripheral equipment that has no input or effect on device calibrations or configurations.

(Added 2001)

(Added 1990) (Amended 1992, 1999, 2000, 2001, 2002, and 2004)

Table S.6.3.b.
Notes for Table S.6.3.a. Marking Requirements

1. Manufacturer's identification and model designation and *model designation prefix.* *
    [*Nonretroactive as of January 1, 2003]
    (Also see G-S.1. Identification.) *[Prefix lettering may be initial capitals, all capitals or all lower case]*
    (Amended 2000)

2. *Serial number [Nonretroactive as of January 1, 1968] and prefix [Nonretroactive as of January 1, 1986].* (Also see G-S.1. Identification.)

3. The device shall be marked with the nominal capacity. *The nominal capacity shall be shown together with the value of the scale division (e.g., 15 × 0.005 kg, 30 × 0.01 lb, or capacity = 15 kg, d = 0.005 kg) in a clear and conspicuous manner and be readily apparent when viewing the reading face of the scale indicator unless already apparent by the design of the device. Each scale division value or weight unit shall be marked on multiple range or multi-interval scales.* *[Nonretroactive as of January 1, 1983]*
    (Amended 2005)

4. *Required only if different from "d."*
    [Nonretroactive as of January 1, 1986]

5. *Required only on Class III, III L, and IIII devices if the temperature range on the NTEP CC is narrower than and within − 10 °C to 40 °C (14 °F to 104 °F). [Nonretroactive as of January 1, 1986]*
    (Amended 1999)

6. *This value may be stated on load cells in units of 1000; e.g., n: 10 is 10 000 divisions.*
    [Nonretroactive as of January 1, 1988]

7. *Denotes compliance for single or multiple load cell applications. It is acceptable to use a load cell with the "S" or Single Cell designation in multiple load cell applications as long as all other parameters meet applicable requirements. A load cell with the "M" or Multiple Cell designation can be used only in multiple load cell applications.* *[Nonretroactive as of January 1, 1988]*
    (Amended 1999)

8. *An indicating element not permanently attached to a weighing element shall be clearly and permanently marked with the accuracy Class of I, II, III, III L, or IIII, as appropriate, and the maximum number of scale divisions, n_{max}, for which the indicator complies with the applicable requirement. Indicating elements that qualify for use in both Class III and III L applications may be marked III/III L and shall be marked with the maximum number of scale divisions for which the device complies with the applicable requirements for each accuracy class.* *[Nonretroactive as of January 1, 1988]*

9. *For vehicle and axle-load scales only. The CLC shall be added to the load-receiving element of any such scale not previously marked at the time of modification.*
[Nonretroactive as of January 1, 1989]

(Amended 2002)

10. Necessary to the weighing system but having no metrological effect, e.g., auxiliary remote display, keyboard, etc.

11. *The markings may be either on the load cell or in an accompanying document; except that, if an accompanying document is provided, the serial number shall appear both on the load cell and in the document. [Nonretroactive as of January 1, 1988] The manufacturer's name or trademark, the model designation, and identifying symbols for the model and serial numbers as required by paragraph G-S.1. Identification shall also be marked both on the load cell and in any accompanying document.*
[Nonretroactive as of January 1, 1991]

12. Required on the indicating element *and the load-receiving element* of vehicle and axle-load scales. *Such marking shall be identified as "concentrated load capacity" or by the abbreviation "CLC."**
[*Nonretroactive as of January 1, 1989]

(Amended 2002)

13. *A scale designed for a special application rather than general use shall be conspicuously marked with suitable words, visible to the operator and to the customer, restricting its use to that application, e.g., postal scale, prepack scale, weight classifier, etc.** When a scale is installed with an operational counting feature, the scale shall be marked on both the operator and customer sides with the statement "The counting feature is not legal for trade," except when a Class I or Class II prescription scale complies with all Handbook 44 requirements applicable to counting features.
[*Nonretroactive as of 1986]

(Amended 1994 and 2003)

14. Required on *livestock** and railway track scales. When marked on vehicle and axle-load scales manufactured before January 1, 1989, it may be used as the CLC. For livestock scales manufactured between January 1, 1989, and January 1, 2003, required markings may be either CLC or section capacity.
[*Nonretroactive as of January 1, 2003]

(Amended 2002)

15. *Required if the direction of loading the load cell is not obvious.*
[Nonretroactive as of January 1, 1988]

16. *Serial number [Nonretroactive as of January 1, 1968] and prefix [Nonretroactive as of January 1, 1986].* (Also see G-S.1. Identification.) Modules without "intelligence" on a modular system (e.g., printer, keyboard module, cash drawer, and secondary display in a point-of-sale system) are not required to have serial numbers.

17. *The accuracy class of a device shall be marked on the device with the appropriate designation as I, II, III, III L, or IIII.*
[Nonretroactive as of January 1, 1986]

18. The nominal capacity shall be conspicuously marked as follows:
    (a) on any scale equipped with unit weights or weight ranges;
    (b) on any scale with which counterpoise or equal-arm weights are intended to be used;
    (c) on any automatic-indicating or recording scale so constructed that the capacity of the indicating or recording element, or elements, is not immediately apparent;
    (d) on any scale with a nominal capacity less than the sum of the reading elements; and
    (e) on the load-receiving element (weighbridge) of vehicle, axle-load, and livestock scales.*
    [*Nonretroactive as of January 1, 1989]

(Amended 1992)

19. *For weighing and load-receiving elements not permanently attached to indicating element or covered by a separate CC.*
[Nonretroactive as of January, 1, 1988]

(Amended 1992)

Table S.6.3.b.
Notes for Table S.6.3.a. Marking Requirements

20. *Combination vehicle/railway track scales must be marked with both the nominal capacity and CLC for vehicle weighing and the nominal capacity and section capacity for railway weighing. All other requirements relating to these markings will apply.*
    [Nonretroactive as of January 1, 2000]

    (Added 1999)

21. *The value of the load cell verification interval (v_{min}) must be stated in mass units. In addition to this information, a device may be marked with supplemental representations of v_{min}.*
    [Nonretroactive as of January 1, 2001]

    (Added 1999)

22. *Combination vehicle/livestock scales must be marked with both the CLC for vehicle weighing and the section capacity for livestock weighing. All other requirements relative to these markings will apply.*
    [Nonretroactive as of January 1, 2003]

    (Added 2002) (Amended 2003)

    Note: The marked section capacity for livestock weighing may be less than the marked CLC for vehicle weighing.

    (Amended 2003)

23. *Required only if a CC has been issued for the device or equipment.*
    [Nonretroactive as of January 1, 2003]

    (G-S.1. Identification (e) Added 2001)

24. *The section capacity shall be prefaced by the words "Section Capacity" or an abbreviation of that term. Abbreviations shall be "Sec Cap" or "Sec C." All capital letters and periods may be used.*
    [Nonretroactive as of January 1, 2005]

    (Added 2004)

S.6.4. Railway Track Scales. – A railway track scale shall be marked with the maximum capacity of each section of the load-receiving element of the scale. Such marking shall be accurately and conspicuously presented on, or adjacent to, the identification or nomenclature plate that is attached to the indicating element of the scale. The nominal capacity marking shall satisfy the following:

(a) For scales manufactured from January 1, 2002, through December 31, 2013:

    (1) the nominal capacity of a scale with more than two sections shall not exceed twice its rated section capacity; and

    (2) the nominal capacity of a two section scale shall not exceed its rated section capacity.

(b) For scales manufactured on or after January 1, 2014, the nominal scale capacity shall not exceed the lesser of:

    (1) the sum of the Weigh Module Capacities as shown in Table S.6.4.M. and Table S.6.4.; or

    (2) the Rated Section Capacity (RSC) multiplied by the Number of Sections (Ns) minus the Number of Dead Spaces (Nd) minus 0.5. As a formula this is stated as:

$$RSC \times (Ns - Nd - 0.5); \text{ or}$$

    (3) 290 300 kg (640 000 lb).

(Amended 1988, 2001, 2002, and 2013)

Table S.6.4.M.
Railway Track Scale – Weigh Module Capacity

Weigh Module Length (meters)	Weigh Module Capacity (kilograms)
< 1.5	36 300
1.5 to < 3.0	72 600
3.0 to < 4.5	108 900
4.5 to < 7.0	145 100
7.0 to < 9.0	168 700
9.0 to < 10.5	192 300
10.5 to < 12.0	234 100
12.0 to < 17.0	257 600
Note: The capacity of a particular module is based on its length as shown above. To determine the "sum of the weigh module capacities" referenced in paragraph S.6.4.(b)(1): (1) determine the length of each individual weigh module in the scale; (2) find its corresponding "weigh module capacity" in the table above; and (3) add all of the individual weigh module capacities."	

(Table Added 2013)

Table S.6.4.
Railway Track Scale – Weigh Module Capacity

Weigh Module Length (feet)	Weigh Module Capacity (pounds)
< 5	80 000
5 to < 10	160 000
10 to < 15	240 000
15 to < 23	320 000
23 to < 29	372 000
29 to < 35	424 000
35 to < 40	516 000
40 to < 56	568 000
Note: The capacity of a particular module is based on its length as shown above. To determine the "sum of the weigh module capacities" referenced in paragraph S.6.4.(b)(1): (1) determine the length of each individual weigh module in the scale; (2) find its corresponding "weigh module capacity" in the table above; and (3) add all of the individual weigh module capacities."	

(Table Added 2013)

S.6.5.   Livestock Scales. – A livestock scale manufactured prior to January 1, 1989, or after January 1, 2003, shall be marked with the maximum capacity of each section of the load-receiving element of the scale. Livestock scales manufactured between January 1, 1989, and January 1, 2003, shall be marked with either the Concentrated Load Capacity (CLC) or the Section Capacity. Such marking shall be accurately and conspicuously presented on, or adjacent to the identification or nomenclature plate that is attached to the indicating element of the scale. *The* *nominal capacity of a scale with more than two sections shall not exceed twice its rated section capacity. The nominal capacity of a two-section scale shall not exceed its rated section capacity.**

[*Nonretroactive as of January 1, 2003]

(Added 2002)

Also see Note 14 in Table S.6.3.b. Notes for Table S.6.3.a.

S.6.6.   Counting Feature, Minimum Individual Piece Weight, and Minimum Sample Piece Count. – A Class I or Class II prescription scale with an operational counting feature shall be marked with the minimum individual piece weight and minimum number of pieces used in the sample to establish an individual piece weight.

(Added 2003)

N. Notes

N.1.   Test Procedures.

N.1.1.   Increasing-Load Test. – The increasing-load test shall be conducted on all scales with the test loads approximately centered on the load-receiving element of the scale, except on a scale having a nominal capacity greater than the total available known test load. When the total test load is less than the nominal capacity, the test load is used to greatest advantage by concentrating it, within prescribed load limits, over the main load supports of the scale.

N.1.2.   Decreasing-Load Test (Automatic Indicating Scales). – The decreasing-load test shall be conducted with the test load approximately centered on the load-receiving element of the scale.

N.1.2.1.   Scales Marked I, II, III, or IIII. – Except for portable wheel load weighers, decreasing-load tests shall be conducted on scales marked I, II, III or IIII and with "n" equal to or greater than 1000 with test loads equal to the maximum test load at each tolerance value. For example, on a Class III scale, at test loads equal to 4000 d, 2000 d, and 500 d; for scales with n less than 1000, the test load shall be equal to one-half of the maximum load applied in the increasing-load test. (Also see Table 6. Maintenance Tolerances.)

(Amended 1998)

N.1.2.2.   All Other Scales. – On all other scales, except for portable wheel load weighers, the decreasing-load test shall be conducted with a test load equal to one-half of the maximum load applied in the increasing-load test.

(Amended 1998)

N.1.3.   Shift Test.

N.1.3.1.   Dairy-Product Test Scales. – A shift test shall be conducted with a test load of 18 g successively positioned at all points on which a weight might reasonably be placed in the course of normal use of the scale.

N.1.3.2.   Equal-Arm Scales. – A shift test shall be conducted with a half-capacity test load centered successively at four points positioned equidistance between the center and the front, left, back, and right edges of each pan as shown in the diagrams below. An equal test load shall be centered on the other pan.

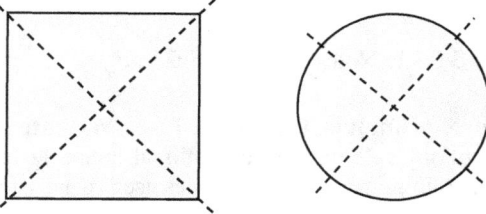

N.1.3.3.   Vehicle Scales, Axle-Load Scales, and Livestock Scales.

   N.1.3.3.1.   Vehicle Scales, Axle-Load Scales, and Combination Vehicle/Livestock Scales.

    (a) **Minimum Shift Test.** – At least one shift test shall be conducted with a minimum test load of 12.5 % of scale capacity, which may be performed anywhere on the load-receiving element using the prescribed test patterns and maximum test loads specified below. (Combination Vehicle/Livestock Scales shall also be tested consistent with N.1.3.3.2. Prescribed Test Pattern and Test Loads for Livestock Scales with More Than Two Sections and Combination Vehicle/Livestock Scales.)

    (Amended 1991, 2000, and 2003)

    (b) **Prescribed Test Pattern and Loading for Vehicle Scales, Axle-Load Scales, and Combination Vehicle/Livestock Scales.** – The normal prescribed test pattern shall be an area of 1.2 m (4 ft) in length and 3.0 m (10 ft) in width or the width of the scale platform, whichever is less. Multiple test patterns may be utilized when loaded in accordance with paragraph (c), (d), or (e) as applicable. An example of a possible test pattern is shown in the diagram below.

    (Amended 1997, 2001, and 2003)

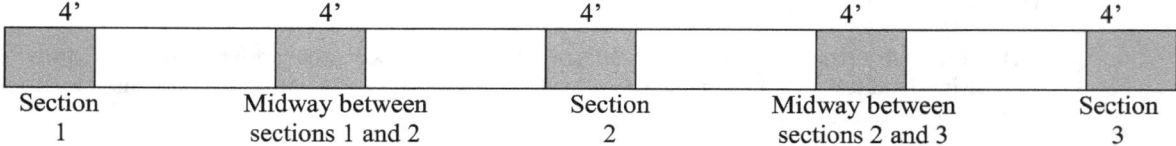

    (c) **Loading Precautions for Vehicle Scales, Axle-Load Scales, and Combination Vehicle/Livestock Scales.** – When loading the scale for testing, one side of the test pattern shall be loaded to no more than half of the concentrated load capacity or test load before loading the other side. The area covered by the test load may be less than 1.2 m (4 ft) × 3.0 m (10 ft) or the width of the scale platform, whichever is less; for test patterns less than 1.2 m (4 ft) in length the maximum loading shall meet the formula: [(wheel base of test cart or length of test load divided by 48 in) × 0.9 × CLC]. The maximum test load applied to each test pattern shall not exceed the concentrated load capacity of the scale. When the test pattern exceeds 1.2 m (4 ft), the maximum test load applied shall not exceed the concentrated load capacity times the largest "r" factor in Table UR.3.2.1. Span Maximum Load for the length of the area covered by the test load. For load-receiving elements installed prior to January 1, 1989, the rated section capacity may be substituted for concentrated load capacity to determine maximum loading. An example of a possible test pattern is shown above.

    (Amended 1997 and 2003)

    (d) **Multiple Pattern Loading.** – To test to the nominal capacity, multiple patterns may be simultaneously loaded in a manner consistent with the method of use.

(e) **Other Designs.** – Special design scales and those that are wider than 3.7 m (12 ft) shall be tested in a manner consistent with the method of use but following the principles described above.

(Amended 1988, 1991, 1997, 2000, 2001, and 2003)

(Amended 2003)

N.1.3.3.2.   Prescribed Test Pattern and Test Loads for Livestock Scales with More Than Two Sections and Combination Vehicle/Livestock Scales. – A minimum test load of 5 000 kg (10 000 lb) or one-half of the rated section capacity, whichever is less, shall be placed, as nearly as possible, successively over each main load support as shown in the diagram below. For livestock scales manufactured between January 1, 1989, and January 1, 2003, the required loading shall be no greater than one-half CLC.

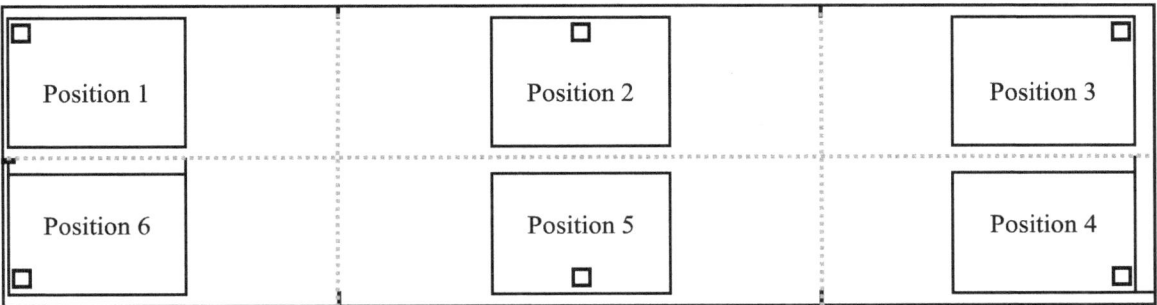

☐ = Load Bearing Point

(Added 2003) (Amended 2016)

N.1.3.3.3.   Prescribed Test Patterns and Test Loads for Two-Section Livestock Scales. – A shift test shall be conducted using the following prescribed test loads and test patterns: 1) When a single field standard weight is used, the prescribed test load shall be applied centrally in the prescribed test pattern; or 2) When multiple field standard weights are used as the prescribed test load, the load shall be applied in a consistent pattern in the shift test positions throughout the test and applied in a manner that does not concentrate the load in a test pattern that is less than when the same load is a single field standard weights on the load-receiving element.

The shift test load shall not exceed one-half the rated section capacity or one-half the rated concentrated load capacity whichever is applicable, using either:

(a) A one-half nominal capacity test load centered as nearly as possible, successively at the center of each quarter of the load-receiving element as shown in N.1.3.7. All Other Scales Except Crane Scales, Hanging Scales, Hopper Scales, Wheel-Load Weighers, and Portable Axle-Load Weighers Figure 1; or

(b) A one-quarter nominal capacity test load centered as nearly as possible, successively over each main load support as shown in N.1.3.7. All Other Scales Except Crane Scales, Hanging Scales, Hopper Scales, Wheel-Load Weighers, and Portable Axle-Load Weighers Figure 2.

(Added 2007) (Amended 2016)

N.1.3.4.   Railway Track Scales Weighing Individual Cars in Single Drafts. – A shift test shall be conducted with at least two different test loads, if available, distributed over, to the right and left of, each pair of main levers or other weighing elements supporting each section of the scale.

N.1.3.5. Monorail Scales, Static Test. – A shift test shall be conducted with a test load equal to the largest load that can be anticipated to be weighed in a given installation, but never less than one-half scale capacity. The load shall be placed successively on the right end, the left end, and the center of the live rail.

(Added 1985)

N.1.3.5.1. Dynamic Monorail Weighing Systems. – Dynamic tests with livestock carcasses or portions of carcasses shall be conducted during normal plant production. No less than 20 test loads using carcasses or portions of carcasses of the type normally weighed shall be used in the dynamic test. If the plant conveyor chain does not space or prevent the carcasses or portions of carcasses from touching one another, dynamic tests shall not be conducted until this condition has been corrected.

All carcasses or portions of carcasses shall be individually weighed statically on either the same scale being tested dynamically or another monorail scale with the same or smaller divisions and in close proximity. (The scale selected for static weighing of the carcasses or portions of carcasses shall first be tested statically with certified test weights that have been properly protected from the harsh environment of the packing plant to ensure they maintain accuracy.)

If the scale being tested is used for weighing freshly slaughtered animals (often referred to as a "hot scale"), care must be taken to get a static weighment as quickly as possible before or following the dynamic weighment to avoid loss due to shrink. If multiple dynamic tests are conducted using the same carcasses or portions of carcasses, static weights shall be obtained before and after multiple dynamic tests. If the carcass or portion of a carcass changes weight between static tests, the amount of weight change shall be taken into account, or the carcass or portion of a carcass shall be disregarded for tolerance purposes.

Note: For a dynamic monorail test, the reference scale shall comply with the principles in the Fundamental Considerations paragraph 3.2. Tolerances for Standards.

(Added 1996) (Amended 1999 and 2007)

N.1.3.6. Vehicle On-Board Weighing Systems. – The shift test for a vehicle on-board weighing system shall be conducted in a manner consistent with its normal use. For systems that weigh as part of the lifting cycle, the center of gravity of the load may be shifted in the vertical direction as well as from side to side. In other cases, the center of gravity may be moved to the extremes of the load-receiving element using loads of a magnitude that reflect normal use (i.e., the load for the shift test may exceed one-half scale capacity), and may, in some cases, be equal to the capacity of the scale. The shift test may be conducted when the weighing system is out of level to the extent that the weighing system remains operational.

(Added 1992)

N.1.3.7. All Other Scales Except Crane Scales, Hanging Scales, Hopper Scales, Wheel-Load Weighers, and Portable Axle-Load Weighers. – A shift test shall be conducted using the following prescribed test loads and test patterns. A single field standard weight used as the prescribed test load shall be applied centrally in the prescribed test pattern. When multiple field standard weights are used as the prescribed test load, the load shall be applied in a consistent pattern in the shift test positions throughout the test and applied in a manner that does not concentrate the load in a test pattern that is less than when that same load is a single field standard weight on the load-receiving element.

(a) For scales with a nominal capacity of 500 kg (1000 lb) or less, a shift test shall be conducted using a one-third nominal capacity test load (defined as test weights in amounts of at least 30 % of scale capacity, but not to exceed 35 % of scale capacity) centered as nearly as possible at the center of each quadrant of the load-receiving element using the prescribed test pattern as shown in Figure 1.

(b) For scales with a nominal capacity greater than 500 kg (1000 lb), a shift test may be conducted by either using a one-third nominal capacity test load (defined as test weights in amounts of at least 30 % of scale capacity, but not to exceed 35 % of scale capacity) centered as nearly as possible at the center of each quadrant of the load-receiving element using the prescribed test pattern as shown in Figure 1, or by using a one-quarter nominal capacity test load centered as nearly as possible, successively, over each corner of the load-receiving element using the prescribed test pattern as shown in Figure 2.

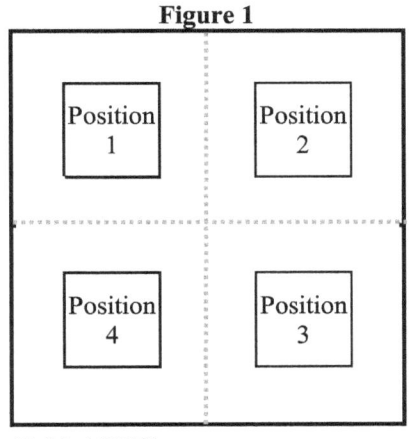

Figure 1

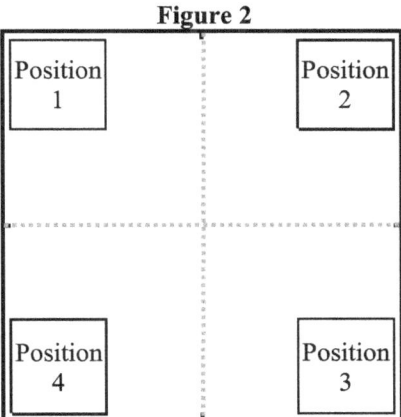

Figure 2

(Added 2003)

(Amended 1987, 2003, and 2007)

N.1.4.   Sensitivity Test. – A sensitivity test shall be conducted on nonautomatic-indicating (weighbeam) scales only, with the weighing device in equilibrium at zero-load and at maximum test load. The test shall be conducted by increasing or decreasing the test load in an amount equal to the applicable value specified in T.2. Sensitivity Requirement (SR) or T.N.6. Sensitivity.

N.1.5.   Discrimination Test. – A discrimination test shall be conducted on all automatic indicating scales with the weighing device in equilibrium at or near zero load and at or near maximum test load, and under controlled conditions in which environmental factors are reduced to the extent that they will not affect the results obtained. For scales equipped with an Automatic Zero-Tracking Mechanism (AZT), the discrimination test may be conducted at a range outside of the AZT range.
[Nonretroactive as of January 1, 1986]

(Added 1985) (Amended 2004)

N.1.5.1.   Digital Device. – On a digital device, this test is conducted from just below the lower edge of the zone of uncertainty for increasing load tests, or from just above the upper edge of the zone of uncertainty for decreasing-load tests.

N.1.6.   RFI Susceptibility Tests, Field Evaluation. – An RFI test shall be conducted at a given installation when the presence of RFI has been verified and characterized if those conditions are considered "usual and customary."

(Added 1986)

N.1.7.   Ratio Test. – A ratio test shall be conducted on all scales employing counterpoise weights and on nonautomatic-indicating equal-arm scales.

N.1.8.   Material Tests. – A material test shall be conducted on all customer-operated bulk weighing systems for recycled materials using bulk material for which the device is used. Insert into the device, in a normal manner, several accurately pre-weighed samples (free of foreign material) in varying amounts approximating average drafts.

N.1.9.   Zero-Load Balance Change. – A zero-load balance change test shall be conducted on all scales after the removal of any test load. The zero-load balance should not change by more than the minimum tolerance applicable. (Also see G-UR.4.2. Abnormal Performance.)

N.1.10. Counting Feature Test. – A test of the counting function shall be conducted on all Class I and Class II prescription scales having an active counting feature used in "legal for trade" applications. The test should verify that the scale will not accept a sample with less than either the minimum sample piece count or the minimum sample weight of 30 e. Counting feature accuracy should be verified at a minimum of two test loads. Verification of the count calculations shall be based upon the weight indication of the test load.

Note:

(1) The minimum sample weight is equal to the marked minimum individual piece weight times the marked minimum sample piece count.

(2) Test load as used in this section refers to actual calibration test weights selected from an appropriate test weight class.

(Added 2003)

N.1.11. Substitution Test. – In the substitution test procedure, material or objects are substituted for known test weights, or a combination of known test weights and previously quantified material or objects, using the scale under test as a comparator. Additional test weights or other known test loads may be added to the known test load to evaluate higher weight ranges on the scale.

(Added 2003)

N.1.12. Strain-Load Test. – In the strain-load test procedure, an unknown quantity of material or objects are used to establish a reference load or tare to which test weights or substitution test loads are added.

(Added 2003)

N.2. Verification (Testing) Standards. – Field standard weights used in verifying weighing devices shall comply with requirements of NIST Handbook 105-Series standards (or other suitable and designated standards) or the tolerances expressed in Fundamental Considerations, paragraph 3.2. (i.e., one-third of the smallest tolerance applied).

(Amended 1986)

N.3. Minimum Test Weights and Test Loads. – The minimum test weights and test loads for in-service tests (except railway track scales) are shown in Table 4. (Also see Footnote 2 in Table 4. Minimum Test Weights and Test Loads.)

(Added 1984) (Amended 1988)

N.3.1. Minimum Test-Weight Load and Tests for Railway Track Scales.

(Amended 1990 and 2012)

N.3.1.1. Initial and Subsequent Tests. – The test-weight load shall be not less than 35 000 kg (80 000 lb). A strain-load test conducted up to the used capacity of the weighing system is recommended.

(Added 1990) (Amended 2012)

N.3.1.2. Interim Test. – An Interim Test may be used to return a railway track scale into temporary service following repairs that could affect the accuracy of the weighing system providing all of the following conditions are met:

(a) a test weight load of not less than 13 500 kg (30 000 lb) shall be used;

(b) a shift (section) test shall be conducted using a test-weight load of not less than 13 500 kg (30 000 lb);

(c) a strain-load test shall be conducted up to at least 25 % of scale capacity;

(d) all test results shall be within applicable tolerances; and

(e) the official with statutory authority shall be immediately notified when scales are repaired and placed in temporary service with an Interim Test. The length of temporary service following repair is at the discretion of the official with statutory authority.

(Added 1990) (Amended 2012)

N.3.1.3.   Enforcement Action for Inaccuracy. – To take enforcement action on a scale that is found to be inaccurate, a minimum test load of 13 500 kg (30 000 lb) must be used.

(Added 1990)

<table>
<tr><td colspan="6" align="center">Table 4.
Minimum Test Weights and Test Loads[1]</td></tr>
<tr><td colspan="3" align="center">Devices in Metric Units</td><td colspan="3" align="center">Devices in U.S. Customary Units</td></tr>
<tr><td rowspan="2">Device Capacity
(kg)</td><td colspan="2" align="center">Minimums
(in terms of device capacity)</td><td rowspan="2">Device Capacity
(lb)</td><td colspan="2" align="center">Minimums
(in terms of device capacity)</td></tr>
<tr><td>Test Weights
(greater of)</td><td>Test
Loads[2]</td><td>Test Weights
(greater of)</td><td>Test
Loads[2]</td></tr>
<tr><td>0 to 150 kg</td><td>100 %</td><td></td><td>0 to 300 lb</td><td>100 %</td><td></td></tr>
<tr><td>151 to 1 500 kg</td><td>25 % or 150 kg</td><td>75 %</td><td>301 to 3 000 lb</td><td>25 % or 300 lb</td><td>75 %</td></tr>
<tr><td>1 501 to 20 000 kg</td><td>12.5 % or 500 kg</td><td>50 %</td><td>3001 to
40 000 lb</td><td>12.5 % or 1 000 lb</td><td>50 %</td></tr>
<tr><td>20 001 kg+</td><td>12.5 % or 5 000 kg</td><td>25 %[3]</td><td>40 001 lb+</td><td>12.5 % or 10 000 lb</td><td>25 %[3]</td></tr>
<tr><td colspan="6">Where practicable:

• Test weights to dial face capacity, 1000 d, or test load to used capacity, if greater than minimums specified.
• During initial verification, a scale should be tested to capacity.</td></tr>
<tr><td colspan="6">[1] If the amount of test weight in Table 4 combined with the load on the scale would result in an unsafe condition, then the appropriate load will be determined by the official with statutory authority.

[2] The term "test load" means the sum of the combination of field standard test weights and any other applied load used in the conduct of a test using substitution test methods. Not more than three substitutions shall be used during substitution testing, after which the tolerances for strain load tests shall be applied to each set of test loads.

[3] The scale shall be tested from zero to at least 12.5 % of scale capacity using known test weights and then to at least 25 % of scale capacity using either a substitution or strain load test that utilizes known test weights of at least 12.5 % of scale capacity. Whenever practical, a strain load test should be conducted to the used capacity of the scale. When a strain load test is conducted, the tolerances apply only to the test weights or substitution test loads.
(Amended 1988, 1989, 1994, and 2003)

Note: GIPSA requires devices subject to their inspection to be tested to at least "used capacity," which is calculated based on the platform area of the scale and a weight factor assigned to the species of animal weighed on the scale. "Used capacity" is calculated using the formula:

Used Scale Capacity = Scale Platform Area x Species Weight Factor

Where species weight factor = 540 kg/m^2 (110 lb/ft^2) for cattle, 340 kg/m^2 (70 lb/ft^2) for calves and hogs, and 240 kg/m^2 (50 lb/ft^2) for sheep and lambs.</td></tr>
</table>

N.3.2.   Field Standard Weight Carts. – Field Standard Weight Carts that comply with the tolerances expressed in Fundamental Considerations, paragraph 3.2. (i.e., one-third of the smallest tolerance applied) may be included as part of the minimum required test load (Also see Table 4. Minimum Test Weights and Test Loads.) for shift tests and other test procedures.

(Added 2004)

N.4.   Coupled-in-Motion Railroad Weighing Systems.[3]

N.4.1.   Weighing Systems Used to Weigh Trains of Less Than Ten Cars. – These weighing systems shall be tested using a consecutive-car test train consisting of the number of cars weighed in the normal operation run over the weighing system a minimum of five times in each mode of operation following the final calibration.

(Added 1990) (Amended 1992)

N.4.2.   Weighing Systems Placed in Service Prior to January 1, 1991, and Used to Weigh Trains of Ten or More Cars. – The minimum test train shall be a consecutive-car test train of no less than ten cars run over the scale a minimum of five times in each mode of operation following final calibration.

(Added 1990) (Amended 1992)

N.4.3.   Weighing Systems Placed in Service on or After January 1, 1991, and Used to Weigh Trains of Ten or More Cars.

(a)   These weighing systems shall be tested using a consecutive-car test train of no less than ten cars run over the scale a minimum of five times in each mode of operation following final calibration; or

(b)   if the official with statutory authority determines it necessary, the As-Used Test Procedures outlined in N.4.3.1. shall be used.

(Added 1990) (Amended 1992)

N.4.3.1.   As-Used Test Procedures – A weighing system shall be tested in a manner that represents the normal method of operation and length(s) of trains normally weighed. The weighing systems may be tested using either a:

(a)   consecutive-car test train of a length typical of train(s) normally weighed; or

(b)   distributed-car test train of a length typical of train(s) normally weighed.

However, a consecutive-car test train of a shorter length may be used, provided that initial verification test results for the shorter consecutive-car test train agree with the test results for the distributed-car or full-length consecutive-car test train as specified in N.4.3.1.1. Initial Verification.

The official with statutory authority shall be responsible for determining the minimum test train length to be used on subsequent tests.

(Added 1990) (Amended 1992)

N.4.3.1.1.   Initial Verification. – Initial verification tests should be performed on any new weighing system and whenever either the track structure or the operating procedure changes. If a consecutive-car test train of length shorter than trains normally weighed is to be used for subsequent verification, the shorter consecutive-car test train results shall be compared either to a distributed-car or to a consecutive-car test train of length(s) typical of train(s) normally weighed.

---

[3] A test weight car that is representative of one of the types of cars typically weighed on the scale under test may be used wherever reference weight cars are specified.

(Added 1991)

The difference between the total train weight of the train(s) representing the normal method of operation and the weight of the shorter consecutive-car test train shall not exceed 0.15 %. If the difference in test results exceeds 0.15 %, the length of the shorter consecutive-car test train shall be increased until agreement within 0.15 % is achieved. Any adjustments to the weighing system based upon the use of a shorter consecutive-car test train shall be offset to correct the bias that was observed between the full-length train test and the shorter consecutive-car test train.

(Added 1990) (Amended 1992 and 1993)

N.4.3.1.2.   Subsequent Verification. – The test train may consist of either a consecutive-car test train with a length not less than that used in initial verification, or a distributed-car test train representing the number of cars used in the normal operation.

(Added 1990)

N.4.3.1.3.   Distributed-Car Test Trains.

(a) The length of the train shall be typical of trains that are normally weighed.

(b) The reference weight cars shall be split into three groups, each group consisting of ten cars or 10 % of the train length, whichever is less.

    (Amended 1991)

(c) The test groups shall be placed near the front, around the middle, and near the end of the train.

(d) Following the final adjustment, the distributed-car test train shall be run over the scale at least three times or shall produce 50 weight values, whichever is greater.

(e) The weighing system shall be tested in each mode of operation.

(Added 1990) (Amended 1992)

N.4.3.1.4.   Consecutive-Car Test Trains.

(a) A consecutive-car test train shall consist of at least ten cars.

(b) If the consecutive-car test train consists of between ten and twenty cars, inclusive, it shall be run over the scale a minimum of five times in each mode of operation following the final calibration.

(c) If the consecutive-car test train consists of more than twenty cars, it shall be run over the scale a minimum of three times in each mode of operation.

(Added 1990) (Amended 1992)

N.5.   Uncoupled-in-Motion Railroad Weighing System. – An uncoupled-in-motion scale shall be tested statically before being tested in motion by passing railroad reference weight cars over the scale. When an uncoupled-in-motion railroad weighing system is tested, the car speed and the direction of travel shall be the same as when the scale is in normal use. The minimum in-motion test shall be three reference weight cars passed over the scale three times. The cars shall be selected to cover the range of weights that are normally weighed on the system and to reflect the types of cars normally weighed.

(Added 1993)

N.6.   Nominal Capacity of Prescription Scales. – The nominal capacity of a prescription scale shall be assumed to be one-half apothecary ounce, unless otherwise marked. (Applicable only to scales not marked with an accuracy class.)

T. Tolerances Applicable to Devices <u>not</u> Marked I, II, III, III L, or IIII

T.1. Tolerance Values.

T.1.1. General. – The tolerances applicable to devices not marked with an accuracy class shall have the tolerances applied as specified in Table T.1.1. Tolerances for Unmarked Scales.

(Amended 1990)

T.1.2. Postal and Parcel Post Scales. – The tolerances for postal and parcel post scales are given in Table T.1.1. Tolerances for Unmarked Scales and Table 5. Maintenance and Acceptance Tolerances for Unmarked Postal and Parcel Post Scales.

(Amended 1990)

Table T.1.1. Tolerances for Unmarked Scales

Type of Device	Subcategory	Minimum Tolerance	Acceptance Tolerance	Maintenance Tolerance	Decreasing-Load Multiplier[1]	Other Applicable Requirements
Vehicle, axle-load, livestock, railway track (weighing statically), crane, and hopper (other than grain hopper)		Class III L, T.N.3.1 (Table 6) and T.N.3.2.			1.0	T.N.2, T.N.3, T.N.4.1., T.N.4.2., T.N.4.3, T.N.4.4., T.N.5, T.N.7.2., *T.N.8.1.4.[4]*, T.N.9.
Grain test scales	n ≤ 10 000	Class III, T.N.3.1. (Table 6) and T.N.3.2.			1.0	*T.N.8.1.4.[4]*, T.N.9.
	n > 10 000	Class II, T.N.3.1. (Table 6) and T.N.3.2.				
Railway track scales weighing in motion		T.N.3.6. except that for T.N.3.6.2. (a), no single error shall exceed four times the maintenance tolerance.			1.0	*T.N.8.1.4.[4]*, T.N.9.
Monorail scales, in-motion		T.N.3.8.			1.0	*T.N.8.1.4.[4]*, T.N.9.
Customer-operated bulk-weighing systems for recycled materials		± 5 % of applied material test load. Average error on 10 or more test loads ≤ 2.5 %.			1.0	*T.N.8.1.4.[4]*, T.N.9.
Wheel-load weighers and portable axle-load scales	Tested individually or in pairs[2]	0.5 d or 50 lb, whichever is greater	1 % of test load	2 % of test load	1.5[3]	*T.N.8.1.4.[4]*, T.N.9.
Prescription scales		0.1 grain (6 mg)	0.1 % of test load	0.1 % of test load	1.5	*T.N.8.1.4.[4]*, T.N.9.
Jewelers' scales	Graduated	0.5 d			1.5	*T.N.8.1.4.[4]*, T.N.9.
	Ungraduated	Sensitivity or smallest weight, whichever is less	0.05 % of test load	0.05 % of test load		
Dairy-product test scale	Loads < 18 g load	0.2 grain	0.2 grain	0.2 grain	1.5	*T.N.8.1.4.[4]*, T.N.9.
	18 g load	0.2 grain	0.3 grain	0.5 grain		
Postal and parcel post scales designed/used to weigh loads < 2 lb	Loads < 2 lb	15 grain, 1 g, 1/32 oz, 0.03 oz, or 0.002 lb	15 grain, 1 g, 1/32 oz, 0.03 oz, or 0.002 lb	15 grain, 1 g, 1/32 oz, 0.03 oz, or 0.002 lb	1.5	*T.N.8.1.4.[4]*, T.N.9.
	Loads ≤ 2 lb	Table 5	Table 5	Table 5		
Other postal and parcel post scales		Table 5	Table 5	Table 5	1.5	T.N.2.5, T.N.4.1., T.N.4.2., T.N.4.3., T.N.5., T.N.7.2., *T.N.8.1.4.[4]*, T.N.9.
All other scales (including grain hopper)	n > 5000	0.5 d or 0.05 % of scale capacity, whichever is less	0.05 % of test load	0.1 % of test load	1.5	T.N.2, T.N.3, T.N.4.1., T.N.4.2., T.N.4.3, T.N.5., T.N.7.2., *T.N.8.1.4.[4]*, T.N.9.
	n ≤ 5000	Class III, T.N.3.1., Table 6 and T.N.3.2.			1.0	

[1] The decreasing load test applies only to automatic indicating scales.

[2] If marked and tested as a pair, the tolerance shall be applied to the sum of the indication.

[3] The decreasing load test does not apply to portable wheel load weighers.

[4] *T.N.8.1.4. Operating Temperature. is nonretroactive and effective for unmarked devices manufactured as of January 1, 1981.*

(Table Added 1990; Amended 1992, 1993, and 2012)

Scale Capacity (lb)	Test Loads (lb)	Maintenance Tolerance (±)		Acceptance Tolerance (±)	
		(oz)	(lb)	(oz)	(lb)
0 to 4, inclusive*	0 to 1, inclusive	$^1/_{32}$	0.002	$^1/_{32}$	0.002
	over 1	$^1/_8$	0.008	$^1/_{16}$	0.004
over 4*	0 to 7, inclusive	$^3/_{16}$	0.012	$^3/_{16}$	0.012
	7+ to 24, inclusive	$^3/_8$	0.024	$^3/_{16}$	0.012
	24+ to 30, inclusive	$^1/_2$	0.030	$^1/_4$	0.015
	over 30	0.1 % of Test Load		0.05 % of Test Load	

Table 5.
Maintenance and Acceptance Tolerances for Unmarked
Postal and Parcel Post Scales

*Also see Table T.1.1. Tolerances for Unmarked Scales for scales designed and/or used to weigh loads less than 2 lb.

T.2.   Sensitivity Requirement (SR).

T.2.1.   Application. – The sensitivity requirement (SR) is applicable to all nonautomatic-indicating scales not marked I, II, III, III L, or IIII, and is the same whether acceptance or maintenance tolerances apply.

T.2.2.   General. – Except for scales specified in paragraphs T.2.3. Prescription Scales through T.2.8. Railway Track Scales: 2 d, 0.2 % of the scale capacity, or 40 lb, whichever is least.

T.2.3.   Prescription Scales. 6 mg (0.1 grain).

T.2.4.   Jewelers' Scales.

  T.2.4.1.   With One-Half Ounce Capacity or Less. – 6 mg (0.1 grain).

  T.2.4.2.   With More Than One-Half Ounce Capacity. – 1 d or 0.05 % of the scale capacity, whichever is less.

T.2.5.   Dairy-Product Test Scales.

  T.2.5.1.   Used in Determining Butterfat Content. – 32 mg (0.5 grain).

  T.2.5.2.   Used in Determining Moisture Content. – 19 mg (0.3 grain).

T.2.6.   Grain Test Scales. – The sensitivity shall be as stated in T.N.6. Sensitivity.
(Amended 1987)

T.2.7.   Vehicle, Axle-Load, Livestock, and Animal Scales.

  T.2.7.1.   Equipped With Balance Indicators. – 1 d.

  T.2.7.2.   Not Equipped With Balance Indicators. – 2 d or 0.2 % of the scale capacity, whichever is less.

T.2.8.   Railway Track Scales. – 3 d or 100 lb, whichever is less.

T.3. Sensitivity Requirement, Equilibrium Change Required.

The minimum change in equilibrium with test loads equal to the values specified in T.2. Sensitivity Requirements (SR) shall be as follows:

(a) **Scale with a Trig Loop but without a Balance Indicator.** – The position of rest of the weighbeam shall change from the center of the trig loop to the top or bottom, as the case may be.

(b) **Scale with a Single Balance Indicator and Having a Nominal Capacity of Less Than 250 kg (500 lb).** – The position of rest of the indicator shall change 1.0 mm (0.04 in) or one division on the graduated scale, whichever is greater.

(c) **Scale with a Single Balance Indicator and Having a Nominal Capacity of 250 kg (500 lb) or Greater.** – The position of rest of the indicator shall change 6.4 mm (0.25 in) or one division on the graduated scale or the width of the central target area, whichever is greater. However, the indicator on a batching scale shall change 3.2 mm (0.125 in) or one division on the graduated scale, whichever is greater.

(d) **Scale with Two Opposite-Moving Balance Indicators.** – The position of rest of the two indicators moving in opposite directions shall change 1.0 mm (0.04 in) with respect to each other.

(e) **Scale with Neither a Trig Loop nor a Balance Indicator.** – The position of rest of the weighbeam or lever system shall change from the horizontal, or midway between limiting stops, to either limit of motion.

T.N.   Tolerances Applicable to Devices Marked I, II, III, III L, and IIII.

T.N.1.   Principles.

T.N.1.1.   Design. – The tolerance for a weighing device is a performance requirement independent of the design principle used.

T.N.1.2.   Accuracy Classes. – Weighing devices are divided into accuracy classes according to the number of scale divisions (n) and the value of the scale division (d).

T.N.1.3.   Scale Division. – The tolerance for a weighing device is related to the value of the scale division (d) or the value of the verification scale division (e) and is generally expressed in terms of d or e.

T.N.2.   Tolerance Application.

T.N.2.1.   General. – The tolerance values are positive (+) and negative (−) with the weighing device adjusted to zero at no load. When tare is in use, the tolerance values are applied from the tare zero reference (zero net weight indication); the tolerance values apply to the net weight indication for any possible tare load using certified test loads.

(Amended 2008)

T.N.2.2.   Type Evaluation Examinations. – For type evaluation examinations, the tolerance values apply to increasing and decreasing load tests within the temperature, power supply, and barometric pressure limits specified in T.N.8.

T.N.2.3.   Subsequent Verification Examinations. – For subsequent verification examinations, the tolerance values apply regardless of the influence factors in effect at the time of the conduct of the examination. (Also see G-N.2. Testing with Nonassociated Equipment.)

T.N.2.4.   Multi-Interval and Multiple Range (Variable Division-Value) Scales. – For multi-interval and multiple range scales, the tolerance values are based on the value of the scale division of the range in use.

T.N.2.5.   Ratio Tests. – For ratio tests, the tolerance values are 0.75 of the applicable tolerances.

T.N.3.   Tolerance Values.

T.N.3.1.   Maintenance Tolerance Values. – The maintenance tolerance values are as specified in Table 6. Maintenance Tolerances.

T.N.3.2.   Acceptance Tolerance Values. – The acceptance tolerance values shall be one-half the maintenance tolerance values.

T.N.3.3.   Wheel-Load Weighers and Portable Axle-Load Weighers of Class IIII. – The tolerance values are two times the values specified in T.N.3.1. Maintenance Tolerance Values and T.N.3.2. Acceptance Tolerance Values.
(Amended 1986)

T.N.3.4.   Crane and Hopper (Other than Grain Hopper) Scales. – The maintenance and acceptance tolerances shall be as specified in T.N.3.1. Maintenance Tolerance Values and T.N.3.2. Acceptance Tolerance Values for Class III L, except that the tolerance for crane and construction materials hopper scales shall not be less than 1 d or 0.1 % of the scale capacity, whichever is less.
(Amended 1986)

Table 6. Maintenance Tolerances (All values in this table are in scale divisions)						
Tolerance in Scale Divisions						
	1		**2**		**3**	**5**
Class	**Test Load**					
I	0 - 50 000		50 001 -	200 000	200 001 +	
II	0 - 5 000		5 001 -	20 000	20 001 +	
III	0 - 500		501 -	2 000	2 001 - 4 000	4 001 +
IIII	0 - 50		51 -	200	201 - 400	401 +
III L	0 - 500		501 -	1 000	(Add 1 d for each additional 500 d or fraction thereof)	

T.N.3.5.   Separate Main Elements: Load Transmitting Element, Indicating Element, Etc. – If a main element separate from a complete weighing device is submitted for laboratory type evaluation, the tolerance for the main element is 0.7 that for the complete weighing device. This fraction includes the tolerance attributable to the testing devices used.
(Amended 2015)

T.N.3.6.   Coupled-In-Motion Railroad Weighing Systems. – The maintenance and acceptance tolerance values for the group of weight values appropriate to the application must satisfy the following conditions:
(Amended 1990 and 1992)

T.N.3.6.1. – For any group of weight values, the difference in the sum of the individual in-motion car weights of the group as compared to the sum of the individual static weights shall not exceed 0.2 %.
(Amended 1990)

T.N.3.6.2. – If a weighing system is used to weigh trains of five or more cars, and if the individual car weights are used, any single weight value within the group must meet the following criteria:

(a)   no single error may exceed three times the static maintenance tolerance;

(b)   not more than 5 % of the errors may exceed two times the static maintenance tolerance; and

(c)   not more than 35 % of the errors may exceed the static maintenance tolerance.
(Amended 1990 and 1992)

T.N.3.6.3. – For any group of weight values wherein the sole purpose is to determine the sum of the group, T.N.3.6.1. alone applies.

(Amended 1990)

T.N.3.6.4. – For a weighing system used to weigh trains of less than five cars, no single car weight within the group may exceed the static maintenance tolerance.

(Amended 1990 and 1992)

T.N.3.7.   Uncoupled-in-Motion Railroad Weighing Systems. – The maintenance and acceptance tolerance values for any single weighment within a group of non-interactive (i.e., uncoupled) loads, the weighment error shall not exceed the static maintenance tolerance.

(Amended 1992)

T.N.3.8.   Dynamic Monorail Weighing System. – Acceptance tolerance shall be the same as the maintenance tolerance shown in Table 6. Maintenance Tolerances. On a dynamic test of twenty or more individual test loads, 10 % of the individual test loads may be in error, each not to exceed two times the tolerance. The error on the total of the individual test loads shall not exceed ± 0.2 %. (Also see Note in N.1.3.5.1. Dynamic Monorail Weighing Systems.) *For equipment undergoing type evaluation, a tolerance equal to one-half the maintenance tolerance values shown in Table 6. Maintenance Tolerances shall apply.* *[Nonretroactive January 1, 2002]*

(Added 1986) (Amended 1999 and 2001)

T.N.3.9.   Materials Test on Customer-Operated Bulk Weighing Systems for Recycled Materials. – The maintenance and acceptance tolerance shall be ± 5 % of the applied materials test load except that the average error on ten or more test materials test loads shall not exceed ± 2.5 %.

(Added 1986)

T.N.3.10. Prescription Scales with a Counting Feature. – In addition to Table 6. Maintenance Tolerances (for weight), the indicated piece count value computed by a Class I or Class II prescription scale counting feature shall comply with the tolerances in Table T.N.3.10. Maintenance and Acceptance Tolerances in Excess and in Deficiency for Count.

Table T.N.3.10. Maintenance and Acceptance Tolerances in Excess and in Deficiency for Count	
Indication of Count	**Tolerance (piece count)**
0 to 100	0
101 to 200	1
201 or more	0.5 %

(Added 2003)

T.N.3.11. Tolerances for Substitution Test. – Tolerances are applied to the scale based on the substitution test load.
(Added 2003)

T.N.3.12. Tolerances for Strain-Load Test. – Tolerances apply only to the test weights or substitution test loads.
(Added 2003)

T.N.4.   Agreement of Indications.

T.N.4.1.   Multiple Indicating/Recording Elements. – In the case of a scale or weighing system equipped with more than one indicating element or indicating element and recording element combination, where the indicators or indicator/recorder combination are intended to be used independently of one another, tolerances shall be applied independently to each indicator or indicator/recorder combination.

(Amended 1986)

T.N.4.2.   Single Indicating/Recording Element. – In the case of a scale or weighing system with a single indicating element or an indicating/recording element combination, and equipped with component parts such as unit weights, weighbeam and weights, or multiple weighbeams that can be used in combination to indicate a weight, the difference in the weight value indications of any load shall not be greater than the absolute value of the applicable tolerance for that load, and shall be within tolerance limits.

(Amended 1986)

T.N.4.3.   Single Indicating Element/Multiple Indications. – In the case of an analog indicating element equipped with two or more indicating means within the same element, the difference in the weight indications for any load other than zero shall not be greater than one-half the value of the scale division (d) and be within tolerance limits.

(Amended 1986)

T.N.4.4.   Shift or Section Tests. – The range of the results obtained during the conduct of a shift test or a section test shall not exceed the absolute value of the maintenance tolerance applicable and each test result shall be within applicable tolerances.

(Added 1986)

T.N.4.5.   Time Dependence. – A time dependence test shall be conducted during type evaluation and may be conducted during field verification, provided test conditions remain constant.

(Amended 1989 and 2005)

T.N.4.5.1.   Time Dependence: Class II, III, and IIII Non-automatic Weighing Instruments. – A non-automatic weighing instrument of Classes II, III, and IIII shall meet the following requirements at constant test conditions. During type evaluation, this test shall be conducted at 20 °C ± 2 °C (68 °F ± 4 °F):

(a) When any load is kept on an instrument, the difference between the indication obtained immediately after placing the load and the indication observed during the following 30 minutes shall not exceed 0.5 e. However, the difference between the indication obtained at 15 minutes and the indication obtained at 30 minutes shall not exceed 0.2 e.

(b) If the conditions in (a) are not met, the difference between the indication obtained immediately after placing the load on the instrument and the indication observed during the following four hours shall not exceed the absolute value of the maximum permissible error at the load applied.

(Added 2005) (Amended 2006 and 2010)

T.N.4.5.2.   Time Dependence: Class III L Non-automatic Weighing Instruments. – A non-automatic weighing instrument of Class III L shall meet the following requirements:

(a) When any load is kept on an instrument, the difference between the indication obtained immediately after placing the load and the indication observed during the following 30 minutes shall not exceed 1.5 e. However, the difference between the indication obtained at 15 minutes and the indication obtained at 30 minutes shall not exceed 0.6 e.

(b) If the conditions in (a) are not met, the difference between the indication obtained immediately after placing the load on the instrument and the indication observed during the following four hours shall not exceed the absolute value of the maximum permissible error at the load applied.

(Added 2005) (Amended 2010)

T.N.4.5.3. Zero Load Return: Non-automatic Weighing Instruments. – A non-automatic weighing instrument shall meet the following requirements at constant test conditions. During type evaluation, this test shall be conducted at 20 °C ± 2 °C (68 °F ± 4 °F). The deviation on returning to zero as soon as the indication has stabilized, after the removal of any load which has remained on the instrument for 30 minutes shall not exceed:

(a) 0.5 e for Class II and IIII devices,

(b) 0.5 e for Class III devices with 4000 or fewer divisions,

(c) 0.83 e for Class III devices with more than 4000 divisions, or

(d) one-half of the absolute value of the applicable tolerance for the applied load for Class III L devices.

For a multi-interval instrument, the deviation shall not exceed 0.83 e_1 (where e_1 is the interval of the first weighing segment of the scale).

On a multiple range instrument, the deviation on returning to zero from Max$_1$ (load in the applicable weighing range) shall not exceed 0.83 e_1 (interval of the weighing range). Furthermore, after returning to zero from any load greater than Max$_1$ (capacity of the first weighing range) and immediately after switching to the lowest weighing range, the indication near zero shall not vary by more than e_1 (interval of the first weighing range) during the following five minutes.
(Added 2010)

T.N.4.6.   Time Dependence (Creep) for Load Cells during Type Evaluation. – A load cell (force transducer) marked with an accuracy class shall meet the following requirements at constant test conditions:

(a) Permissible Variations of Readings. – With a constant maximum load for the measuring range (D$_{max}$) between 90 % and 100 % of maximum capacity (E$_{max}$), applied to the load cell, the difference between the initial reading and any reading obtained during the next 30 minutes shall not exceed the absolute value of the maximum permissible error (mpe) for the applied load. (Also see Table T.N.4.6. Maximum Permissible Error (mpe) for Load Cells During Type Evaluation.) The difference between the reading obtained at 20 minutes and the reading obtained at 30 minutes shall not exceed 0.15 times the absolute value of the mpe. (Also see Table T.N.4.6. Maximum Permissible Error (mpe) for Load Cells During Type Evaluation)

(b) Apportionment Factors. – The mpe for creep shall be determined from Table T.N.4.6. Maximum Permissible Error (mpe) for Load Cells During Type Evaluation using the following apportionment factors (p$_{LC}$):

p$_{LC}$ = 0.7 for load cells marked with S (single load cell applications),
p$_{LC}$ = 1.0 for load cells marked with M (multiple load cell applications), and
p$_{LC}$ = 0.5 for Class III L load cells marked with S or M.

(Added 2005, Amended 2006)

Table T.N.4.6.
Maximum Permissible Error (mpe)* for Load Cells During Type Evaluation

	mpe in Load Cell Verifications Divisions (v) = p_{LC} × Basic Tolerance in v		
Class	p_{LC} × 0.5 v	p_{LC} × 1.0 v	p_{LC} × 1.5 v
I	0 - 50 000 v	50 001 v - 200 000 v	200 001 v +
II	0 - 5 000 v	5 001 v - 20 000 v	20 001 v +
III	0 - 500 v	501 v - 2 000 v	2 001 v +
IIII	0 - 50 v	51 v - 200 v	201 v +
III L	0 - 500 v	501 v - 1 000 v	(Add 0.5 v to the basic tolerance for each additional 500 v or fraction thereof up to a maximum load of 10 000 v)

v represents the load cell verification interval
p_{LC} represents the apportionment factors applied to the basic tolerance
p_{LC} = 0.7 for load cells marked with S (single load cell applications)
p_{LC} = 1.0 for load cells marked with M (multiple load cell applications)
p_{LC} = 0.5 for Class III L load cells marked with S or M
* mpe = p_{LC} × Basic Tolerance in load cell verifications divisions (v)

(Table Added 2005) (Amended 2006)

T.N.4.7.   Creep Recovery for Load Cells During Type Evaluation. – The difference between the initial reading of the minimum load of the measuring range (D_{min}) and the reading after returning to minimum load subsequent to the maximum load (D_{max}) having been applied for 30 minutes shall not exceed:

(a) 0.5 times the value of the load cell verification interval (0.5 v) for Class II and IIII load cells;

(b) 0.5 times the value of the load cell verification interval (0.5 v) for Class III load cells with 4000 or fewer divisions;

(c) 0.83 times the value of the load cell verification interval (0.83 v) for Class III load cells with more than 4000 divisions; or

(d) 2.5 times the value of the load cell verification interval (2.5 v) for Class III L load cells.

(Added 2006) (Amended 2009 and 2011)

T.N.5.   Repeatability. – The results obtained from several weighings of the same load under reasonably static test conditions shall agree within the absolute value of the maintenance tolerance for that load, and shall be within applicable tolerances.

T.N.6.   Sensitivity. – This section is applicable to all nonautomatic-indicating scales marked I, II, III, III L, or IIII.

T.N.6.1.   Test Load.

(a) The test load for sensitivity for nonautomatic-indicating vehicle, axle-load, livestock, and animal scales shall be 1 d for scales equipped with balance indicator, and 2 d or 0.2 % of the scale capacity, whichever is less, for scales not equipped with balance indicators.

(b) For all other nonautomatic-indicating scales, the test load for sensitivity shall be 1 d at zero and 2 d at maximum test load.

T.N.6.2.   Minimum Change of Indications. – The addition or removal of the test load for sensitivity shall cause a minimum permanent change as follows:

T.N.6.2.   Minimum Change of Indications. – The addition or removal of the test load for sensitivity shall cause a minimum permanent change as follows:

(a)   for a scale with trig loop but without a balance indicator, the position of the weighbeam shall change from the center to the outer limit of the trig loop;

(b)   for a scale with balance indicator, the position of the indicator shall change one division on the graduated scale, the width of the central target area, or the applicable value as shown below, whichever is greater:

Scale of Class I or II: 1 mm (0.04 in),
Scale of Class III or IIII with a maximum capacity of 30 kg (70 lb) or less: 2 mm (0.08 in),
Scale of Class III, III L, or IIII with a maximum capacity of more than 30 kg (70 lb): 5 mm (0.20 in);

(c)   for a scale without a trig loop or balance indicator, the position of rest of the weighbeam or lever system shall change from the horizontal or midway between limiting stops to either limit of motion.

(Amended 1987)

T.N.7.   Discrimination.

T.N.7.1.   Analog Automatic Indicating (i.e., Weighing Device with Dial, Drum, Fan, etc.). – A test load equivalent to 1.4 d shall cause a change in the indication of at least 1.0 d. (Also see N.1.5. Discrimination Test.)

T.N.7.2.   Digital Automatic Indicating. – A test load equivalent to 1.4 d shall cause a change in the indicated or recorded value of at least 2.0 d. This requires the zone of uncertainty to be not greater than three-tenths of the value of the scale division. (Also see N.1.5.1. Digital Device.)

T.N.8.   Influence Factors. – The following factors are applicable to tests conducted under controlled conditions only, provided that:

(a)   types of devices approved prior to January 1, 1986, and manufactured prior to January 1, 1988, need not meet the requirements of this section;

(b)   new types of devices submitted for approval after January 1, 1986, shall comply with the requirements of this section; and

(c)   all devices manufactured after January 1, 1988, shall comply with the requirements of this section.

(Amended 1985)

T.N.8.1.   Temperature. – Devices shall satisfy the tolerance requirements under the following temperature conditions:

T.N.8.1.1.   If not specified in the operating instructions for Class I or II scales, or if not marked on the device for Class III, III L, or IIII scales, the temperature limits shall be: − 10 °C to 40 °C (14 °F to 104 °F).

T.N.8.1.2.   If temperature limits are specified for the device, the range shall be at least that specified in Table T.N.8.1.2. Temperature Range by Class.

Table T.N.8.1.2. Temperature Range by Class	
Class	**Temperature Range**
I	5 °C (9 °F)
II	15 °C (27 °F)
III, III L, and IIII	30 °C (54 °F)

T.N.8.1.3.   Temperature Effect on Zero-Load Balance. – The zero-load indication shall not vary by more than:

    (a) three divisions per 5 °C (9 °F) change in temperature for Class III L devices; or

    (b) one division per 5 °C (9 °F) change in temperature for all other devices.

(Amended 1990)

T.N.8.1.4.   Operating Temperature. – Except for Class I and II devices, an indicating or recording element shall not display nor record any usable values until the operating temperature necessary for accurate weighing and a stable zero balance condition have been attained.

T.N.8.2.   Barometric Pressure. – Except for Class I scales, the zero indication shall not vary by more than one scale division for a change in barometric pressure of 1 kPa over the total barometric pressure range of 95 kPa to 105 kPa (28 in to 31 in of Hg).

T.N.8.3.   Electric Power Supply.

    T.N.8.3.1.   Power Supply, Voltage and Frequency.

        (a) Weighing devices that operate using alternating current must perform within the conditions defined in paragraphs T.N.3. Tolerance Values through T.N.7. Discrimination, inclusive, when tested over the range of − 15 % to + 10 % of the marked nominal line voltage(s) at 60 Hz, or the voltage range marked by the manufacturer, at 60 Hz.

        (Amended 2003)

        (b) Battery operated instruments shall not indicate nor record values outside the applicable tolerance limits when battery power output is excessive or deficient.

    T.N.8.3.2.   Power Interruption. – A power interruption shall not cause an indicating or recording element to display or record any values outside the applicable tolerance limits.

T.N.9.   Radio Frequency Interference (RFI) and Other Electromagnetic Interference Susceptibility. – The difference between the weight indication due to the disturbance and the weight indication without the disturbance shall not exceed one scale division (d); or the equipment shall:

    (a) blank the indication; or

    (b) provide an error message; or

    (c) the indication shall be so completely unstable that it cannot be interpreted, or transmitted into memory or to a recording element, as a correct measurement value.

The tolerance in T.N.9. Radio Frequency Interference (RFI) and Other Electromagnetic Interference Susceptibility is to be applied independently of other tolerances. For example, if indications are at allowable basic tolerance error limits when the disturbance occurs, then it is acceptable for the indication to exceed the applicable basic tolerances during the disturbance.

(Amended 1997)

UR.   User Requirements

UR.1.   Selection Requirements. – Equipment shall be suitable for the service in which it is used with respect to elements of its design, including but not limited to, its capacity, number of scale divisions, value of the scale division or verification scale division, minimum capacity, and computing capability.[4]

UR.1.1.   General.

(a)   For devices marked with a class designation, the typical class or type of device for particular weighing applications is shown in Table 7a. Typical Class or Type of Device for Weighing Applications.

(b)   For devices not marked with a class designation, Table 7b. Applicable to Devices not Marked with a Class Designation applies.

Table 7a. Typical Class or Type of Device for Weighing Applications	
Class	**Weighing Application or Scale Type**
I	Precision laboratory weighing
II	Laboratory weighing, precious metals and gem weighing, grain test scales
III	All commercial weighing not otherwise specified, grain test scales, retail precious metals and semi-precious gem weighing, grain-hopper scales, animal scales, postal scales, vehicle on-board weighing systems with a capacity less than or equal to 30 000 lb, and scales used to determine laundry charges
III L	Vehicle scales, vehicle on-board weighing systems with a capacity greater than 30 000 lb, axle-load scales, livestock scales, railway track scales, crane scales, and hopper (other than grain hopper) scales
IIII	Wheel-load weighers and portable axle-load weighers used for highway weight enforcement
Note: A scale with a higher accuracy class than that specified as "typical" may be used.	

(Amended 1985, 1986, 1987, 1988, 1992, 1995, and 2012)

---

[4] Purchasers and users of scales such as railway track, hopper, and vehicle scales should be aware of possible additional requirements for the design and installation of such devices.

(Footnote Added 1995)

| Table 7b. Applicable to Devices not Marked with a Class Designation ||
Scale Type or Design	Maximum Value of d
Retail Food Scales, 50 lb capacity and less	1 oz
Animal Scales	1 lb

Grain Hopper Scales
    Capacity up to and including 50 000 lb
    Capacity over 50 000 lb
	
	10 lb (not greater than 0.05 % of capacity)
	20 lb

Crane Scales	not greater than 0.2 % of capacity

Vehicle and Axle-Load Scales Used in Combination
    Capacity up to and including 200 000 lb
    Capacity over 200 000 lb
	
	20 lb
	50 lb

Railway Track Scales
    With weighbeam
    Automatic indicating
	
	20 lb
	100 lb

Scales with capacities greater than 500 lb except otherwise specified	0.1 % capacity (but not greater than 50 lb)
Wheel-Load Weighers	0.25 % capacity (but not greater than 50 lb)
Note: For scales not specified in this table, G-UR.1.1. and UR.1. apply.	

(Added 1985) (Amended 1989)

UR.1.2.   Grain Hopper Scales. – Hopper scales manufactured as of January 1, 1986, that are used to weigh grain shall be Class III and have a minimum of 2000 scale divisions.

(Amended 2012)

UR.1.3.   *Value of the Indicated and Recorded Scale Division.* – *The value of the scale division as recorded shall be the same as the division value indicated.*
[Nonretroactive as of January 1, 1986]

(Added 1985) (Amended 1999)

UR.1.3.1.   Exceptions. – The provisions of UR.1.3. Value of the Indicated and Recorded Scale Division shall not apply to:

(a) Class I scales, or

(b) Dynamic monorail weighing systems when the value of d is less than the value of e.

(Added 1999)

UR.1.4.   Grain-Test Scales: Value of the Scale Divisions. – The scale division for grain-test scales shall not exceed 0.2 g for loads through 500 g, and shall not exceed 1 g for loads above 500 g through 1000 g.

(Added 1992)

UR.1.5.   *Recording Element, Class III L Railway Track Scales.* – *Class III L Railway Track Scales must be equipped with a recording element.*
[Nonretroactive as of January 1, 1996]

(Added 1995)

UR.2.   Installation Requirements.

UR.2.1.   Supports. – A scale that is portable and that is being used on a counter, table, or the floor shall be so positioned that it is firmly and securely supported.

UR.2.2.   Suspension of Hanging Scale. – A hanging scale shall be freely suspended from a fixed support when in use.

UR.2.3.   Protection From Environmental Factors. – The indicating elements, the lever system or load cells, and the load-receiving element of a permanently installed scale, and the indicating elements of a scale not intended to be permanently installed, shall be adequately protected from environmental factors such as wind, weather, and RFI that may adversely affect the operation or performance of the device.

UR.2.4.   Foundation, Supports, and Clearance. – The foundation and supports of any scale installed in a fixed location shall be such as to provide strength, rigidity, and permanence of all components, and clearance shall be provided around all live parts to the extent that no contacts may result when the load-receiving element is empty, nor throughout the weighing range of the scale. An in-motion railway track scale is not required to provide clearance using rail gaps to separate the live rail portion of the weighing/load-receiving element from that which is not live if the scale is designed to be installed and operated using continuous rail. *On vehicle and livestock scales, the clearance between the load-receiving elements and the coping at the bottom edge of the platform shall be greater than at the top edge of the platform.**
[*Nonretroactive as of January 1, 1973]
(Amended 2014)

UR.2.5.   Access to Weighing Elements. – Adequate provision shall be made for ready access to the pit of a vehicle, livestock, animal, axle-load, or railway track scale for the purpose of inspection and maintenance. Any of these scales without a pit shall be installed with adequate means for inspection and maintenance of the weighing elements.
(Amended 1985)

UR.2.6.   Approaches.

UR.2.6.1.   Vehicle Scales. – On the entrance and exit end(s) of a vehicle scale, there shall be a straight approach as follows:

(a)   the width at least the width of the platform,

(b)   the length at least one-half the length of the platform but not required to be more than 12 m (40 ft), and

(c)   not less than 3 m (10 ft) of any approach adjacent to the platform shall be in the same plane as the platform. Any slope in the remaining portion of the approach shall ensure (1) ease of vehicle access, (2) ease for testing purposes, and (3) drainage away from the scale.

In addition to (a), (b), and (c), scales installed in any one location for a period of six months or more shall have not less than 3 m (10 ft) of any approach adjacent to the platform constructed of concrete or similar durable material to ensure that this portion remains smooth and level and in the same plane as the platform; however, grating of sufficient strength to withstand all loads equal to the concentrated load capacity of the scale may be installed in this portion.
[Nonretroactive as of January 1, 1976]
(Amended 1977, 1983, 1993, 2006, and 2010)

UR.2.6.2.   Axle-Load Scales. – At each end of an axle-load scale there shall be a straight paved approach in the same plane as the platform. The approaches shall be the same width as the platform and of sufficient length to insure the level positioning of vehicles during weight determinations.

UR.2.7.   Stock Racks. – A livestock or animal scale shall be equipped with a suitable stock rack, with gates as required, which shall be securely mounted on the scale platform. Adequate clearances shall be maintained around the outside of the rack.

UR.2.8.   Hoists. – On vehicle scales equipped with means for raising the load-receiving element from the weighing element for vehicle unloading, means shall be provided so that it is readily apparent to the scale operator when the load-receiving element is in its designed weighing position.

UR.2.9.   Provision for Testing Dynamic Monorail Weighing Systems. *– Provisions shall be made at the time of installation of a dynamic monorail weighing systems for testing in accordance with N.1.3.5.1. Dynamic Monorail Weighing Systems (a rail around or other means for returning the test carcasses to the scale being tested). [Nonretroactive as of January 1, 1998]*

(Added 1997) (Amended 1999)

UR.3.   Use Requirements.

UR.3.1.   Recommended Minimum Load. – A recommended minimum load is specified in Table 8 since the use of a device to weigh light loads is likely to result in relatively large errors.

Table 8. Recommended Minimum Load		
Class	Value of Scale Division (d or e*)	Recommended Minimum Load (d or e*)
I	equal to or greater than 0.001 g	100
II	0.001 g to 0.05 g, inclusive	20
	equal to or greater than 0.1 g	50
III	All**	20
III L	All	50
IIII	All	10

*For Class I and II devices equipped with auxiliary reading means (i.e., a rider, a vernier, or a least significant decimal differentiated by size, shape or color), the value of the verification scale division "e" is the value of the scale division immediately preceding the auxiliary means. For Class III and IIII devices the value of "e" is specified by the manufacturer as marked on the device; "e" must be less than or equal to "d."

**A minimum load of 10 d is recommended for a weight classifier marked in accordance with a statement identifying its use for special applications.

(Amended 1990)

UR.3.1.1.   Minimum Load, Grain Dockage Determination. – When determining the quantity of foreign material (dockage) in grain, the weight of the sample shall be equal to or greater than 500 scale divisions.

(Added 1985)

UR.3.2.   Maximum Load. – A scale shall not be used to weigh a load of more than the nominal capacity of the scale.

UR.3.2.1.   Maximum Loading for Vehicle Scales. – A vehicle scale shall not be used to weigh loads exceeding the maximum load capacity of its span as specified in Table UR.3.2.1. Span Maximum Load.

(Added 1996)

Distance in Feet Between the Extremes of any Two or More Consecutive Axles	Ratio of CLC to Maximum Load ("r" factor) Carried on Any Group of Two or More Consecutive Axles.							
	2 axles	3 axles	4 axles	5 axles	6 axles	7 axles	8 axles	9 axles
4[1]	1.000		INSTRUCTIONS:					
5[1]	1.000		1. Determine the scale's CLC.					
6[1]	1.000							
7[1]	1.000		2. Count the number of axles on the vehicle in a given span and determine the distance in feet between the first and last axle in the span.					
8 and less[1]	1.000	1.000						
More than 8[1]	1.118	1.235	3. Multiply the CLC by the corresponding multiplier in the table.*					
9	1.147	1.257	4. The resulting number is the scale's maximum concentrated load for a single span based on the vehicle configuration.					
10	1.176	1.279						
11	1.206	1.301	* note and formula on next page.					
12	1.235	1.324	1.471	1.632				
13	1.265	1.346	1.490	1.651				
14	1.294	1.368	1.510	1.669				
15	1.324	1.390	1.529	1.688	1.853			
16	1.353	1.412	1.549	1.706	1.871			
17	1.382	1.434	1.569	1.724	1.888			
18	1.412	1.456	1.588	1.743	1.906			
19	1.441	1.478	1.608	1.761	1.924			
20	1.471	1.500	1.627	1.779	1.941			
21	1.500	1.522	1.647	1.798	1.959			
22	1.529	1.544	1.667	1.816	1.976			
23	1.559	1.566	1.686	1.835	1.994			
24	1.588	1.588	1.706	1.853	2.012	2.176		
25	1.618	1.610	1.725	1.871	2.029	2.194		
26		1.632	1.745	1.890	2.047	2.211		
27		1.654	1.765	1.908	2.065	2.228		
28		1.676	1.784	1.926	2.082	2.245	2.412	
29		1.699	1.804	1.945	2.100	2.262	2.429	
30		1.721	1.824	1.963	2.118	2.279	2.445	
31		1.743	1.843	1.982	2.135	2.297	2.462	
32		1.765	1.863	2.000	2.153	2.314	2.479	2.647
33			1.882	2.018	2.171	2.331	2.496	2.664
34			1.902	2.037	2.188	2.348	2.513	2.680
35			1.922	2.055	2.206	2.365	2.529	2.697
36			2.000[2]	2.074	2.224	2.382	2.546	2.713
37			2.000[2]	2.092	2.241	2.400	2.563	2.730
38			2.000[2]	2.110	2.259	2.417	2.580	2.746
39			2.000	2.129	2.276	2.434	2.597	2.763
40			2.020	2.147	2.294	2.451	2.613	2.779
41			2.039	2.165	2.312	2.468	2.630	2.796
42			2.059	2.184	2.329	2.485	2.647	2.813
43			2.078	2.202	2.347	2.502	2.664	2.829
44			2.098	2.221	2.365	2.520	2.681	2.846
45			2.118	2.239	2.382	2.537	2.697	2.862
46			2.137	2.257	2.400	2.554	2.714	2.879
47			2.157	2.276	2.418	2.571	2.731	2.895
48			2.176	2.294	2.435	2.588	2.748	2.912
49			2.196	2.313	2.453	2.605	2.765	2.928
50			2.216	2.331	2.471	2.623	2.782	2.945

Table UR.3.2.1. Span Maximum Load

Table UR.3.2.1.
Span Maximum Load

Distance in Feet Between the Extremes of any Two or More Consecutive Axles	Ratio of CLC to Maximum Load ("r" factor) Carried on Any Group of Two or More Consecutive Axles.							
	2 axles	3 axles	4 axles	5 axles	6 axles	7 axles	8 axles	9 axles
51			2.235	2.349	2.488	2.640	2.798	2.961
52			2.255	2.368	2.506	2.657	2.815	2.978
53			2.275	2.386	2.524	2.674	2.832	2.994
54			2.294	2.404	2.541	2.691	2.849	3.011
55			2.314	2.423	2.559	2.708	2.866	3.028
56			2.333	2.441	2.576	2.725	2.882	3.044
57			2.353[3]	2.460	2.594	2.742	2.899	3.061
58				2.478	2.612	2.760	2.916	3.077
59				2.496	2.629	2.777	2.933	3.094
60				2.515	2.647	2.794	2.950	3.110

*Note: This table was developed based upon the following formula. Values may be rounded in some cases for ease of use.

$$W = r \times 500\left[\left(\frac{LN}{N-1}\right) + 12N + 36\right]$$

[1] Tandem Axle Weight.
[2] Exception – These values in the third column correspond to the maximum loads in which the inner bridge dimensions of 36, 37, and 38 ft are considered to be equivalent to 39 ft. This allows a weight of 68 000 lb on axles 2 through 5.
[3] Corresponds to the Interstate Gross Weight Limit.

UR.3.3.   Single-Draft Vehicle Weighing. – A vehicle or a coupled-vehicle combination shall be commercially weighed on a vehicle scale only as a single draft. That is, the total weight of such a vehicle or combination shall not be determined by adding together the results obtained by separately and not simultaneously weighing each end of such vehicle or individual elements of such coupled combination. However, the weight of:

(a) a coupled combination may be determined by uncoupling the various elements (tractor, semitrailer, trailer), weighing each unit separately as a single draft, and adding together the results; or

(b) a vehicle or coupled-vehicle combination may be determined by adding together the weights obtained while all individual elements are resting simultaneously on more than one scale platform.

Note: This paragraph does not apply to highway-law-enforcement scales and scales used for the collection of statistical data.
(Added 1992)

UR.3.4.   Wheel-Load Weighing.

UR.3.4.1.   Use in Pairs. – When wheel-load weighers or portable axle-load weighers are to be regularly used in pairs, both weighers of each such pair shall be appropriately marked to identify them as weighers intended to be used in combination.

UR.3.4.2.   Level Condition. – A vehicle of which either an axle-load determination or a gross-load determination is being made utilizing wheel-load weighers or portable axle-load weighers, shall be in a reasonably level position at the time of such determination.

UR.3.5.   Special Designs. – A scale designed and marked for a special application (such as a prepackaging scale or prescription scale with a counting feature) shall not be used for other than its intended purpose.[5]

(Amended 2003)

UR.3.6.   Wet Commodities. – Wet commodities not in watertight containers shall be weighed only on a scale having a pan or platform that will drain properly.

(Amended 1988)

UR.3.7.   Minimum Load on a Vehicle Scale. – A vehicle scale shall not be used to weigh net loads smaller than:

(a)   10 d when weighing scrap material for recycling or weighing refuse materials at landfills and transfer stations; and

(b)   50 d for all other weighing.

As used in this paragraph, scrap materials for recycling shall be limited to ferrous metals, paper (including cardboard), textiles, plastic, and glass.

(Amended 1988, 1992, and 2006)

UR.3.8.   Minimum Load for Weighing Livestock. – A scale with scale divisions greater than 2 kg (5 lb) shall not be used for weighing net loads smaller than 500 d.

(Amended 1989)

UR.3.9.   Use of Manual Weight Entries. – Manual gross or net weight entries are permitted for use in the following applications only when:

(a)   a point-of-sale system interfaced with a scale is giving credit for a weighed item;

(b)   an item is pre-weighed on a legal for trade scale and marked with the correct net weight;

(c)   a device or system is generating labels for standard weight packages;

(d)   postal scales or weight classifiers are generating manifests for packages to be picked up at a later time; or

(e)   livestock and vehicle scale systems generate weight tickets to correct erroneous tickets.

(Added 1992) (Amended 2000 and 2004)

UR.3.10.   Dynamic Monorail Weighing Systems. – When the value of d is different from the value of e, the commercial transaction must be based on e.

(Added 1999)

---

[5] Prepackaging scales and prescription scales with a counting feature (and other commercial devices) used for putting up packages in advance of sale are acceptable for use in commerce only if all appropriate provisions of NIST Handbook 44 are met. Users of such devices must be alert to the legal requirements relating to the declaration of quantity on a package. Such requirements are to the effect that, on the average, the contents of the individual packages of a particular commodity comprising a lot, shipment, or delivery must contain at least the quantity declared on the label. The fact that a prepackaging scale may overregister, but within established tolerances, and is approved for commercial service is not a legal justification for packages to contain, on the average, less than the labeled quantity.

(Amended 2003)

UR.3.11. Minimum Count. – A prescription scale with an operational counting feature shall not be used to count a quantity of less than 30 pieces weighing a minimum of 90 e.

(Added 2003)

Note: The minimum count as defined in this paragraph refers to the use of the device in the filling of prescriptions and is different from the minimum sample piece count as defined in S.1.2.3. and as required to be marked on the scale by S.6.6.

(Note Added 2004)

UR.3.12. Correct Stored Piece Weight. – For prescription scales with a counting feature, the user is responsible for maintaining the correct stored piece weight. This is especially critical when a medicine has been reformulated or comes from different lots.

(Added 2003)

UR.4. Maintenance Requirements.

UR.4.1. Balance Condition. – The zero-load adjustment of a scale shall be maintained so that, with no load on the load-receiving element and with all load-counterbalancing elements of the scale (such as poises, drop weights, or counterbalance weights) set to zero, the scale shall indicate or record a zero balance condition. A scale not equipped to indicate or record a zero-load balance shall be maintained in balance under any no-load condition.

UR.4.2. Level Condition. – If a scale is equipped with a level-condition indicator, the scale shall be maintained in level.

UR.4.3. Scale Modification. – The dimensions (e.g., length, width, thickness, etc.) of the load receiving element of a scale shall not be changed beyond the manufacturer's specifications, nor shall the capacity of a scale be increased beyond its design capacity by replacing or modifying the original primary indicating or recording element with one of a higher capacity, except when the modification has been approved by a competent engineering authority, preferably that of the engineering department of the manufacturer of the scale, and by the weights and measures authority having jurisdiction over the scale.

(Amended 1996)

UR.5. Coupled-in-Motion Railroad Weighing Systems. – A coupled-in-motion weighing system placed in service on or after January 1, 1991, should be tested in the manner in which it is operated, with the locomotive either pushing or pulling the cars at the designed speed and in the proper direction. The cars used in the test train should represent the range of gross weights that will be used during the normal operation of the weighing system. Except as provided in N.4.2. Weighing Systems Placed in Service Prior to January 1, 1991, and Used to Weigh Trains of Ten or More Cars and N.4.3.(a) Weighing Systems Placed in Service on or After January 1, 1991, and Used to Weigh Trains of Ten or More Cars, normal operating procedures should be simulated as nearly as practical. Approach conditions for a train length in each direction of the scale site are more critical for a weighing system used for individual car weights than for a unit-train-weights-only facility, and should be considered prior to installation.

(Added 1990) (Amended 1992)

Scales Code Index

Table of Contents

Section 2.21. Belt-Conveyor Scale Systems

A. Application

A.1.   General. – This code applies to belt-conveyor scale systems and weigh-belt systems used for the weighing of bulk materials.

(Amended 2015)

A.2.   Exceptions. – The code does not apply to:

(a)   devices used for discrete weighing while moving on conveyors;

(b)   devices that measure quantity on a time basis;

(c)   checkweighers; or

(d)   controllers or other auxiliary devices except as they may affect the weighing performance of the belt-conveyor scale.

A.3.   Additional Code Requirements. – In addition to the requirements of this code, Belt-Conveyor Scale Systems shall meet the requirements of Section 1.10. General Code.

S.   Specifications

S.1.   Design of Indicating and Recording Elements.

S.1.1.   General. – A belt-conveyor scale shall be equipped with a primary indicating element in the form of a master weight totalizer *and shall also be equipped with a recording element, and a rate of flow indicator and recorder (which may be analog).* * An auxiliary indicator shall not be considered part of the master weight totalizer.

[*Nonretroactive as of January 1, 1986]

(Amended 1986)

S.1.2.   Units. – A belt-conveyor scale shall indicate and record weight units in terms of pounds, tons, long tons, metric tons, or kilograms. The value of a scale division (d) expressed in a unit of weight shall be equal to:

(a)   1, 2, or 5; or

(b)   a decimal multiple or submultiples of 1, 2, or 5.

S.1.3.   Value of the Scale Division.

S.1.3.1.   For Scales Installed After January 1, 1986. – *The value of the scale division shall not be greater than 0.125 % ($^1/_{800}$) of the minimum totalized load.*
[Nonretroactive as of January 1, 1986]

(Added 1985)(Amended 2009)

S.1.3.2.   For Scales Installed Before January 1, 1986. – The value of the scale division shall not be greater than $^1/_{1200}$ of the rated capacity of the device. However, provision shall be made so that compliance with the requirements of the zero-load test as prescribed in N.3.1. Zero Load Tests may be readily and accurately determined in 20 minutes of operation.

S.1.4.   Recording Elements and Recorded Representations. – *The value of the scale division of the recording element shall be the same as that of the indicating element.*

   a)   *The belt-conveyor scale system shall record the unit of measurement (i.e., kilograms, tonnes, pounds, tons, etc.), the date, and the time.*

   b)   *The belt-conveyor scale system shall record the initial indication and the final indication of the master weight totalizer and the quantity.**

All of the information in (a) and (b) must be recorded for each delivery.*
[Nonretroactive as of January 1, 1986]
[*Nonretroactive as of January 1, 1994]

(Amended 1993)

   S.1.4.1.   *The belt-conveyor scale system shall be capable of recording the results of automatic or semi-automatic zero load tests.***
   [**Nonretroactive as of January 1, 2004]

   (Added 2002)

S.1.5.   Rate of Flow Indicators and Recorders. - *A belt-conveyor scale shall be equipped with a rate of flow indicator and an analog or digital recorder. Permanent means shall be provided to produce an audio or visual signal when the rate of flow is equal to or less than 20 % and when the rate of flow is equal to or greater than 100 % of the rated capacity of the scale. The type of alarm (audio or visual) shall be determined by the individual installation.*
[Nonretroactive as of January 1, 1986]

(Amended 1989 and 2004)

S.1.6.   Advancement of Primary Indicating or Recording Elements. – The master weight totalizer shall advance only when the belt conveyor is in operation and under load.
(Amended 1989)

S.1.7.   Master Weight Totalizer. – *The master weight totalizer shall not be resettable without breaking a security means.*
[Nonretroactive as of January 1, 1986]

S.1.8.   Power Loss. – *In the event of a power failure of up to 24 hours, the accumulated measured quantity on the master weight totalizer of an electronic digital indicator shall be retained in memory during the power loss.*
[Nonretroactive as of January 1, 1986]

(Amended 1989)

S.1.9.   Zero-Ready Indicator. – *A belt-conveyor scale shall be equipped with a zero-ready indicator that produces an audio or visual signal when the zero balance is within ± 0.12 % of the rated capacity of the scale during an unloaded belt condition. The type of indication (audio or visual) shall be determined by the individual installation.*
[Nonretroactive as of January 1, 2014]

(Added 2012)

S.2.   Design of Weighing Elements. – A belt-conveyor scale system shall be designed to combine automatically belt travel with belt load to provide a determination of the weight of the material that has passed over the scale.

S.2.1.   Speed Measurement. – A belt-conveyor scale shall be equipped with a belt speed or travel sensor that will accurately sense the belt speed or travel whether the belt is empty or loaded.

S.2.2. **Adjustable Components.** – An adjustable component that can affect the performance of the device (except as prescribed in S.3.1. Design of Zero-Setting Mechanism) shall be held securely in adjustment.
(Amended 1998)

S.2.3. **Overload Protection.** – The load-receiving elements shall be equipped with means for overload protection of not less than 150 % of rated capacity. The accuracy of the scale in its normal loading range shall not be affected by overloading.

S.3. Zero Setting.

S.3.1. **Design of Zero-Setting Mechanism.** – Automatic and semiautomatic zero-setting mechanisms shall be so constructed that the resetting operation is carried out only after a whole number of belt revolutions and the completion of the setting or the whole operation is indicated. *An audio or visual signal shall be given when the automatic and semiautomatic zero-setting mechanisms reach the limit of adjustment of the zero-setting mechanism.**
(Amended 1999 and 2002)

Except for systems that record the zero load reference at the beginning and end of a delivery, the range of the zero-setting mechanism shall not be greater than ± 2 % of the rated capacity of the scale without breaking the security means. *For systems that record the zero-load reference at the beginning and end of a delivery, the range of zero-setting mechanism shall not be greater than ± 5 % without breaking the security means.***
[*Nonretroactive as of January 1, 1990]
[**Nonretroactive as of January 1, 2004]
(Amended 1989 and 2002)

> ***S.3.1.1.*** ***Automatic Zero-Setting Mechanism.*** *– The automatic zero-setting mechanism shall indicate or record any change in the zero reference.*
> *[Nonretroactive as of January 1, 2010]*
>
> (Added 2009)

S.3.2. ***Sensitivity at Zero Load (For Type Evaluation).*** *– When a system is operated for a time period equal to the time required to deliver the minimum test load and with a test load calculated to indicate two scale divisions applied directly to the weighing element, the totalizer shall advance not less than one or more than three scale divisions. An alternative test of equivalent sensitivity, as specified by the manufacturer, shall also be acceptable. [Nonretroactive as of January 1, 1986]*

S.4. **Marking Requirements.** – Belt-conveyor scale systems and weigh-belt systems shall be marked with the following: (Also see also G-S.1. Identification.)

(a) the rated capacity in units of weight per hour (minimum and maximum);

(b) the value of the scale division;

(c) the belt speed in terms of feet (or meters) per minute at which the belt will deliver the rated capacity, or the maximum and minimum belt speeds at which the conveyor system will be operated for variable speed belts;

(d) the load in terms of pounds per foot or kilograms per meter (determined by material tests); and

(e) *the operational temperature range if other than − 10 °C to 40 °C (14 °F to 104 °F).*
   [Nonretroactive as of January 1, 1986]
(Amended 2015)

S.5. ***Provision for Sealing.*** *– A device shall be designed using the format set forth in Table S.5. with provision(s) for applying a security seal that must be broken, or for using other approved means of providing security (e.g. data*

change audit trail available at the time of inspection), before any change that affects the metrological integrity of the device can be made to any electronic mechanism.
[Nonretroactive as of January 1, 1999]

(Added 1998)

Table S.5. Categories of Device and Methods of Sealing	
Categories of Devices	**Methods of Sealing**
Category 1: *No remote configuration capability.*	*Seal by physical seal or two event counters: one for calibration parameters and one for configuration parameters.*
Category 3: *Remote configuration capability.*	*An event logger is required in the device; it must include an event counter (000 to 999), the parameter ID, the date and time of the change, and the new value of the parameter. A printed copy of the information must be available through the device or through another on-site device. The event logger shall have a capacity to retain records equal to 10 times the number of sealable parameters in the device, but not more than 1000 records are required. (**Note:** Does not require 1000 changes to be stored for each parameter.)*

[Nonretroactive as of January 1, 1999]

(Table Added 1998)

N. Notes

N.1. General. – Belt-conveyor scales are capable of weighing bulk material accurately. (Also see Tolerances.) However, their performance can be detrimentally affected by the conditions of the installation. (Also see User Requirements.) The performance of the equipment is not to be determined by averaging the results of the individual tests. The results of all tests shall be within the tolerance limits.

(Amended 2002)

N.1.1. Official Test. – An official test of a belt-conveyor scale system shall include tests specified in N.3.1. Zero Load Tests, N.3.2. Material Tests, and, if applicable, N.3.3. Simulated Load Tests.

(Amended 2006)

N.1.2. Simulated Test. – Simulated loading conditions as recommended by the manufacturer and approved by the official with statutory authority may be used to properly monitor the system operational performance between official tests, but shall not be used for official certification.

(Amended 1991)

N.2. Conditions of Tests. – A belt-conveyor scale shall be tested after it is installed on the conveyor system with which it is to be used and under such environmental conditions as may normally be expected. Each test shall be conducted with test loads no less than the minimum test load. Before each test run, the inspector shall check the zero setting and adjust as necessary.

(Amended 1986, 2004, and 2009)

N.2.1.   Initial Verification. – A belt-conveyor scale system or a weigh-belt system shall be tested using a minimum of two test runs as indicated in Table N.2.1. Initial Verification.

Results of the individual test runs in each pair of tests shall not differ by more than the absolute value of the tolerance as specified in T.2. Tolerance Values, Repeatability Tests. All tests shall be within the tolerance as specified in T.1. Tolerance Values.

Test runs may also be conducted at any other rate of flow that may be used at the installation. A minimum of four test runs may be conducted at only one flow rate if evidence is provided that the system is used at a constant speed/constant loading setting and that rate does not vary in either direction by an amount more than 10 % of the normal flow rate that can be developed at the installation for at least 80 % of the time.

Table N.2.1. Initial Verification		
Device Configuration	**Minimum of Two Test Runs at Each of the Following Settings**	**Total Tests (Minimum)**
Constant Belt Speed and Variable Loading	- Belt Loading: high (normal) - Belt Loading: medium (intermediate) - Belt Loading: low (35 %)	6
Variable Belt Speed and Constant Loading	- Belt Speed: maximum - Belt Speed: medium - Belt Speed: minimum	6
Variable Belt Speed and Variable Loading	- Belt Speed: maximum; Belt Loading: high (normal) - Belt Speed: maximum; Belt Loading: medium (intermediate) - Belt Speed: maximum; Belt Loading: low (35 %) - Belt Speed: minimum; Belt Loading: high (normal) - Belt Speed: minimum; Belt Loading: medium (intermediate) - Belt Speed: minimum; Belt Loading: low (35 %)	12

1. Use the device configurations in the left-hand column to identify the scale being tested.
2. Perform two test runs (minimum) at each of the settings shown in the center column.
3. The following terminology applies to "Belt Loading":
   - Low: 35 % of the maximum rated capacity of the system.
   - Medium: an intermediate rate between the high and low settings.
   - High: maximum (normal use) operational rate.

(Table Added 2015)

(Added 2004) (Amended 2009 and 2015)

N.2.2.   Subsequent Verification. – Subsequent testing shall include testing at the normal use flow rate and other flow rates used at the installation. The official with statutory authority may determine that testing only at the normal use flow rate is necessary for subsequent verifications if evidence is provided that the system is used to operate:

(a) at no less than 70 % of the maximum rated capacity for at least 80 % of the time (excluding time that the belt is unloaded); or

(b) with a normal use flow rate that does not vary by more than 10 % of the maximum rated capacity.

Example: If a belt-conveyor scale system has a maximum rated capacity of 200 tons per hour (tph), and the normal use flow rate is 150 tph (75 % of the maximum rated capacity), no testing at additional flow rates is

required provided the flow rates remain above 140 tph for more than 80 % of the time. If the same device were operating with a normal use flow rate of 130 tph, it is operating at 65 % of the maximum rated capacity. In this case, testing at flow rates in addition to the normal use flow rate would be required if the normal use flow rate varies by more than 20 tph (10 % of the maximum rated capacity).
(Added 2004)

N.2.3. Minimum Test Load.

N.2.3.1.   Minimum Test Load, Weigh-Belt Systems. – The minimum test load shall not be less than the largest of the following values:

(a) 800 scale divisions;

(b) the load obtained at maximum flow rate in one revolution of the belt; or

(c) at least one minute of operation.
(Amended 2015)

N.2.3.2.   Minimum Test Load, All Other Belt-Conveyor Scale Systems. – Except for applications where a normal weighment is less than 10 minutes, the minimum test load shall not be less than the largest of the following values:

(a) 800 scale divisions;

(b) the load obtained at maximum flow rate in one revolution of the belt; or

(c) at least 10 minutes of operation.

For applications where a normal weighment is less than 10 minutes (e.g., belt-conveyor scale systems used exclusively to issue net weights for material conveyed by individual vehicles and railway track cars) the minimum test load shall be the normal weighment that also complies with N.2.3.2.(a) and (b).

The official with statutory authority may determine that a smaller minimum totalized load down to 2 % of the load totalized in one hour at the maximum flow rate may be used for subsequent tests, provided that:

1.   the smaller minimum totalized load is greater than the quantities specified in N.2.3.2.(a) and (b); and

2.   consecutive official testing with the minimum totalized loads described in N.2.3.2.(a), (b), or (c) and the smaller minimum test load has been conducted that demonstrates the system complies with applicable tolerances for repeatability, acceptance, and maintenance.
(Added 2004) (Amended 2008 and 2015)

N.3. Test Procedures.

N.3.1.   Zero-Load Tests. – A zero-load test shall be conducted to establish that the belt scale system (including the conveyor) is capable of holding a stable, in-service zero.
(Amended 1989 and 2002)

N.3.1.1.   Determination of Zero. – A zero-load test is a determination of the error in zero, expressed as an internal reference, a percentage of the full-scale capacity, or a change in a totalized load over a whole number of complete belt revolutions. A zero-load test shall be performed as follows:

(a) For belt-conveyor scales with electronic integrators, the test must be performed over a period of at least three minutes and with a whole number of complete belt revolutions.

(b) For belt-conveyor scales with mechanical integrators, the test shall be performed with no less than three complete revolutions or 10 minutes of operation, whichever is greater.

(c) For weigh belt systems, the test must be performed over a period of at least one minute and at least one complete revolution of the belt.

(Added 2002) (Amended 2015)

N.3.1.2.   Test of Zero Stability. – The conveyor system shall be operated to warm up the belt and the belt scale shall be zero adjusted as required. A series of zero-load tests shall be carried out immediately before conducting the simulated load or materials test until the three consecutive zero-load tests each indicate an error which does not exceed ± 0.06 % of the totalized load at full scale capacity for the duration of the test. No adjustments can be made during the three consecutive zero-load test readings.

(Added 2002) (Amended 2004 and 2009)

N.3.1.3.   Check for Consistency of the Conveyor Belt along Its Entire Length. – During a zero-load test with any operational low-flow lock-out disabled, the absolute value of the difference between the maximum and minimum totalizer readings indicated on the totalizer during any complete revolution of the belt shall not exceed 0.12 % of the minimum test load.

Note: The end value of the zero-load test must meet the ± 0.06 % requirement referenced in the "Test for Zero Stability."

(Added 2002) (Amended 2004 and 2011)

N.3.2.   Material Tests. – Material tests should be conducted using actual belt loading conditions. These belt loading conditions shall include, but are not limited to conducting materials tests using different belt loading points, all types and sizes of products weighed on the scale, at least one other belt speed, and in both directions of weighing.

On subsequent verifications, at least two individual tests shall be conducted. The results of all these tests shall be within the tolerance limits.

Either pass a quantity of pre-weighed material over the belt-conveyor scale in a manner as similar as feasible to actual loading conditions, or weigh all material that has passed over the belt-conveyor scale. Means for weighing the material test load will depend on the capacity of the belt-conveyor scale and availability of a suitable scale for the test. To assure that the test load is accurately weighed and determined, the following precautions shall be observed:

(a) The containers, whether railroad cars, trucks, or boxes, must not leak, and shall not be overloaded to the point that material will be lost.

(b) The actual empty or tare weight of the containers shall be determined at the time of the test. Stenciled tare weight of railway cars or trucks shall not be used. Gross and tare weights shall be determined on the same scale.

(c) When a pre-weighed test load is passed over the scale, the belt-loading hopper shall be examined before and after the test to assure that the hopper is empty and that only the material of the test load has passed over the scale.

(d) Where practicable, a reference scale should be tested within 24 hours preceding the determination of the weight of the test load used for a belt-conveyor scale material test.

A reference scale which is not "as found" within maintenance tolerance should have its accuracy re-verified after the belt-conveyor test with a suitable known weight load if the "as found" error of the belt-conveyor scale material test exceeds maintenance tolerance values.*

(e) If any suitable known weight load other than a certified test weight load is used for re-verification of the reference scale accuracy, its weight shall be determined on the reference scale after the reference scale certification and before commencing the belt scale material test.*

(f) The test shall not be conducted if the weight of the test load has been affected by environmental conditions.

*Note: Even if the reference scale is within maintenance tolerance it may require adjusting to be able to meet paragraph N.3.2.1. Accuracy of Material.

(Amended 1986, 1989, 1998, 2000, 2002, and 2009)

N.3.2.1.   Accuracy of Material. – The quantity of material used to conduct a material test shall be weighed on a reference scale to an accuracy within 0.1 %. Scales typically used for this purpose include Class III and III L scales or a scale without a class designation as described in Handbook 44, Section 2.20., Table T.1.1. Tolerances for Unmarked Scales.

(Added 1989) (Amended 1991, 1993, 1998 and 2000)

N.3.3.   Simulated Load Tests.

(a) As required by the official with statutory authority, simulated load tests as recommended by the manufacturer are to be conducted between material tests to monitor the system's operational performance, but shall not be used for official certification.

(Amended 1991)

(b) A simulated load test consisting of at least three consecutive test runs shall be conducted as soon as possible, but not more than 12 hours after the completion of the material test, to establish the factor to relate the results of the simulated load test to the results of the material tests.

(Added 1990)

(c) The results of the simulated load test shall repeat within 0.1 %.

(Added 1990)

(Amended 1989 and 1990)

T. Tolerances

T.1.   Tolerance Values.[1] – Maintenance and acceptance tolerances on materials tests, relative to the weight of the material, shall be ± 0.25 % of the test load.

(Amended 1993)

T.1.1.   Tolerance Values – Test of Zero Stability. – Immediately after material has been weighed over the belt-conveyor scale during the conduct of any material test run, the zero-load test shall be repeated. The change

---

[1] The variables and uncertainties included in the relative tolerance represent only part of the variables that affect the accuracy of the material weighed on belt-conveyor scales. If this tolerance was based on an error analysis beginning with mass standards through all of the test processes and following the principle expressed in Section 3.2. of the Fundamental Considerations in Appendix A, the tolerance would be 0.5 %.

(Added 1993)

in the accumulated or subtracted weight during the zero-load test shall not exceed 0.12 % of the totalized load at full scale capacity for the duration of that test. If the range of zero adjustments during a complete (official) verification test exceeds 0.18 % of the totalized load at full scale capacity for the duration of the zero-load test, the official with statutory authority may establish an interval for zero-load testing during normal operation.
(Added 2004) (Amended 2009)

T.2.   Tolerance Values, Repeatability Tests. – The variation in the values obtained during the conduct of materials tests shall not be greater than 0.25 % ($^1/_{400}$).

T.3.   Influence Factors. – The following factors are applicable to tests conducted under controlled conditions only, provided that:

(a)   types of devices approved prior to January 1, 1986, and manufactured prior to January 1, 1988, need not meet the requirements of this section;

(b)   new types of devices submitted for approval after January 1, 1986, shall comply with the requirements of the section; and

(c)   all devices manufactured after January 1, 1988, shall comply with the requirements of this section.

T.3.1.   Temperature. – Devices shall satisfy the tolerance requirements at temperatures from − 10 °C to 40 °C (14 °F to 104 °F).

T.3.1.1.   Effect on Zero-Load Balance. – The zero-load indication shall not change by more than 0.035 % of the rated capacity of the scale (without the belt) for a change in temperature of 10 °C (18 °F) at a rate not to exceed 5 °C (9 °F) per hour.
(Amended 2004)

T.3.1.2.   Temperature Limits. *– If a temperature range other than − 10 °C to 40 °C (14 °F to 104 °F) is specified for the device, the range shall be at least 30 °C (54 °F).*
[Nonretroactive as of January 1, 1990]
(Added 1989)

T.3.2.   Power Supply, Voltage, and Frequency. – A belt-conveyor scale system shall satisfy the tolerance requirements over a range of 100 V to 130 V or 200 V to 250 V as appropriate and over a frequency range of 59.5 Hz to 60.5 Hz.

UR.   User Requirements

UR.1.   Installation Requirements.

UR.1.1.   Protection from Environmental Factors. – The indicating elements, the lever system or load cells, and the load-receiving element of a belt-conveyor scale shall be adequately protected from environmental factors such as wind, moisture, dust, weather, and radio frequency interference (RFI) and electromagnetic interference (EMI) that may adversely affect the operation or performance of the device.

UR.1.2.   Conveyor Installation. – The design and installation of the conveyor leading to and from the belt-conveyor scale is critical with respect to scale performance. Installation shall be in accordance with the scale manufacturer's instructions and the following:

(a) **Installation - General.** – A belt-conveyor scale shall be so installed that neither its performance nor operation will be adversely affected by any characteristic of the installation, including but not limited to, the foundation, supports, covers, or any other equipment.
(Amended 2002)

(b) Live Portions of Scale. – All live portions of the scale shall be protected with appropriate guard devices and clearances, as recommended by the scale manufacturer, to prevent accidental interference with the weighing operation. (Also see UR.3.1. Scale and Conveyor Maintenance.)
(Amended 2004)

(c) Storage of Simulated Load Equipment. – Suitable protection shall be provided for storage of any simulated load equipment.

(d) Take-up Device. – Any take-up device shall provide constant and consistent tension for the belt under all operating conditions.
(Amended 2014)

(e) Scale Location and Training Idlers. – The scale shall be so installed that the first weigh idler of the scale is at least 6 m (20 ft) or five idler spaces, whichever is greater, from loading point, skirting, head or tail pulley, or convex curve in the conveyor. Any training idler shall be located at least 18 m (60 ft) from the centerline of the weigh span of the scale. Training idlers shall not be restrained at any time in order to force belt alignment.
(Amended 1998)

(f) Concave Curve. – If there is a concave curve in the conveyor, before or after the scale, the scale shall be installed so that the belt is in contact with all the idler rollers at all times for at least 6 m (20 ft) or five idler spaces, whichever is greater, before and after the scale.[2] A concave curve shall start no closer than 12 m (40 ft) from the scale to the tangent point of the concave curve.
(Amended 1998)

(g) Tripper and Movable Pulleys. – There shall be no tripper or movable head pulleys in the conveyor.

(h) Conveyor Orientation. – The conveyor may be horizontal or inclined, but, if inclined, the angle shall be such that slippage of material along the belt does not occur.

(i) Conveyor Stringers. – Conveyor stringers at the scale and for not less than 6 m (20 ft) before and beyond the scale shall be continuous or securely joined and of sufficient size and so supported as to eliminate relative deflection between the scale and adjacent idlers when under load. The conveyor stringers should be so designed that the deflection between any two adjacent idlers within the weigh area does not exceed 0.6 mm (0.025 in) under load.

(j) Identification of Scale Area. – The scale area and five idlers on both ends of the scale shall be of a contrasting color, or other suitable means shall be used to distinguish the scale from the remainder of the conveyor installation, and the scale shall be readily accessible.
(Amended 1998)

(k) Belt Composition and Maintenance. – In a loaded or unloaded condition, the belt shall make constant contact with horizontal and wing rollers of the idlers in the scale area. Splices shall not cause any undue disturbance in scale operation. (Also see N.3. Test Procedures.)
(Amended 1998, 2000, 2001, and 2015)

---

[2] Installing the belt scale five-idler spaces from the tail pulley or the infeed skirting will be in the area of least belt tension on the conveyor and should produce the best accuracy. The performance of a belt-conveyor scale may be adversely affected by a concave curve in the conveyor that is located between the loading point and the scale. Therefore, whenever possible, a belt-conveyor scale should not be installed with a concave curve in the conveyor between the loading point and the scale.
(Amended 1995 and 1998)

(l) **Uniformity of Belt Loading and Flow.** – The conveyor loading mechanism shall be designed to provide uniform belt loading. The distance from the loading point to the scale shall allow for adequate settling time of the material on the belt before it is weighed. Feeding mechanisms shall have a positive closing or stopping action so that material leakage does not occur. Feeders shall provide an even flow over the scale through the full range of scale operation. Sufficient impact idlers shall be provided in the conveyor under each loading point to prevent deflection of the belt during the time material is being loaded.

(m) **Belt Alignment.** – The belt shall not extend beyond the edge of the outermost roller of any carry side (top) roller in any area of the conveyor nor touch the conveyor structure on the return (bottom) side of the conveyor.

(Amended 1998 and 2008)

(Amended 2002, 2012, 2013, 2014, and 2015)

UR.1.3.   Material Test. – *A belt-conveyor scale shall be installed so that a material test can be conveniently conducted.*

[Nonretroactive as of January 1, 1981]

UR.1.4.   Belt Travel (Speed or Velocity). – The belt travel sensor shall be so positioned that it accurately represents the travel of the belt over the scale for all flow rates between the maximum and minimum values. The belt travel sensor shall be so designed and installed that there is no slip.

(Amended 2012)

UR.2.   Use Requirements.

UR.2.1.   Rate of Operation. – A belt-conveyor scale system shall be operated between 20 % and 100 % of its rated capacity.

(Amended 2004)

UR.2.2.   Minimum Totalized Load. – Delivered quantities of less than the minimum test load shall not be considered a valid weighment.

UR.2.3.   Security Means. – When a security means has been broken, it shall be reported to the official with statutory authority.

(Amended 1991)

UR.2.4.   Loading. – The feed of material to the scale shall be controlled to assure that, during normal operation, the material flow is in accordance with the manufacturer's recommendation for rated capacity.

UR.2.5.   Diversion or Loss of Measured Product. – There shall be no operation(s) or condition(s) of use that result in loss or diversion that adversely affects the quantity of measured product.

(Added 2005)

UR.2.6.   Retention of Maintenance, Test, and Analog or Digital Recorder Information. – Records of calibration and maintenance, including conveyor alignment, analog or digital recorder, zero-load test, and material test data shall be maintained on site for at least the three concurrent years as a history of scale performance. Copies of any report as a result of a test or repair shall be mailed to the official with statutory authority as required. The current date and correction factor(s) for simulated load equipment shall be recorded and maintained in the scale cabinet.

(Added 2002)

(Amended 2012)

UR.3.   Maintenance Requirements – Scale and Conveyor Maintenance. – Weighing systems and idlers shall be maintained and serviced in accordance with manufacturer's instructions and the following:

(a) **Zero Balance.** – The zero balance condition of a belt-conveyor scale shall be maintained such that, prior to beginning any commercial transaction, with no load on the belt, the zero balance condition is within $\pm$ 0.12 % of the scale's rated capacity.

(Added 2012)

(b) **Scale Clearance.** – The scale and area surrounding the scale shall be kept clean of debris or other foreign material that can detrimentally affect the performance of the system.

(c) **Weighed Material.** – There shall be provisions to ensure that weighed material does not adhere to the belt and return to the scale system area.

(Added 2004)

(d) **Simulated and Zero-Load Test Intervals.** – Zero-load tests and simulated load or material tests shall be conducted at periodic intervals between official tests and after a repair or mechanical adjustment to the conveyor system in order to provide reasonable assurance that the device is performing correctly. The minimum interval for periodic zero-load tests and simulated load tests shall be established by the official with statutory authority or according to manufacturer recommendations.

The actions to be taken as a result of the zero-load test are shown in the following table.

Change in Zero (Δ 0)	Actions to be Taken
If the change in zero is less than $\pm$ 0.25 % (Δ 0 < 0.25 %)	Perform zero adjustment and proceed to simulated load test.
If the change in zero is $\pm$ 0.25 % to $\pm$ 0.5 % (0.25 % $\leq \Delta$ 0 $\leq$ 0.5 %)	Inspect the conveyor and weighing area for compliance with UR.1. Installation Requirements and repeat the zero-load test.
If the change in zero is greater than $\pm$ 0.5 % (Δ 0 > 0.5 %)	Inspect the conveyor and weighing area for compliance with UR.1. Installation Requirements, repeat the zero-load test, and reduce the interval between zero-load tests.

The action to be taken as a result of the simulated load or material tests is shown in the following table.

Change in Factor (Reference) Established in N.3.3.(b) [Δ N.3.3.(b)]	Action to be Taken
If the error is less than 0.25 % (Δ N.3.3.(b) < 0.25 %)	No Action
If the error is at least 0.25 % but not more than 0.6 % (0.25 % ≤ Δ N.3.3.(b) ≤ 0.6 %)	Inspect the conveyor and weighing area for compliance with UR.1. Installation Requirements and, after compliance is verified, repeat the test. If the result of that test remains greater than ± 0.25 %, a span correction shall be made and the official with statutory authority notified. (Amended 1991)
If the error is greater than 0.6 % but does not exceed 0.75 % (0.6 % < Δ N.3.3.(b) ≤ 0.75 %)	Inspect the conveyor and weighing area for compliance with UR.1. Installation Requirements and, after compliance is verified, repeat the test. If the result of that test remains greater than ± 0.25 %, a span correction shall be made, the official with statutory authority shall be notified, and an official test shall be conducted. (Amended 1991)
If the error is greater than 0.75 % (Δ N.3.3.(b) > 0.75 %)	An official test is required. (Amended 1987)

(Amended 2002 and 2009)

    (e) Scale Alignment. – Alignment checks shall be conducted in accordance with the manufacturer's recommendation. A material test is required after any realignment.

    (Amended 1986, 2000, and 2015)

    (f) Simulated Load Equipment. – Simulated load equipment shall be clean and properly maintained.

    (g) Zero Load Reference Information. – When zero load reference information is recorded for a delivery, the information must be based upon zero load tests performed as a minimum both immediately before and immediately after the totalized load.

    (Added 2002)

(Amended 1986, 2000, 2002, 2004, 2009, 2012, and 2015)

UR.4.   Compliance. – Prior to initial verification, the scale manufacturer or installer shall certify to the owner that the scale meets code requirements. Prior to initial verification and each subsequent verification, the scale owner or his agent shall notify the official with statutory authority in writing that the belt-conveyor scale system is in compliance with this specification and ready for material testing.

(Amended 1991)

THIS PAGE INTENTIONALLY LEFT BLANK

Table of Contents

Section 2.22.   Automatic Bulk Weighing Systems[1]

A. Application

A.1.   General. – This code applies to automatic bulk weighing systems, that is, weighing systems adapted to the automatic weighing of a commodity in successive drafts of predetermined amounts automatically recording the no-load and loaded weight values and accumulating the net weight of each draft.
(Amended 1987)

A.2.   Additional Code Requirements. – In addition to the requirements of this code, Automatic Bulk Weighing Systems shall meet the requirements of Section 1.10. General Code.

S. Specifications

S.1.   Design of Indicating and Recording Elements and Recorded Representations.

    S.1.1.   Zero Indication. – Provisions shall be made to indicate and record a no-load reference value and, if the no-load reference value is a zero value indication, to indicate and record an out-of-balance condition on both sides of zero.

        S.1.1.1.   Digital Zero Indication. – A digital zero indication shall represent a balance condition that is within ± ½ the value of the scale division.

    S.1.2.   Value of Scale Division (d). – *The value of the scale division (d), expressed in a unit of weight, shall be equal to:*

        (a)   1, 2, or 5; or

        (b)   a decimal multiple or submultiple of 1, 2, or 5; or

        (c)   a binary submultiple of a unit of weight.

    Examples: Scale divisions may be 0.01, 0.02, or 0.05; 0.1, 0.2, or 0.5; 1, 2, or 5; 10, 20, or 50; or ½, ¼, ⅛, 1/16, etc.
[*Nonretroactive as of January 1, 1986*]
(Amended 1987)

    S.1.3.   Capacity Indication and Recorded Representation. – An indicating or recording element shall not indicate or record any values when the gross load is in excess of 105 % of the capacity of the system.

    S.1.4.   Weighing Sequence. – For systems used to receive (weigh in), the no-load reference value shall be determined and recorded only at the beginning of each weighing cycle. For systems used to deliver (weigh out), the no-load reference value shall be determined and recorded only after the gross load reference value for each weighing cycle has been indicated and recorded.

    S.1.5.   Recording Sequence. – Provision shall be made so that all weight values are indicated until the completion of the recording of the indicated value.

---

[1] (Title amended 1986)

S.1.6.   Provision for Sealing Adjustable Components on Electronic Devices. – Provision shall be made for applying a security seal in a manner that requires the security seal to be broken before an adjustment can be made to any component affecting the performance of the device.

S.2.   Design of Balance and Damping Mechanism.

S.2.1.   Zero-Load Adjustment. – The weighing system shall be equipped with manual or semiautomatic means by which the zero-load balance or no-load reference value indication may be adjusted. Automatic zero-tracking and automatic zero-setting mechanisms are prohibited.
(Amended 2010)

S.2.1.1.   Manual. – A manual zero-load or no-load reference value setting mechanism shall be operable or accessible only by a tool outside of or entirely separate from this mechanism or enclosed in a cabinet.

S.2.1.2.   Semiautomatic. – A semiautomatic zero-load or no-load reference value setting mechanism shall meet the provisions of S.2.1.1. or shall be operable only when:

(a) the indication is stable within ± 3 scale divisions; and

(b) cannot be operated during a weighing operation.

S.2.2.   Damping Means. – A system shall be equipped with effective means necessary to bring the indications quickly to a readable, stable equilibrium. Effective means shall also be provided to permit the recording of weight values only when the indication is stable within plus or minus three scale divisions for devices with 10 000 scale divisions, or plus or minus one division for devices with less than 10 000 scale divisions.

S.3.   Interlocks and Gate Control.

S.3.1.   Gate Position. – Provision shall be made to clearly indicate to the operator the position of the gates leading directly to and from the weigh hopper.

S.3.2.   Interlocks. – Each automatic bulk weighing system shall have operating interlocks to provide for the following:

(a) Product cannot be cycled and weighed if the weight recording element is disconnected or subjected to a power loss.

(b) The recording element cannot print a weight if either of the gates leading directly to or from the weigh hopper is open.

(c) A "low paper" sensor, when provided, is activated.

(d) The system will operate only in the proper sequence in all modes of operation.

(e) When an overfill alarm is activated, the system shall indicate and record an overfill condition.
(Amended 1993)

S.3.3.   Overfill Sensor.

(a) The weigh hopper shall be equipped with an overfill sensor which will cause the feed gate to close, activate an alarm, and inhibit weighing until the overfill condition has been corrected.
(Added 1993)

(b) *If the system is equipped with a lower garner or surge bin, that garner shall also be equipped with an overfill sensor which will cause the gate of the weigh hopper to remain open, activate an alarm, and inhibit weighing until the overfill condition has been corrected.*
[Nonretroactive as of January 1, 1998]

(Amended 1997)

S.4.   Design of Weighing Elements.

S.4.1.   Antifriction Means. – At all points at which a live part of the mechanism may come into contact with another part in the course of normal usage, frictional effects shall be reduced to a minimum by means of suitable antifriction means, opposing surfaces and points being properly shaped, finished, and hardened.

S.4.2.   Adjustable Components. – An adjustable component, such as a potentiometer, shall be held securely in adjustment and, except for a component for adjusting level or a no-load reference value, shall not be adjustable from the outside of the device.

S.4.3.   Multiple Load-Receiving Elements. – A system with a single indicating or recording element, or a combination indicating recording element, that is coupled to two or more load-receiving elements with independent weighing systems, shall be provided with means to prohibit the activation of any load-receiving element (or elements) not in use, and shall be provided with automatic means to indicate clearly and definitely which load-receiving element (or elements) is in use.

S.4.4.   Venting. – All weighing systems shall be vented so that any internal or external pressure will not affect the accuracy or operation of the system.

S.5.   Marking Requirements. (Also see Section 1.10. General Code paragraph G-S.1. Identification.)

S.5.1.   Capacity and Value of the Scale Division. – The capacity of the weighing system and the value of the scale division shall be clearly and conspicuously marked on the indicating element near the weight value indications.

S.5.2.   Weighing Elements. – On a weighing element not permanently attached to an indicating element, there shall be clearly and permanently marked for the purposes of identification, the name, initials, or trademark of the manufacturer, the manufacturer's designation that positively identifies the pattern or design, and the nominal capacity.

S.5.3.   Temperature Limits. – *Unless the temperature range is − 10 °C to + 40 °C (14 °F to 104 °F), the temperature range shall be marked on the device.*
[Nonretroactive as of January 1, 1986]

(Added 1985)

S.5.4.   Accuracy Class.

(a) *All systems used to weigh grain shall be marked Class III.**

(b) *All other systems shall be marked either Class III or III L.**

*(*Also see Section 2.20. Scales Code for the parameters for these accuracy classes for scales. The specific requirements for automatic bulk weighing systems apply to these devices when there is a conflict between the Scales Code and the Automatic Bulk Weighing Systems Code.)*
[Nonretroactive as of January 1, 1986]

(Added 1985) (Amended 1992)

N. Notes

N.1. Testing Procedures.

N.1.1. Test Weights. – The increasing load test shall be conducted using test weights equal to at least 10 % of the capacity of the system:

(a) on automatic grain bulk-weighing systems installed after January 1, 1984; and

(b) on other automatic bulk-weighing systems installed after January 1, 1986.
(Amended 1987)

N.1.2. Increasing-Load Test. – An increasing-load test consisting of substitution and strain-load tests shall be conducted up to the used capacity of the weighing system.
(Amended 1987)

N.1.3. Decreasing-Load Test. – A decreasing-load test shall be conducted on devices used to weigh out.
(Added 1986)

N.1.4. Zero-Balance or No-Load Reference Value Change Test. – A test for change of zero-balance or no-load reference value shall be conducted on all scales after the removal of any test load. The change shall not be more than the minimum tolerance applicable.

N.1.5. Discrimination Test. – A discrimination test shall be conducted on all automatic indicating scales with the weighing device in equilibrium at zero-load and at maximum test load, and under controlled conditions in which environmental factors are reduced to the extent that they will not affect the results obtained. [Nonretroactive as of January 1, 1986]

N.1.5.1. Digital Device. – On a digital device, this test is conducted from just below the lower edge of the zone of uncertainty for increasing-load tests, or from just above the upper edge of the zone of uncertainty for decreasing-load tests.
(Added 1987)

N.2. Verification (Testing) Standards. – Standard weights and masses used in verifying weighing devices shall comply with requirements of NIST Handbook 105-1 (Class F) or the tolerances expressed in Appendix A, Fundamental Considerations, paragraph 3.2. (i.e., one-third of the smallest tolerance applied).

T. Tolerances

T.1. Tolerance Application. – Tolerance values shall be applied to all indications and recorded representations of a weighing system.

T.1.1. To Errors of Underregistration and Overregistration. – The tolerances hereinafter prescribed shall be applied equally to errors of underregistration and errors of overregistration.

T.1.2. To Increasing-Load Tests. – Basic tolerances shall be applied.

T.1.3. To Decreasing-Load Tests. – Basic tolerances shall be applied to systems used to weigh out.
(Added 1986)

T.1.4. To Tests Involving Digital Indications or Representations. – To the tolerances that would otherwise be applied, there shall be added an amount equal to one-half the value of the scale division. This does not apply to digital indications or recorded representations that have been corrected for rounding using error weights.
(Added 1986)

T.2.   Minimum Tolerance Values. – The minimum tolerance value shall not be less than half the value of the scale division.

   T.2.1.   For Systems Used to Weigh Construction Materials. – The minimum maintenance and acceptance tolerance shall be 0.1 % of the weighing capacity of the system, or the value of the scale division, whichever is less.

   (Added 1986)

T.3.   Basic Tolerance Values.

   T.3.1.   Acceptance Tolerance. – The basic acceptance tolerance shall be one-half the basic maintenance tolerance.

   T.3.2.   For Systems Used to Weigh Grain. – The basic maintenance tolerance shall be 0.1 % of test load.

   T.3.3.   For All Other Systems. – The basic maintenance tolerance shall be 0.2 % of test load.

   (Amended 1986)

T.4.   Time Dependence. – At constant test conditions, the indication 20 seconds after the application of a load and the indication after one hour shall not differ by more than the absolute value of the applicable tolerance for the applied load.
[Nonretroactive and enforceable as of January 1, 1987]

(Added 1986)

T.5.   Repeatability. – The results obtained by several weighings of the same load under reasonably static test conditions shall agree within the absolute value of the maintenance tolerance for that load, and shall be within applicable tolerances.

(Added 1986)

T.6.   Discrimination, Digital Automatic Indicating Scales. – A test load equivalent to 1.4 d shall cause a change in the indicated or recorded value of at least 2.0 d. This requires the zone of uncertainty to be not greater than 0.3 times the value of the scale division.

(Added 1985)

T.7.   Influence Factors. – The following factors are applicable to tests conducted under controlled conditions only, provided that:

   (a)   types of devices approved prior to January 1, 1986, and manufactured prior to January 1, 1988, need not meet the requirements of this section; and

   (b)   new types of devices submitted for approval after January 1, 1986, shall comply with the requirements of this section; and

   (c)   all devices manufactured after January 1, 1988, shall comply with the requirements of this section.
[Nonretroactive as of January 1, 1986]

   T.7.1.   Temperature. – Devices shall satisfy the tolerance requirements under the following temperature conditions:

      T.7.1.1.   If not marked on the device, the temperature limits shall be: − 10 °C to 40 °C (14 °F to 104 °F).

      T.7.1.2.   If temperature limits are specified for the device, the range shall be at least 30 °C (54 °F).

      T.7.1.3.   Temperature Effect on Zero-Load Balance. – The zero-load indicator shall not vary by more than one division per 5 °C (9 °F) change in temperature.

T.7.1.4.   Operating Temperature. – *An indicating or recording element shall not display or record any usable values until the operating temperature necessary for accurate weighing and a stable zero-balance condition has been attained.*
[Nonretroactive as of January 1, 1986]

T.7.2.   Barometric Pressure. – *The zero indication shall not vary by more than one scale division for a change in barometric pressure of 1 kPa over the total barometric range of 95 kPa to 105 kPa (28 in to 31 in of mercury).*
[Nonretroactive as of January 1, 1986]

T.7.3.   Electric Power Supply.

T.7.3.1.   Power Supply, Voltage, and Frequency.

(a) Weighing devices that operate using alternating current must perform within the conditions defined in paragraphs T.2. through T.7., inclusive over the line voltage range of 100 V to 130 V or 200 V to 250 V rms as appropriate and over the frequency range of 59.5 Hz to 60.5 Hz.

(b) Battery-operated instruments shall not indicate nor record values outside the applicable tolerance limits when battery power output is excessive or deficient.

T.7.3.2.   Power Interruption. – *A power interruption shall not cause an indicating or recording element to display or record any values outside the applicable tolerance limits.*
[Nonretroactive as of January 1, 1986]

(Added 1985)

UR.   User Requirements

UR.1.   Selection Requirements.

UR.1.1.   For Systems used to Weigh Grain. – *The number of scale divisions of a weighing system shall not be less than 2 000 nor greater than 10 000 divisions.*
[Nonretroactive as of January 1, 1984]

(Amended 1986 and 1992)

U.R.1.2.   For Systems used to Weigh Commodities other than Grain. – *The number of scale divisions shall not be less than 500 nor greater than 10 000.*
[Nonretroactive as of January 1, 1987]

(Added 1986)

UR.2.   Installation Requirements.

UR.2.1.   Protection from Environmental Factors. – The indicating elements, the lever system or load cells, the load-receiving element, and any permanently installed test weights shall be adequately protected from environmental factors such as wind, weather, and RFI that may adversely affect the operation or performance of the system.

UR.2.2.   Foundation, Supports, and Clearance. – The foundation and supports of any system shall be such as to provide strength, rigidity, and permanence of all components, and clearance shall be provided around all live parts so that no contact can result before or during operation of the system.

UR.3.   Loading Requirements.

UR.3.1.   For Systems Used to Weigh Grain. – A system shall not be used to weigh drafts less than 40 % of the weighing capacity of the system except for a final partial draft. Loads shall not normally be retained on the weighing element for a period longer than a normal weighing cycle.

(Amended 1986)

UR.3.2.   For Systems Used to Weigh Commodities Other than Grain. *– A system shall not be used to weigh drafts less than 20 % of the weighing capacity of the system except for a final partial draft. Loads shall not normally be retained on the weighing element for a period longer than a normal weighing cycle.*
[Nonretroactive as of January 1, 1987]

(Added 1986)

UR.4.   System Modification. – The weighing system shall not be modified except when the modification has been approved by a competent engineering authority, preferably that of the engineering department of the manufacturer of the scale, and the official with statutory authority having jurisdiction over the scale.

(Amended 1991)

THIS PAGE INTENTIONALLY LEFT BLANK

Table of Contents

THIS PAGE INTENTIONALLY LEFT BLANK

Section 2.23. Weights

A. Application

A.1.   General. – This code applies to commercial weights; that is, weights used in connection with commercial weighing devices.

A.2.   Exceptions. – This code does not apply to test weights or to other "standards" of mass.

A.3.   Additional Code Requirements. – In addition to the requirements of this code, Weights shall meet the requirements of Section 1.10. General Code.

S. Specifications

S.1.   Material. – The material used for weights shall be as follows:

(a) Weights of 6 g or 100 gr and larger shall be made of a metal, or a metal alloy, not softer than brass.

(b) Weights of less than 6 g or 100 gr may be made of aluminum, but shall not be made of iron or of unplated steel, except stainless steel.

S.2.   Design.

S.2.1.   Surface. – The surface of a weight shall be smooth and shall not be coated with thick, soft, or brittle material. A weight of more than 2 g or 30 gr or shall not have sharp edges, points, or corners.

S.2.2.   Ring. – A ring on a weight shall not be split or removable.

S.3.   Adjusting Material. – Adjusting material shall be securely positioned and shall not project beyond the surface of the weight.

S.4.   Marking Requirements.

S.4.1.   General. – A weight shall be marked to show clearly its nominal value, which shall include identification of the unit; however, the nominal value of a weight of 30 gr or 2 g, or less, may be designated by dots, lines, figures, distinctive shape, or other appropriate means.

S.4.2.   Apothecaries' Weights. – On apothecaries' dram, ounce, and pound weights, the letters "ap" shall be used in combination with the nominal value and the appropriate abbreviation of or symbol for the unit.

S.4.3.   Troy Weights. – On troy ounce and pound weights, the letter "t" shall be used in combination with the nominal value and the appropriate symbol of the unit.

S.4.4.   Metric Weights. – On metric weights, the symbols "kg," "g," and "mg" shall be used in combination with the nominal value of kilograms, grams, and milligrams, respectively.

S.4.5.   Carat Weights. – On carat weights, the letter "c" shall be used in combination with the nominal value.

S.4.6.   Counterpoise Weight. – A counterpoise weight shall be marked to show clearly both its nominal value and the value it represents when used on the multiplying-lever scale for which it is intended.

N. Notes

N.1. **Testing Procedures.** – Commercial weights should be tested on a precision balance using standard weights, the errors of which, when used without correction, do not exceed $\frac{1}{3}$ of the smallest tolerance to be applied. (Also see Appendix A, Fundamental Considerations, paragraphs 3.2. Tolerance for Standards and 3.3. Accuracy of Standards.)

T. Tolerances

T.1. **In Excess and In Deficiency.** – The tolerances hereinafter prescribed shall be applied equally to errors in excess and errors in deficiency.

T.2. **On Avoirdupois Weights.** – The maintenance tolerances shall be as shown in Table 1. Maintenance Tolerance for Avoirdupois Weights. Acceptance tolerances shall be one-half the maintenance tolerances.

Table 1. Maintenance Tolerance for Avoirdupois Weights						
Maintenance Tolerance						
Nominal Value	**Equal-Arm Weights**		**Counterpoise Weights**			
			For scales with multiples of less than 1000		**For scales with multiples of 1000 or over**	
oz	grains	mg	grains	mg	grains	mg
1/64	0.1	6				
1/32	0.3	19				
1/16	0.4	26				
1/8	0.5	32				
1/4	1.0	65				
1/2	1.5	97	1.0	65		
1	1.7	110	1.0	65		
2	2.0	130	1.0	65		
3	2.0	130	1.5	97		
4	3.0	190	1.5	97	1.0	65
5	3.5	230	1.5	97	1.0	65
6	3.5	230	1.5	97		
8	4.0	260	2.0	130	1.5	97
10	4.0	260	2.5	160	2.0	130
12	5.0	320	2.5	160	2.0	130
Lb	**Grains**	**mg**	**grains**	**mg**	**grains**	**mg**
1	5.0	320	3.0	190	2.5	160
2	7.0	450	6.0	390	4.0	260
3	9.0	580	9.0	580	5.0	320
4	11.0	710	11.0	710	6.0	390
5	15.0	970	12.0	780	6.5	420
6	17.0	1190				
7	19.0	1200				
8	21.0	1400	15.0	970	9.0	580
9	23.0	1500				
10	25.0	1600	18.0	1160	10.0	650
15	28.0	1800				
20	30.0	1900				
25	35.0	2300				
30	40.0	2600				
40	45.0	2900				
50	50.0	3200				

T.3.   On Metric Weights. – The maintenance tolerances shall be as shown in Table 2. Maintenance Tolerances for Metric Weights. Acceptance tolerances shall be one-half the maintenance tolerances.

T.4.   On Carat Weights. – The maintenance tolerances shall be as shown in Table 2. Maintenance Tolerances for Metric Weights. Acceptance tolerances shall be one-half the maintenance tolerances.

Table 2. Maintenance Tolerances for Metric Weights			
Nominal Value (mg)	Maintenance Tolerance (mg)	Nominal Value (g)	Maintenance Tolerance (mg)
5 or less	0.1	1	4
10	0.3	2	6
20	0.4	3	8
30	0.6	5	10
50	0.8	10	15
100	1.0	20	20
200	1.5	30	30
300	2.0	50	40
500	3.0	100	70
		200	100
		300	150
		500	175
Nominal Value (kg)	Maintenance Tolerance (mg)	Nominal Value (carats)	Maintenance Tolerance (mg)
1	250	0.25*	0.6
2	400	0.5**	1.0
3	500	1.0	1.5
5	800	2.0	2.0
10	1000	3.0	3.0
20	1500	5.0	4.0
		10.0	6.0
		20.0	10.0
		30.0	12.0
		50.0	15.0
		100.0	25.0
		*25 points or less **50 points	

T.5.   On Apothecaries and Troy Weights. – The maintenance tolerances shall be as shown in Table 3. Maintenance Tolerances for Apothecaries' and Troy Weights. Acceptance tolerances shall be one-half the maintenance tolerances.

Nominal Value	Maintenance Tolerance		Nominal Value	Maintenance Tolerance	
Table 3.					
Maintenance Tolerances for Apothecaries' and Troy Weights					
grains	**grains**	**mg**	**oz**	**grains**	**mg**
1	0.01	0.6	1	0.4	25.0
2	0.02	1.3	2	0.6	40.0
3	0.03	2.0	3	1.0	65.0
5	0.03	2.0	4	1.5	100.0
10	0.04	2.5	5	1.6	105.0
20	0.06	4.0			
scruples	**grains**	**mg**	**oz**	**grains**	**mg**
1	0.06	4.0	6	1.8	115.0
2	0.10	6.5	7	1.9	125.0
			8	2.0	130.0
			9	2.1	135.0
			10	2.2	145.0
dr	**grains**	**mg**	**oz**	**grains**	**mg**
0.5	0.07	4.5	11	2.4	155.0
1.0	0.10	6.5	12	2.5	160.0
2.0	0.20	13.0	20	2.9	190.0
3.0	0.30	20.0	30	3.7	240.0
4.0	0.40	25.0	50	5.4	350.0
5.0	0.50	30.0			
6.0	0.60	40.0			
dwt	**grains**	**mg**	**oz**	**grains**	**mg**
1	0.06	4.0	100	7.7	500.0
2	0.10	6.5	200	12.3	800.0
3	0.15	10.0	300	15.4	1 000.0
4	0.20	13.0	500	23.1	1 500.0
5	0.30	20.0	1 000	38.6	2 500.0
10	0.40	25.0			

THIS PAGE INTENTIONALLY LEFT BLANK

Table of Contents

Section 2.24. Automatic Weighing Systems

A. Application

A.1. General. – This code applies to devices used to automatically weigh pre-assembled discrete loads or single loads or loose materials in applications where automatic weighing systems[1] are used or employed in the determination of quantities, things, produce, or articles for distribution, for purchase, offered or submitted for sale, for distribution, purchase, or in computing any basic charge or payment for services rendered on the basis of weight, and in packaging plants subject to regulation by the USDA. Some weigh-labelers and checkweighers may also include a scale that is incorporated in a conveyor system that weighs packages in a static or non-automatic weighing mode.[2]

This includes:

(a) Automatic weigh-labelers;

(b) Combination automatic and non-automatic weigh-labelers;

(c) Automatic checkweighers;

(d) Combination automatic and non-automatic checkweighers; and

(e) Automatic gravimetric filling machines that weigh discrete loads or single loads of loose materials and determine package and production lot compliance with net content representations.

(Amended 1997 and 2004)

A.2. Exceptions. – This code does not apply to:

(a) Belt-Conveyor Scale Systems;

(b) Railway Track Scales;

(c) Monorail Scales;

(d) Automatic Bulk-Weighing Systems;

(e) Devices that measure quantity on a time basis;

(f) Controllers or other auxiliary devices except as they may affect the weighing performance; or

---

[1] An automatic weighing system does not require the intervention of an operator during the weighing process. The necessity to give instructions to start a process or to release a load or the function of the instrument (static, dynamic, set-up, etc.) is not relevant in deciding the category of automatic or non-automatic instruments.
(Added 2004)

[2] Prepackaging scales (and other commercial devices) used for putting up packages in advance of sale are acceptable for use in commerce if all appropriate provisions of Handbook 44 are met. Users of such devices must be alert to the legal requirements relating to the declaration of quantity on a package. Such requirements are to the effect that, on the average, the contents of the individual packages of a particular commodity comprising a lot, shipment, or delivery must contain at least the quantity declared on the label. The fact that a scale or other commercial device may overregister, but within established tolerances, and is approved for commercial service is not a legal justification for packages to contain, on the average, less than the labeled quantity.
(Added 2004)

(g) Automatic gravimetric filling machines and other automatic weighing systems employed in determining the weight of a commodity in a plant or business with a separate quantity control program (e.g., a system of statistical process control) using suitable weighing instruments and measurement standards traceable to national standards to determine production lot compliance with net content representations.[3]

(Added 2004)

A.3.   Additional Code Requirements. – In addition to the requirements of this code, Automatic Weighing Systems shall meet the requirements of Section 1.10. General Code.

S. Specifications

S.1.   Design of Indicating and Recording Elements and of Recorded Representations.

S.1.1.   Zero Indication.

(a) A weigh-labeler shall be equipped with an indicating or recording element. Additionally, a weigh-labeler equipped with an indicating or recording element shall either indicate or record a zero-balance condition and an out-of-balance condition on both sides of zero.

(Amended 2004)

(b) An automatic checkweigher may be equipped with an indicating or recording element.

(c) A zero-balance condition may be indicated by other than a continuous digital zero indication, provided that effective automatic means is provided to inhibit a weighing operation or to return to a continuous digital indication when the device is in an out-of-balance condition.

S.1.1.1.   Digital Indicating Elements.

(a) A digital zero indication shall represent a balance condition that is within ± ½ scale division.

(b) A digital indicating device shall either automatically maintain a "center of zero" condition to ± ¼ scale division or less, or have an auxiliary or supplemental "center-of-zero" indicator that defines a zero-balance condition to ± ¼ scale division or less.

(c) Verification of the accuracy of the center of zero indication to ± ¼ scale division or less during automatic operation is not required on automatic checkweighers.

(Amended 2004)

S.1.2.   Value of Division Units. – The value of a division d expressed in a unit of weight shall be equal to:

(a) 1, 2, or 5; or

(b) a decimal multiple or submultiple of 1, 2, or 5.

The requirement that the value of the scale division be expressed only as 1, 2, or 5, or a decimal multiple or submultiple of only 1, 2, or 5 does not apply to net weight indications and recorded representations that are calculated from gross and tare weight indications where the scale division of the gross weight is different from the scale division of the tare weight(s) on multi-interval or multiple range scales. For example, a multiple range or multi-interval scale may indicate and record tare weights in a lower weighing range (WR) or weighing segment (WS), gross weights in the higher weighing range or weighing segment, and net weights as follows:

---

[3] See NIST Handbook 130, "Uniform Laws and Regulations in the Area of Legal Metrology and Engine Fuel Quality," Interpretations and Guidelines, paragraph 2.6.11. Good Quantity Control Practices.

55 kg	Gross Weight	(WR2 d = 5 kg)		10.05 lb	Gross Weight	(WS2 d = 0.05 lb)
– 4 kg	Tare Weight	(WR1 d = 2 kg)		– 0.06 lb	Tare Weight	(WS1 d = 0.02 lb)
= 51 kg	Net Weight	(Mathematically Correct)		= 9.99 lb	Net Weight	(Mathematically Correct)

(Amended 2008)

S.1.2.1.   Weight Units. – Except for postal scales, indicating and recording elements for shipping and postal applications, and scales used to print standard pack labels, a device shall indicate weight values using only a single unit of measure.

(Amended 2004)

S.1.3.   Provision for Sealing.

(a) **Automatic Weighing Systems, Except Automatic Checkweighers.** – A device shall be designed with provision(s) as specified in Table S.1.3. Categories of Device and Methods of Sealing for applying a security seal that must be broken, or for using other approved means of providing security (e.g., data change audit trail available at the time of inspection), before any change that detrimentally affects the metrological integrity of the device can be made to any electronic mechanism.

(b) **For Automatic Checkweighers.** – Security seals are not required in applications where it would prohibit an authorized user from having access to the calibration functions of a device.

Table S.1.3. Categories of Device and Methods of Sealing	
Categories of Device	**Methods of Sealing**
Category 1: No Remote configuration capability.	Seal by physical seal or two event counters: one for calibration parameters and one for configuration parameters.
Category 2: Remote configuration capability, but access is controlled by physical hardware. The device shall clearly indicate that it is in the remote configuration mode and record such message if capable of printing in this mode.	The hardware enabling access for remote communication must be at the device and sealed using a physical seal or two event counters: one for calibration parameters and one for configuration parameters.
Category 3: Remote configuration capability access may be unlimited or controlled through a software switch (e.g., password).	An event logger is required in the device; it must include an event counter (000 to 999), the parameter ID, the date and time of the change, and the new value of the parameter. A printed copy of the information must be available through the device or through another on-site device. The event logger shall have a capacity to retain records equal to 10 times the number of sealable parameters in the device, but not more than 1000 records are required. (**Note:** Does not require 1000 changes to be stored for each parameter.)

S.1.4.   Automatic Calibration. – A device may be fitted with an automatic or a semi-automatic calibration mechanism. This mechanism shall be incorporated inside the device. After sealing, neither the mechanism nor the calibration process shall facilitate fraud.

S.1.5.   Adjustable Components. – Adjustable components shall be held securely in adjustment and, except for a zero-load balance mechanism, shall be located within the housing of the element.

S.2.   Design of Zero and Tare Mechanisms.

S.2.1.   Zero Load Adjustment.

S.2.1.1.   Automatic Zero-Tracking Mechanism. – Except for automatic checkweighers, under normal operating conditions the maximum load that can be "rezeroed," when either placed on or removed from the platform all at once, shall be 1.0 scale division.

Except for an initial zero-setting mechanism, an automatic zero adjustment outside these limits is prohibited. (Amended 2004 and 2010)

S.2.1.2.   Initial Zero-Setting Mechanism. – Except for automatic checkweighers, an initial zero-setting mechanism shall not zero a load in excess of 20 % of the maximum capacity of the automatic weighing system unless tests show that the scale meets all applicable tolerances for any amount of initial load compensated by this device within the specified range.

S.2.2.   Tare. – On any automatic weighing system (except for multi-interval scales or multiple range scales when the value of tare is determined in a lower weighing range or weighing segment) the value of the tare division shall be equal to the value of the scale division. The tare mechanism shall operate only in a backward direction (i.e., in a direction of underregistration) with respect to the zero-load balance condition of the automatic weighing system. A device designed to automatically clear any tare value shall also be designed to prevent the automatic clearing of tare until a complete transaction has been indicated.
(Amended 2008)

Note: On a computing automatic weighing system, this requires the input of a unit price, the display of the unit price, and a computed positive total price at a readable equilibrium. Other devices require that a transaction or lot run be completed.
(Note Amended 2004)

S.3.   Verification Scale Interval.

S.3.1.   Multiple Range and Multi-Interval Automatic Weighing System. – The value of e shall be equal to the value of d.

S.3.2.   Load Cell Verification Interval Value. – The relationship of the value for the load cell verification scale interval, v_{min}, to the scale division d for a specific scale installation shall be:

$$v_{min} \leq \frac{d}{\sqrt{N}}$$, where N is the number of load cells in the scale.

Note: When the value of the scale division d differs from the verification scale division e for the scale, the value of e must be used in the formula above.

S.3.3. – For automatic checkweighers, the value of e shall be specified by the manufacturer and may be larger than d, but in no case can e be more than ten times the value of d.

S.4.   Weight Indicators, Weight Displays, Reports, and Labels.

S.4.1.   Additional Digits in Displays. – Auxiliary digital displays that provide additional digits for use during performance evaluation may be included on automatic checkweighers. However, in cases where these indications are not valid for determining the actual weight of a package (e.g., only appropriate for use in statistical process control programs by users) they shall be clearly and distinctly differentiated from valid weight displays by indicating them to the user.

For example, the additional digits may be differentiated by color, partially covered by placing crosshatch overlays on the display, or made visible only after the operator presses a button or turns a key to set the device in a mode which enables the additional digits.

S.4.2.   Damping. – An indicating element equipped with other than automatic recording elements shall be equipped with effective means to permit the recording of weight values only when the indication is stable within plus or minus one scale division. The values recorded shall be within applicable tolerances.

S.4.3.   Over Capacity Indication. – An indicating or recording element shall not display nor record any values when the scale capacity is exceeded by nine scale divisions.

S.4.4.   Label Printer. – A device that produces a printed ticket to be used as the label for a package shall print all values digitally and of such size, style of type, and color as to be clear and conspicuous on the label.

   S.4.4.1.   Label Printing. – If an automatic checkweigher prints a label containing weight information that will be used in a commercial transaction, it must conform to all of the requirements specified for weigh-labelers so that the printed ticket meets appropriate requirements.

S.5.   Accuracy Class.

   S.5.1.   Marking. – Weigh-labelers and automatic checkweighers shall be Class III devices and shall be marked accordingly, except that a weigh-labeler marked Class IIIS may be used in package shipping applications.
   (Amended 1997)

S.6.   Parameters for Accuracy Classes. – The number of divisions for device capacity is designated by the manufacturer and shall comply with parameters shown in Table S.6. Parameters for Accuracy Classes.

Table S.6. Parameters for Accuracy Classes			
		Number of Divisions (n)	
Class	**Value of the Verification Division (e)**	**Minimum**	**Maximum**
SI Units			
III	0.1 to 2 g, inclusive	100	10 000
	equal to or greater than 5 g	500	10 000
U.S. Customary Units			
III	0.0002 lb to 0.005 lb, inclusive	100	10 000
	0.005 oz to 0.125 oz, inclusive	100	10 000
	equal to or greater than 0.01 lb	500	10 000
	equal to or greater than 0.25 oz	500	10 000
IIIS	greater than 0.01 lb	100	1 000
	greater than 0.25 oz	100	1 000
For Class III devices, the value of e is specified by the manufacturer as marked on the device; d shall not be smaller than 0.1 e. e shall be differentiated from d by size, shape, or color.			

(Amended 2004)

S.7.   Marking Requirements. – [Also see G-S.1. Identification, G-S.4. Interchange or Reversal of Parts, G-S.6. Marking Operational Controls, Indications, and Features, G-S.7. Lettering, G-UR.2.1.1. Visibility of Identification, and UR.3.3. Special Designs]

   S.7.1.   Location of Marking Information. – Automatic weighing systems which are not permanently attached to an indicating element, and for which the load-receiving element is the only part of the weighing/load-receiving element visible after installation, may have the marking information required in Section 1.10. General Code,

G-S.1. Identification, and Section 2.24. Automatic Weighing Systems Code, Table S.7.a. Marking Requirements and Table S.7.b. Notes for Table S.7.a. located in an area that is accessible only through the use of a tool; provided that the information is easily accessible (e.g., the information may appear on the junction box under an access plate). The identification information for these automatic weighing systems shall be located on the weighbridge (load-receiving element) near the point where the signal leaves the weighing element, or beneath the nearest access cover.

Table S.7.a. Marking Requirements					
	Weighing Equipment				
To Be Marked With ⇩	Weighing, load-receiving, and indicating element in same housing	Indicating element not permanently attached to weighing and load-receiving element	Weighing and load-receiving element not permanently attached to indicating element	Load cell with CC (10)	Other equipment or device (9)
Manufacturer's ID (1)	x	x	x	x	x
Model Designation (1)	x	x	x	x	x
Serial Number and Prefix (2)	x	x	x	x	x (13)
Certificate of Conformance (CC) Number (16)	x	x	x	x	x (16)
Accuracy Class (14)	x	x (8)	x	x	
Nominal Capacity (3)(15)	x	x	x		
Value of Division, d (3)	x	x			
Value of e (4)	x	x			
Temperature Limits (5)	x	x	x	x	
Special Application (11)	x	x	x		
Maximum Number of Scale Divisions, n_{max} (6)		x (8)	x	x	
Minimum Verification Division, (e_{min})			x		
"S" or "M" (7)				x	
Direction of Loading (12)				x	
Minimum Dead Load				x	
Maximum Capacity (Max)	x			x	
Minimum Capacity (Min)	x				
Safe Load Limit				x	
Load Cell Verification Interval (v_{min})				x	
Maximum Belt Speed (m/sec or m/min)	x		x		
Note: Also see Table S.7.b. for applicable parenthetical notes.					

(Amended 1999)

Table S.7.b.
Notes for Table S.7.a.

1. Manufacturer's identification and model designation. (Also see G-S.1. Identification)

2. Serial number and prefix. Also see G-S.1. Identification)

3. The nominal capacity and value of the automatic weighing system division shall be shown together (e.g., 50 000 × 5 kg, or 30 × 0.01 lb) adjacent to the weight display when the nominal capacity and value of the automatic weighing system division are not immediately apparent. Each division value or weight unit shall be marked on variable-division value or division-unit automatic weighing systems.

4. Required only if different from d.

5. Required only on automatic weighing systems if the temperature range on the NTEP CC is narrower than and within − 10 °C to 40 °C (14 °F to 104 °F).
    (Amended 2007)

6. This value may be stated on load cells in units of 1000; (e.g., n_{max} 10 is 10 000 divisions.)

7. Denotes compliance for single or multiple load cell applications.

8. An indicating element not permanently attached to a weighing element shall be clearly and permanently marked with the accuracy Class III, or IIIS and the maximum number of divisions, n_{max}.

9. Necessary to the weighing system but having no metrological effect (e.g., auxiliary remote display, keyboard, etc.).

10. The markings may be either on the load cell or in an accompanying document; except that, if an accompanying document is provided, the serial number shall appear both on the load cell and in the document. The manufacturer's name or trademark, the model designation, and identifying symbol for the serial number shall also be marked both on the load cell and in any accompanying document.

11. An automatic weighing system designed for a special application rather than general use shall be conspicuously marked with suitable words visible to the operator and customer restricting its use to that application.

12. Required if the direction of loading the load cell is not obvious.

13. Serial number and prefix (Also see G-S.1. Identification) modules without "intelligence" on a modular system (e.g., printer, keyboard module, cash drawer, and secondary display in a point-of-sale system) are not required to have serial numbers.

14. The accuracy class of a device shall be marked on the device with the appropriate designation.

15. The nominal capacity shall be conspicuously marked on any automatic-indicating or recording automatic weighing system so constructed that the capacity of the indicating or recording element, or elements, is not immediately apparent.

16. Required only if a CC has been issued for the equipment.

S.7.2.   Marking Required on Components of Automatic Weighing Systems. – The following components of automatic weighing systems shall be marked as specified in Tables S.7.a. Marking Requirements and S.7.b. Notes for Table S.7.2.a.:

    (a) Main elements and components when not contained in a single enclosure for the entire automatic weighing system;

    (b) Load cells for which Certificates of Conformance (CC) have been issued under the National Type Evaluation Program; and

    (c) Other equipment necessary to a weighing system but having no metrological effect on the weighing system.

N. Notes

N.1.   Test Requirements for Automatic Weighing Systems.

    N.1.1.   Test Pucks and Packages.

        (a) Test pucks and packages shall be:

            (1) representative of the type, size, and weight ranges to be weighed on a device; and

            (2) stable while in motion, hence the length and width of a puck or package should be greater than its height.

        (b) For type evaluation the manufacturer shall supply the test pucks or packages for each range of test loads.
(Amended 1997)

    N.1.2.   Accuracy of Test Pucks or Packages. – The error in any test puck or package shall not exceed one-fourth (¼) of the acceptance tolerance. If packages are used to conduct field tests on automatic weighing systems, the package weights shall be determined on a reference scale or balance with an inaccuracy that does not exceed one-fifth (⅕) of the smallest tolerance that can be applied to the device under test.

    N.1.3.   Verification (Testing) Standards. – Field standard weights shall comply with requirements of NIST Handbook 105-1, "Specifications and Tolerances for Field Standard Weights (Class F)" or the tolerances expressed in Fundamental Considerations, paragraph 3.2. (i.e., one-third of the smallest tolerance applied).

    N.1.4.   Radio Frequency Interference (RFI) and Other Electromagnetic Interference Susceptibility, Field Evaluation. – An RFI test shall be conducted at a given installation when the presence of RFI has been verified and characterized if those conditions are considered "usual and customary."
(Added 2004)

    N.1.5.   Tests Loads. – A performance test shall consist of four separate test runs conducted at different test loads according to Table N.1.5. Test Loads.

Table N.1.5. Test Loads
At or near minimum capacity
At or near maximum capacity
At two (2) critical points between minimum and maximum capacity
Test may be conducted at other loads if the device is intended for use at other specific capacities

N.1.6.   Influence Factor Testing. – Influence factor testing shall be conducted statically.

N.2.   Test Procedures - Weigh-Labelers. – If the device is designed for use in a non-automatic weighing mode, it shall be tested in the non-automatic mode according to NIST Handbook 44, Section 2.20. Scales Code.

Note: If the device is designed for only automatic weighing, it shall only be tested in the automatic weighing mode.
(Amended 2004)

N.2.1.   Non-Automatic Tests.

N.2.1.1.   Increasing-Load Test. – The increasing-load test shall be conducted with the test loads approximately centered on the load-receiving element of the scale.

N.2.1.2.   Decreasing-Load Test. – The decreasing-load test shall be conducted with the test loads approximately centered on the load-receiving element of the scale.

N.2.1.3.   Shift Test. – To determine the effect of off-center loading, a test load equal to one-half (½) maximum capacity shall be placed in the center of each of the four points equidistant between the center and front, left, back, and right edges of the load receiver.

N.2.1.4.   Discrimination Test. – A discrimination test shall be conducted with the weighing device in equilibrium at zero-load and at maximum test load, and under controlled conditions in which environmental factors are reduced to the extent that they will not affect the results obtained. This test is conducted from just below the lower edge of the zone of uncertainty for increasing-load tests, or from just above the upper edge of the zone of uncertainty for decreasing-load tests.

N.2.1.5.   Zero-Load Balance Change. – A zero-load balance change test shall be conducted on all automatic weighing systems after the removal of any test load. The zero-load balance should not change by more than the minimum tolerance applicable. (Also see G-UR.4.2. Abnormal Performance)
(Amended 2004)

N.2.2.   Automatic Test Procedures.

N.2.2.1.   Tests Non-Automatic. – If the automatic weighing system is designed to operate non-automatically, and is used in that manner, during normal use operation, it shall be tested non-automatically using mass standards. The device shall not be tested non-automatically if it is used only in the automatic mode.

N.2.2.2.   Automatic Tests. – The device shall be tested at the normal operating speed using packages. Test runs should be conducted using at least two test loads distributed over its normal weighing range (e.g., near the lowest and highest ranges in which the device is typically operated.) Each test load should be run a minimum of ten consecutive times.
(Amended 2004)

N.3.   Test Procedures - Automatic Checkweigher.

N.3.1.   Tests Non-Automatic. – If the scale is designed to operate non-automatically during normal user operation, it shall be tested non-automatically according to paragraphs N.2.1.1. Increasing Load Test through N.2.1.5. Zero-Balance Change.
(Amended 2004)

N.3.2.   Automatic Tests. – The device shall be tested at the highest speed in each weight range using standardized test pucks or packages. Test runs shall be conducted using two test loads. The number of consecutive test weighments shall be as specified in Table N.3.2. Number of Sample Weights per Test for Automatic Checkweighers.

(Amended 2004)

Table N.3.2. Number of Sample Weights per Test for Automatic Checkweighers		
Weighing Range m = mass of test load	Number of Sample Weights per Test	
	Field	Type Evaluation
20 divisions ≤ m ≤ 10 kg 20 divisions ≤ m ≤ 22 lb	30	60
10 kg < m ≤ 25 kg 22 lb < m ≤ 55 lb	16	32
25 kg < m ≤ 100 kg 55 lb < m ≤ 220 lb	10	20
100 kg (220 lb) < m	10	10

T. Tolerances

T.1. Principles.

T.1.1. Design. – The tolerance for a weighing device is a performance requirement independent of the design principle used.

T.1.2. Scale Division. – The tolerance for a weighing device is related to the value of the scale division (d) or the value of the verification scale division (e) and is generally expressed in terms of d or e. The random tolerance for automatic checkweighers is expressed in terms of Maximum Allowable Variance (MAV).

T.2. Tolerance Application.

T.2.1. General. – The tolerance values are positive (+) and negative (−) with the weighing device adjusted to zero at no load. When tare is in use, the tolerance values are applied from the tare zero reference (zero net weight indication); the tolerance values apply to the net weight indication for any possible tare load using certified test loads.

(Amended 2008)

T.2.2. Type Evaluation Examinations. – For type evaluation examinations, the tolerance values apply to increasing and decreasing load tests within the temperature and power supply limits specified in T.7. Influence Factors.

(Amended 2004)

T.2.3. Subsequent Verification Examinations. – For subsequent verification examinations, the tolerance values apply regardless of the influence factors in effect at the time of the conduct of the examination. (Also see G-N.2. Testing with Nonassociated Equipment.)

(Added 2007)

T.2.4. Multiple Range and Multi-Interval Automatic Weighing System. – For multiple range and multi-interval devices, the tolerance values are based on the value of the scale division of the range in use.

T.3.   Tolerance Values.

Table T.3. Class III - Tolerance in Divisions (e)		
Test Load in Divisions	**Tolerance in Divisions**	
Class III	**Acceptance**	**Maintenance**
0 - 500	± 0.5	± 1
501 - 2000	± 1.0	± 2
2001 - 4000	± 1.5	± 3
4001 +	± 2.5	± 5

T.3.1.   Tolerance Values – Class III Weigh-Labeler. (Also see Section T.3.2. Class IIIS Weigh-Labelers.)

T.3.1.1.   Non-automatic Tests. – Tolerance values shall be as specified in Table T.3. Class III - Tolerance in Divisions (e).
(Amended 2004)

T.3.1.2.   Automatic Tests. – Acceptance tolerance values shall be the same as maintenance tolerance values specified in Table T.3. Class III - Tolerance in Divisions (e).
(Amended 2004)

T.3.2.   Tolerance Values - Class IIIS Weigh-labelers in Package Shipping Applications.
(Added 1997)

T.3.2.1.   Non-automatic Tests. – Tolerance values shall be as specified in Table T.3.2.1. Non-automatic Tolerances for Class IIIS Weigh-labelers.
(Amended 2004)

T.3.2.2.   Automatic Tests. – Tolerance values specified in Table T.3.2.2. Automatic Tolerances for Class IIIS Weigh-labelers shall be applied.
(Amended 2004)

Table T.3.2.1. Non-automatic Tolerances for Class IIIS Weigh-labelers		
Test Load in Divisions	**Tolerance in Divisions**	
Class IIIS	**Acceptance**	**Maintenance**
0 - 50	± 0.5	± 1
51 - 200	± 1.0	± 2
201 - 1000	± 1.5	± 3

(Added 1997) (Amended 2004)

Table T.3.2.2. Automatic Tolerances for Class IIIS Weigh-labelers		
Test Load in Divisions	**Tolerance in Divisions**	
Class IIIS	**Acceptance**	**Maintenance**
0 - 50	± 1.5	± 2
51 - 200	± 2.0	± 3
201 - 1000	± 2.5	± 4

(Added 1997) (Amended 2004)

T.3.3.   Tolerance Values. – Automatic Checkweighers.

T.3.3.1.   Laboratory Tests for Automatic Checkweighers.

T.3.3.1.1.   Non-Automatic Tests. – The acceptance tolerance values specified in Table T.3. Class III - Tolerance in Divisions (e), shall be applied.
(Amended 2004)

T.3.3.1.2. Automatic Tests.

(a) The systematic error for each test run shall be within the acceptance tolerances specified in Table T.3. Class III - Tolerance in Divisions (e) for the test loads specified in Table N.1.5. Test Loads.

(Amended 2004)

(b) The standard deviation of the results shall not exceed one-ninth ($\frac{1}{9}$) of the MAV for specific package weights (which means that three standard deviations cannot exceed one-third ($\frac{1}{3}$) of the MAV value) as required in the latest edition of NIST Handbook 133, "Checking the Net Contents of Packaged Goods." This value does not change regardless of whether acceptance or maintenance tolerances are being applied to the device under test.

(Amended 2004)

(1) For U.S. Department of Agriculture (USDA) inspected meat and poultry products packaged at a plant subject to inspection by the USDA Food Safety and Inspection Service, use NIST Handbook 133, Appendix A. Tables, Table 2-9, U.S. Department of Agriculture, Meat and Poultry, Groups and Lower Limits for Individual Packages;

(2) for all other packages with a labeled net quantity in terms of weight, use NIST Handbook 133, Appendix A. Tables, Table 2-5, Maximum Allowable Variations (MAVs) for Packages Labeled by Weight; or

(3) for all packages with a labeled net quantity in terms of liquid or dry volume use NIST Handbook 133, Appendix A. Tables, Table 2-6, Maximum Allowable Variations (MAVs) for Packages Labeled by Liquid or Dry Volume.

(Amended 2004)

T.3.3.2. Field Tests for Automatic Checkweighers.

T.3.3.2.1. Non-Automatic Test. – The tolerance values shall be as specified in Table T.3. Class III – Tolerance in Divisions (e).

(Amended 2004)

T.3.3.2.2. Automatic Test.

(a) The systematic error requirement is not applied in a field test.

(b) The standard deviation of the test results shall not exceed one-ninth ($\frac{1}{9}$) of the MAV for specific package weights (which means that three standard deviations cannot exceed one-third ($\frac{1}{3}$) of the MAV value) as required in the latest Edition of NIST Handbook 133. This value does not change regardless of whether acceptance or maintenance tolerances are being applied to the device under test.

(Amended 2004)

(1) For USDA inspected meat and poultry products packaged at a plant subject to inspection by the USDA Food Safety and Inspection Service, use NIST Handbook 133, Appendix A, Tables, Table 2-9, U.S. Department of Agriculture, Meat and Poultry, Groups and Lower Limits for Individual Packages;

(2) for all other packages with a labeled net quantity in terms of weight, use NIST Handbook 133, Appendix A. Tables, Table 2-5, Maximum Allowable Variations (MAVs) for Packages Labeled by Weight; or

(3) for all packages with a labeled net quantity in terms of liquid or dry volume use NIST Handbook 133, Appendix A. Tables, Table 2-6. Maximum Allowable Variations (MAVs) for Packages Labeled by Liquid or Dry Volume.

T.4.   Agreement of Indications. – In the case of a weighing system equipped with more than one indicating element or indicating element and recording element combination, the difference in the weight value indications of any load shall not be greater than the absolute value of the applicable tolerance for that load, and shall be within tolerance limits.

T.5.   Repeatability. – The results obtained from several weighings of the same load under reasonably constant test conditions shall agree within the absolute value of the maintenance tolerance for that load, and shall be within applicable tolerances.

(Amended 2004)

T.6.   Discrimination. – A test load equivalent to 1.4 d shall cause a change in the indicated or recorded value of at least 2.0 d. This requires the zone of uncertainty to be not greater than 0.3 d (See N.2.1.4. Discrimination Test).

(Amended 2004)

T.7.   Influence Factors. – The following factors are applicable to tests conducted under controlled conditions only.

T.7.1.   Temperature. – Devices shall satisfy the tolerance requirements under the following temperature conditions:

T.7.1.1. if not specified in the operating instructions or if not marked on the device, the temperature limits shall be: − 10 °C to 40 °C (14 °F to 104 °F).

T.7.1.2. if temperature limits are specified for the device, the range shall be at least 30 °C (54 °F).

T.7.1.3.   Temperature Effect on Zero-Load Balance. – The zero-load indication shall not vary by more than one division per 5 °C (9 °F) change in temperature.

T.7.1.4.   Operating Temperature. – The indicating or recording element shall not display nor record any usable values until the operating temperature necessary for accurate weighing and a stable zero balance condition have been attained.

T.7.2.   Electric Power Supply.

T.7.2.1.   Range of Voltages.

(a) Automatic weighing systems that operate using alternating current must perform within the conditions defined in paragraphs T.3. Tolerance Values through T.6. Discrimination, inclusive, when tested over the range of − 15 % to + 10 % of the marked nominal line voltage(s) at 60 Hz, or the voltage range marked by the manufacturer, at 60 Hz.

(b) Automatic weighing systems that operate using DC current must perform within the conditions defined in paragraphs T.3. Tolerance Values through T.6. Discrimination, inclusive, when tested over the range from minimum operating voltage[4] to + 20 % of the voltage marked on the instrument (nominal voltage).

(c) Battery-operated electronic automatic weighing systems with external or plug-in power supply (AC or DC) shall either continue to function correctly or not indicate any weight values if the voltage is

---

[4] The minimum operating voltage is defined as the lowest possible operating voltage before the automatic weighing system no longer indicates nor records weight values.

(Added 2004)

below the manufacturer's specified value, the latter being larger than or equal to the minimum operating voltage.[4]

Note: This requirement applies only to metrologically significant voltage supplies.

(Amended 2001)

(Amended 2004)

T.7.2.2.   Power Interruption. – A power interruption shall not cause an indicating or recording element to display or record any values outside the applicable tolerance limits.

T.8.   Radio Frequency Interference (RFI) and Other Electromagnetic Interference Susceptibility. – The difference between the weight indication with the disturbance and the weight indication without the disturbance (also see N.1.4. Radio Frequency Interference (RFI) and Other Electromagnetic Interference Susceptibility, Field Evaluation) shall not exceed one scale division (d) or the equipment shall:

(a) blank the indication;

(b) provide an error message; or

(c) the indication shall be so completely unstable that it could not be interpreted, or transmitted into memory or to a recording element, as a correct measurement value.

(Amended 2004)

UR.   User Requirements

UR.1.   Selection Requirements. – Equipment shall be suitable for the service in which it is used with respect to elements of its design, including but not limited to, its capacity, number of scale divisions, value of the scale division or verification scale division, minimum capacity, and computing capability.

UR.1.1.   General. – Automatic Weighing Systems shall be designated by the manufacturer for that service.

UR.1.2.   Value of the Indicated and Recorded Scale Division. – The value of the division as recorded shall be the same as the division value indicated.

UR.2.   Installation Requirements.

UR.2.1.   Protection from Environmental Factors. – The indicating elements, the lever system or load cells, and the load-receiving element of a permanently installed scale, and the indicating elements of a scale not intended to be permanently installed, shall be adequately protected from environmental factors such as wind, weather, and RFI that may adversely affect the operation or performance of the device.

UR.2.2.   Foundation, Supports, and Clearance. – The foundation and supports of any scale installed in a fixed location shall be such as to provide strength, rigidity, and permanence of all components, and clearance shall be provided around all live parts to the extent that no contacts may result when the load-receiving element is empty, nor throughout the weighing range of the scale.

UR.2.3.   Entry and Departure from Weighing Area. – The belt or other conveyance that introduces the weighed load to the weighing zone and that carries the weighed load away from the weighing zone shall be maintained per the manufacturer's recommendations.

UR.3.   Use Requirements.

UR.3.1.   Minimum Load. – The minimum load shall be as specified by the manufacturer, but not less than twenty divisions since the use of a device to weigh light loads is likely to result in relatively large errors.

UR.3.1.1.   Minimum Load for Class IIIS Weigh-Labelers. – The minimum load shall be as specified by the manufacturer, but not less than ten divisions since the use of a device to weigh light loads is likely to result in relatively large errors.

(Added 1997)

UR.3.2.   Maximum Load. – An automatic weighing system shall not be used to weigh a load of more than its maximum capacity.

(Amended 2004)

UR.3.3.   Special Designs. – An automatic weighing system designed and marked for a special application shall not be used for other than its intended purpose.

UR.3.4.   Use of Manual Gross Weight Entries. – Manual entries are permitted only when a device or system is generating labels for standard weight packages.

UR.4.   Maintenance Requirements.

UR.4.1.   Balance Condition. – If an automatic weighing system is equipped with a zero-load display, the zero-load adjustment of an automatic weighing system shall be maintained so that the device indicates or records a zero balance condition.

UR.4.2.   Level Condition. – If an automatic weighing system is equipped with a level-condition indicator, the automatic weighing system shall be maintained in level.

UR.4.3.   Automatic Weighing System Modification. – The length or the width of the load-receiving element of an automatic weighing system shall not be increased beyond the manufacturer's design dimension, nor shall the capacity of an automatic weighing system be increased beyond its design capacity by replacing or modifying the original primary indicating or recording element with one of a higher capacity, except when the modification has been approved by competent engineering authority, preferably that of the engineering department of the manufacturer of the automatic weighing system, and by the weights and measures authority having jurisdiction over the automatic weighing system.

THIS PAGE INTENTIONALLY LEFT BLANK

Table of Contents

Section 2.25.   Weigh-In-Motion Systems
Used for Vehicle Enforcement Screening – Tentative Code

This tentative code has a trial or experimental status and is not intended to be enforced. The requirements are designed for study prior to the development and adoption of a final code. Officials wanting to conduct an official examination of a device or system are advised to see paragraph G-A.3. Special and Unclassified Equipment.

(Tentative Code Added 2015)

A. Application

A.1.   General. – This code applies to systems used to weigh vehicles, while in motion, for the purpose of screening and sorting the vehicles based on the vehicle weight to determine if a static weighment is necessary.

A.2.   Exception. – This code does not apply to weighing systems intended for the collection of statistical traffic data.

A.3.   Additional Code Requirements. – In addition to the requirements of this code, weigh-in-motion screening systems shall meet the requirements of Section 1.10. General Code.

S. Specifications

S.1.   Design of Indicating and Recording Elements and of Recorded Representations.

S.1.1.   Ready Indication. – The system shall provide a means of verifying that the system is operational and ready for use.

S.1.2.   Value of System Division Units. – The value of a system division "d" expressed in a unit of weight shall be equal to:

(a)   1, 2, or 5; or

(b)   a decimal multiple or submultiple of 1, 2, or 5.

Examples: divisions may be 10, 20, 50, 100; or 0.01, 0.02, 0.05; or 0.1, 0.2, 0.5, etc.

S.1.2.1.   Units of Measure. – The system shall indicate weight values using only a single unit of measure.

S.1.3.   Maximum Value of Division. – The value of the system division "d" for a Class A, weight-in-motion system shall not be greater than 50 kg (100 lb).

S.1.4.   Value of Other Units of Measure.

S.1.4.1.   Speed. – Vehicle speeds shall be measured in miles per hour or kilometers per hour.

S.1.4.2.   Axle-Spacing (Length). – The center-to-center distance between any two successive axles shall be measured in:

(a)   meters and decimal submultiples of a meter;

(b)   feet and inches; or

(c) feet and decimal submultiples of a foot.

S.1.4.3.   Vehicle Length. – If the system is capable of measuring the overall length of the vehicle, the length of the vehicle shall be measured in feet and/or inches, or meters.

S.1.5.   Capacity Indication. – An indicating or recording element shall not display nor record any values greater than 105 % of the specified capacity of the load receiving element.

S.1.6.   Identification of a Fault. – Fault conditions shall be presented to the operator in a clear and unambiguous means. The following fault conditions shall be identified:

(a) Vehicle speed is below the minimum or above the maximum speed as specified.

(b) The maximum number of vehicle axles as specified has been exceeded.

(c) A change in vehicle speed greater than that specified has been detected.

S.1.7.   Recorded Representations.

S.1.7.1.   Values to be Recorded. – At a minimum, the following values shall be printed and/or stored electronically for each vehicle weighment:

(a) transaction identification number;

(b) lane identification (required if more than one lane at the site has the ability to weigh a vehicle in motion);

(c) vehicle speed;

(d) number of axles;

(e) weight of each axle;

(f) identification and weight of axle groups;

(g) axle spacing;

(h) total vehicle weight;

(i) all fault conditions that occurred during the weighing of the vehicle;

(j) violations, as identified in paragraph S.2.1. Violation Parameters, which occurred during the weighing of the vehicle; and

(k) time and date.

S.1.8.   Value of the Indicated and Recorded System Division. – The value of the system's division "(d)," as recorded, shall be the same as the division value indicated.

S.2.   System Design Requirements.

S.2.1.   Violation Parameters. – The instrument shall be capable of accepting user-entered violation parameters for the following items:

(a) single axle weight limit;

(b) axle group weight limit;

(c) gross vehicle weight limit; and

(d) bridge formula maximum.

The instrument shall display and/or record violation conditions when these parameters have been exceeded.

S.3. Design of Weighing Elements.

S.3.1. Multiple Load-Receiving Elements. – An instrument with a single indicating or recording element, or a combination indicating-recording element, that is coupled to two or more load-receiving elements with independent weighing systems, shall be provided with means to prohibit the activation of any load-receiving element (or elements) not in use, and shall be provided with automatic means to indicate clearly and definitely which load receiving element (or elements) is in use.

S.4. Design of Weighing Devices, Accuracy Class.

S.4.1. Designation of Accuracy. – Weigh-in-motion systems meeting the requirements of this code shall be designated as accuracy Class A.

Note: This does not preclude higher accuracy classes from being proposed and added to this Code in the future when it can be demonstrated that weigh-in-motion systems grouped within those accuracy classes can achieve the higher level of accuracy specified for those devices.

S.5. Marking Requirements. – In addition to the marking requirements in G-S.1. Identification (except G.S.1.(e)), the system shall be marked with the following information:

(a) accuracy class;

(b) value of the system division "d";

(c) operational temperature limits;

(d) number of instrumented lanes (not required if only one lane is instrumented);

(e) minimum and maximum vehicle speed;

(f) maximum number of axles per vehicle;

(g) maximum change in vehicle speed during weighment; and

(h) minimum and maximum load.

S.5.1.   Location of Marking Information. – The marking information required in Section 1.10. General Code, G-S.1. Identification and Section 2.25. Weigh-in-Motion Systems, S.5. Marking Requirements shall be visible after installation. The information shall be marked on the system or recalled from an information screen.

N. Notes

N.1.   Test Procedures.

N.1.1.   Selection of Test Vehicles. – All dynamic testing associated with the procedures described in each of the subparagraphs of N.1.5 shall be performed with a minimum of two test vehicles.

(a)   The first test vehicle may be a two-axle, six-tire, single-unit truck; that is, a vehicle with two axles with the rear axle having dual wheels. The vehicle shall have a maximum gross vehicle weight of 10 000 lb.

(b)   The second test vehicle shall be a five-axle, single-trailer truck with a maximum gross vehicle weight of 80 000 lb.

Note: Consideration should be made for testing the systems using vehicles which are typical to the system's daily operation.

N.1.1.1.   Weighing of Test Vehicles. – All test vehicles shall be weighed on a reference scale before being used to conduct the dynamic tests.

N.1.1.2.   Determining Reference Weights for Axle, Axle Groups, and Gross Vehicle Weight. – The reference weights shall be the average weight value of a minimum of three static weighments of all single axles, axle groups, and gross vehicle weight.

Note: The axles within an axle group are not considered single axles.

N.1.2.   Test Loads.

N.1.2.1.   Static Test Loads. – All static test loads shall use certified test weights.

N.1.2.2.   Dynamic Test Loads. – Test vehicles used for dynamic testing shall be loaded to 85 % to 95 % of their legal maximum Gross Vehicle Weight. The "load" shall be non-shifting and shall be positioned to present as close as possible, an equal side-to-side load.

N.1.3.   Reference Scale. – Each reference vehicle shall be weighed statically on a multiple platform vehicle scale comprised of three individual weighing/load-receiving elements, each an independent scale. The three individual weighing/load receiving elements shall be of such dimension and spacing to facilitate:

(a)   the single-draft weighing of all reference test vehicles;

(b)   the simultaneous weighing of each single axle and axle group of the reference test vehicles on different individual elements of the scale; and

(c)   gross vehicle weight determined by summing the values of the different reference axle and reference axle groups of a test vehicle.

The scale shall be tested immediately prior to using it to establish reference test loads and in no case more than 24 hours prior. To qualify for use as a suitable reference scale, it must meet NIST Handbook 44, Class III L maintenance tolerances.

N.1.3.1.   Location of a Reference Scale. – The location of the reference scale must be considered since vehicle weights will change due to fuel consumption.

N.1.4.   Test Speeds. – All dynamic tests shall be conducted within 20 % below or at the posted speed limit.

N.1.5.   Test Procedures.

N.1.5.1.   Dynamic Load Test. – The dynamic test shall be conducted using the test vehicles defined in N.1.1. Selection of Test Vehicles. The test shall consist of a minimum of 20 runs for each test vehicle at the speed as stated in N.1.4. Test Speeds.

At the conclusion of the dynamic test there will be a minimum of 20 weight readings for each single axle, axle group, and gross vehicle weight of the test vehicle. The tolerance for each weight reading shall be based on the percentage values specified in Table T.2.2. Tolerances for Accuracy Class A.

N.1.5.2.   Vehicle Position Test. – During the conduct of the dynamic testing, ensure the vehicle stays within the defined roadway along the width of the sensor. The test shall be conducted with 10 runs with the vehicle centered along the width of the sensor; 5 runs with the vehicle on the right side along the width of the sensor; and 5 runs with the vehicle on the left side along the width of the sensor. Only gross vehicle weight is used for this test and the tolerance for each weighment shall be based on the tolerance value specified in T.2.3. Tolerance Value for Vehicle Position Test.

N.1.5.3.   Axle Spacing Test. – The axle spacing test is a review of the displayed and/or recorded axle spacing distance of the test vehicles. The tolerance value for each distance shall be based on the tolerance value specified in T.2.4. Tolerance Value for Axle Spacing.

T.   Tolerances

T.1.   Principles.

T.1.1.   Design. – The tolerance for a weigh-in-motion system is a performance requirement independent of the design principle used.

T.2.   Tolerance Values for Accuracy Class A.

T.2.1.   Tests Involving Digital Indications or Representations. – To the tolerances that would otherwise be applied in paragraphs T.2.2. Tolerance Value for Dynamic Load Test and T.2.3. Tolerance Value for Vehicle Position Test, there shall be added an amount equal to one-half the value of the scale division to account for the uncertainty of digital rounding.

T.2.2.   Tolerance Values for Dynamic Load Test. – The tolerance values applicable during dynamic load testing are as specified in Table T.2.2.

Table T.2.2. Tolerances for Accuracy Class A	
Load Description*	**Tolerance as a Percentage of Applied Test Load**
Axle Load	± 20 %
Axle Group Load	± 15 %
Gross Vehicle Weight	± 10 %
* No more than 5 % of the weighments in each of the load description subgroups shown in this table shall exceed the applicable tolerance.	

T.2.3.   Tolerance Value for Vehicle Position Test. – The tolerance value applied to each gross vehicle weighment is ± 10 % of the applied test load.

T.2.4.   Tolerance Value for Axle Spacing. – The tolerance value applied to each axle spacing measurement shall be ± 0.15 m (0.5 ft).

T.3.   Influence Factors. – The following factor is applicable to tests conducted under controlled conditions only.

T.3.1.   Temperature. – Systems shall satisfy the tolerance requirements under all operating temperature unless a limited operating temperature range is specified by the manufacturer.

T.4.   Radio Frequency Interference (RFI) and Other Electromagnetic Interference Susceptibility. – The difference between the weight indication due to the disturbance and the weight indication without the disturbance shall not exceed the tolerance value as stated in Table T.2.2. Tolerances for Accuracy Class A.

UR.   User Requirements

UR.1.   Selection Requirements. – Equipment shall be suitable for the service in which it is used with respect to elements of its design, including but not limited to, its capacity, number of scale divisions, value of the scale division, or verification scale division and minimum capacity.

UR.1.1.   General. – The typical class or type of device for particular weighing applications is shown in Table 1. Typical Class or Type of Device for Weighing Applications.

Table 1. Typical Class or Type of Device for Weighing Applications	
Class	**Weighing Application**
A	Screening and sorting of vehicles based on axle, axle group, and gross vehicle weight.
Note: A WIM system with a higher accuracy class than that specified as "typical" may be used.	

UR.2.   User Location Conditions and Maintenance. – The system shall be installed and maintained as defined in the manufacturer's recommendation.

UR.2.1.   System Modification. – The dimensions (e.g., length, width, thickness, etc.) of the load receiving element of a system shall not be changed beyond the manufacturer's specifications, nor shall the capacity of a scale be increased beyond its design capacity by replacing or modifying the original primary indicating or recording element with one of a higher capacity, except when the modification has been approved by a competent engineering authority, preferably that of the engineering department of the manufacturer of the system, and by the weights and measures authority having jurisdiction over the system.

UR.2.2.   Foundation, Supports, and Clearance. – The foundation and supports shall be such as to provide strength, rigidity, and permanence of all components.

On load-receiving elements, which use moving parts for determining the load value, clearance shall be provided around all live parts to the extent that no contacts may result when the load-receiving element is empty, nor throughout the weighing range of the system.

UR.2.3.   Access to Weighing Elements. – If necessary, adequate provision shall be made for inspection and maintenance of the weighing elements.

UR.3.   Maximum Load. – A system shall not be used to weigh a load of more than the marked maximum load of the system.

Appendix D. Definitions

The specific code to which the definition applies is shown in the [brackets] at the end of the definition. Definitions for the General Code [1.10] apply to all codes in NIST Handbook 44.

A

axle. – The axis oriented transversely to the nominal direction of vehicle motion, and extending the full width of the vehicle, about which the wheel(s) at both ends rotate. [2.25]

axle-group load. – The sum of all tire loads of the wheels on a group of adjacent axles; a portion of the gross-vehicle weight. [2.25]

axle load. – The sum of all tire loads of the wheels on an axle; a portion of the gross-vehicle weight. [2.25]

axle spacing. – The distance between the centers of any two axles. When specifying axle spacing, the axels used also need to be identified. [2.25]

S

single-axle load. – The load transmitted to the road surface by the tires lying on the same longitudinal axis (that axis transverse to the movement of the vehicle and about which the wheels rotate). [2.25]

T

tandem-axle load. – The load transmitted to the road surface by the tires of two single-axles lying on the same longitudinal axis (that axis transverse to the movement of the vehicle and about which the wheels rotate). [2.25]

triple-axle load. – The load transmitted to the road surface by the tires of three single-axles lying on the same longitudinal axis (that axis transverse to the movement of the vehicle and about which the wheels rotate). [2.25]

W

weigh-in-motion (WIM). – A process of estimating a moving vehicle's gross weight and the portion of that weight that is carried by each wheel, axle, or axle group, or combination thereof, by measurement and analysis of dynamic vehicle tire forces. [2.25]

weigh-in-motion screening scale. – A weigh-in-motion system used to identify potentially overweight vehicles. [2.25]

wheel weight. – The weight value of any single or set of wheels on one side of a vehicle on a single axle. [2.25]

WIM System. – A set of sensors and supporting instruments that measure the presence of a moving vehicle and the related dynamic tire forces at specified locations with respect to time; estimate tire loads; calculate speed, axle spacing, vehicle class according to axle arrangement, and other parameters concerning the vehicle; and process, display, store, and transmit this information. This standard applies only to highway vehicles. [2.25]

Section 3

Table of Contents

Note: In this section of Handbook 44, the reference temperature for the temperature compensation of refined petroleum products is shown as "15 °C (60 °F)." Although these values are not exact equivalents, they reflect industry usage when the SI and U.S. customary units are used in measurements.

THIS PAGE INTENTIONALLY LEFT BLANK

Table of Contents

Section 3.30.   Liquid-Measuring Devices

A. Application

A.1.   General. – This code applies to:

(a) devices used for the measurement of liquids, including liquid fuels and lubricants; and

(b) wholesale devices used for the measurement and delivery of agri-chemical liquids such as fertilizers, feeds, herbicides, pesticides, insecticides, fungicides, and defoliants.
(Added 1985)

A.2.   Exceptions. – This code does not apply to:

(a) meters mounted on vehicle tanks (Also see Section 3.31. Code for Vehicle-Tank Meters.);

(b) devices used for dispensing liquefied petroleum gases (Also see Section 3.32. Code for Liquefied Petroleum Gas and Anhydrous Ammonia Liquid-Measuring Devices.);

(c) devices used for dispensing other liquids that do not remain in a liquid state at atmospheric pressures and temperatures;

(d) water meters;

(e) devices used solely for dispensing a product in connection with operations in which the amount dispensed does not affect customer charges; or

(f) mass flow meters. (Also see Section 3.37. Code for Mass Flow Meters.)
(Added 1994)

A.3.   Additional Code Requirements. – In addition to the requirements of this code, liquid-measuring devices shall meet the requirements of Section 1.10. General Code.

S. Specifications

S.1.   Indicating and Recording Elements and Recorded Representations.

S.1.1.   General. – A liquid-measuring device:

(a) shall be equipped with a primary indicating element; and

(b) may be equipped with a primary recording element.

S.1.2.   Units. – A liquid-measuring device shall indicate, and record if the device is equipped to record, its deliveries in liters, gallons, quarts, pints, fluid ounces, or binary-submultiples or decimal subdivisions of the liter or gallon.
(Amended 1987, 1994, and 2006)

S.1.2.1.   Retail Motor-Fuel Devices. – Deliveries shall be indicated and recorded, if the device is equipped to record, in liters or gallons and decimal subdivisions or fractional equivalents thereof.
(Added 1979)

S.1.2.2.   Agri-Chemical Liquid Devices.

S.1.2.2.1.   Liquid Measure. – Deliveries shall be indicated and recorded in liters or gallons and decimal subdivisions or fractional equivalents thereof.

S.1.2.3.   Value of Smallest Unit. – The value of the smallest unit of indicated delivery, and recorded delivery if the device is equipped to record, shall not exceed the equivalent of:

(a) 0.5 L (0.1 gal) on devices with a maximum rated flow rate of 750 L/min (200 gal/min) or less;

(b) 5 L (1 gal) on devices with a maximum rated flow of more than 750 L/min (200 gal/min); or

(c) 5 L (1 gal) on meters with a rated maximum flow rate of 375 L/min (100 gal/min) or more used for jet fuel aviation refueling systems.
(Added 2007)

This requirement does not apply to manually operated devices equipped with stops or stroke-limiting means.
(Amended 1983, 1986, and 2007)

S.1.3.   Advancement of Indicating and Recording Elements. – It shall not be possible to advance primary indicating and recording elements except by the mechanical operation of the device. Clearing a device by advancing its elements to zero is permitted, but only if:

(a) once started, the advancement movement cannot be stopped until zero is reached; and

(b) in the case of indicating elements only, such elements are automatically obscured until the elements reach the correct zero position.

S.1.4.   Graduations.

S.1.4.1.   Length. – Graduations shall be varied in length so that they may be conveniently read.

S.1.4.2.   Width. – In a series of graduations, the width of:

(a) every graduation shall be at least 0.2 mm (0.008 in) but not greater than the minimum clear interval between graduations; and

(b) main graduations shall be not more than 50 % greater than the width of subordinate graduations.

S.1.4.3.   Clear Interval Between Graduations. – The clear interval between graduations shall be not less than 1.0 mm (0.04 in). If the graduations are not parallel, the measurement shall be made:

(a) along the line of movement of the tip of the index of the indicator as it passes over the graduations; or

(b) if the indicator extends over the entire length of the graduations, at the point of widest separation of the graduations.

S.1.5.   Indicators.

S.1.5.1.   Symmetry. – The portion of the index of an indicator associated with the graduations shall be symmetrical with respect to the graduations.

S.1.5.2.   Length.

(a) If the indicator and the graduations are in different planes, the index of the indicator shall extend to each graduation with which it is to be used.

(b) If the indicator is in the same plane as the graduations, the distance between the index of the indicator and the ends of the graduations, measured along the line of the graduations, shall be not more than 1.0 mm (0.04 in).

S.1.5.3.   Width.

(a) *The index of an indicator shall not be wider than the width of the narrowest graduation.*
[Nonretroactive as of January 1, 2002]
(Amended 2000)

(b) If the index of an indicator extends over the entire length of a graduation, it shall be of uniform width throughout the portion that coincides with the graduation.

S.1.5.4.   Clearance. – If the indicator and the graduations are in different planes, the clearance between the index of an indicator and the plane of the graduations shall be no greater than 1.5 mm (0.06 in).

S.1.5.5.   Parallax. – Parallax effects shall be reduced to the practical minimum.

S.1.6.   Additional Operating Requirements, Retail Devices (Except Slow-flow Meters).

S.1.6.1.   Indication of Delivery. – The device shall automatically show on its face the initial zero condition and the quantity delivered (up to the nominal capacity). However, the following requirements shall apply:

For electronic devices manufactured prior to January 1, 2006, the first 0.03 L (or 0.009 gal) of a delivery and its associated total sales price need not be indicated.

For electronic devices manufactured on or after January 1, 2006, the measurement, indication of delivered quantity, and the indication of total sales price shall be inhibited until the fueling position reaches conditions necessary to ensure that the delivery starts at zero.
[Nonretroactive as of January 1, 2006]
(Added 2005)
(Amended 1982 and 2005)

S.1.6.2.   Provisions for Power Loss.

S.1.6.2.1.   Transaction Information. – *In the event of a power loss, the information needed to complete any transaction in progress at the time of the power loss (such as the quantity and unit price, or sales price) shall be determinable for at least 15 minutes at the dispenser or at the console if the console is accessible to the customer.*
[Nonretroactive as of January 1, 1983]

S.1.6.2.2.   User Information. – *The device memory shall retain information on the quantity of fuel dispensed and the sales price totals during power loss.*
[Nonretroactive as of January 1, 1983]

S.1.6.3.   Return to Zero.

(a) The primary indicating elements, and primary recording elements if the device is equipped to record, shall be readily returnable to a definite zero indication. However, a key-lock operated or other self-operated device may be equipped with cumulative indicating or recording elements, provided that it is also equipped with a zero-return indicating element.

(b) It shall not be possible to return primary indicating elements, or primary recording elements beyond the correct zero position.

(c) Primary indicating elements shall not be resettable to zero during a delivery.

(Amended 1972 and 2016)

S.1.6.4.   Display of Unit Price and Product Identity.

S.1.6.4.1.   Unit Price.

(a) A computing or money-operated device shall be able to display on each face the unit price at which the device is set to compute or to dispense.

(b) *Except for dispensers used exclusively for fleet sales, other price contract sales, and truck refueling (e.g., truck stop dispensers used only to refuel trucks), whenever a grade, brand, blend, or mixture is offered for sale from a device at more than one unit price, then all of the unit prices at which that product is offered for sale shall meet the following conditions:*

   (1) *For a system that applies a discount prior to the delivery, all unit prices shall be displayed or shall be capable of being displayed on the dispenser through a deliberate action of the customer prior to the delivery of the product. It is not necessary that all of the unit prices for all grades, brands, blends, or mixtures be simultaneously displayed prior to the delivery of the product.*
   [Effective and Nonretroactive as of January 1, 1991]

   (2) For a system that offers post-delivery discounts on fuel sales, display of pre-delivery unit price information is exempt from (b)(1), provided the system complies with S.1.6.8. Recorded Representations for Transactions Where a Post-Delivery Discount(s) is Provided.

   (Added 2012)

   Note: When a product is offered at more than one unit price, display of the unit price information may be through the deliberate action of the customer: 1) using controls on the device; 2) through the customer's use of personal or vehicle-mounted electronic equipment communicating with the system; or 3) verbal instructions by the customer.

   (Added 2012)

(Amended 1989, 1997, and 2012)

S.1.6.4.2.   Product Identity.

(a) A device shall be able to conspicuously display on each side the identity of the product being dispensed.

(b) A device designed to dispense more than one grade, brand, blend, or mixture of product also shall be able to display on each side the identity of the grade, brand, blend, or mixture being dispensed.

S.1.6.5.   Money-Value Computations.

(a) *A computing device shall compute the total sales price at any single-purchase unit price (i.e., excluding fleet sales, other price contract sales, and truck stop dispensers used only to refuel trucks) for which the product being measured is offered for sale at any delivery possible within*

either the measurement range of the device or the range of the computing elements, whichever is less.
[Effective and Nonretroactive as of January 1, 1991]

(b) The analog sales price indicated for any delivered quantity shall not differ from a mathematically computed price (quantity × unit price = total sales price) by an amount greater than the value in Table 1. Money-Value Divisions and Maximum Allowable Variations for Money-Value Computations on Mechanical Analog Computers.

(Amended 1984, 1989, and 1993)

S.1.6.5.1.   Money-Value Divisions, Analog. – The values of the graduated intervals representing money values on a computing type device shall be no greater than those in Table 1. Money-Value Divisions and Maximum Allowable Variations for Money-Value Computations on Mechanical Analog Computers.

(Amended 1991)

Table 1. Money-Value Divisions and Maximum Allowable Variations for Money-Value Computations on Mechanical Analog Computers					
Unit Price		Money-Value Division	Maximum Allowable Variation	
From	To and including		Design Test	Field Test
0	$0.25/liter or $1.00/gallon	1¢	± 1¢	± 1¢
$0.25/liter or $1.00/gallon	$0.75/liter or $3.00/gallon	1¢ or 2¢	± 1¢	± 2¢
$0.75/liter or $3.00/gallon	$2.50/liter or $10.00/gallon	1¢ or 2¢	± 1¢	± 2¢
$0.75/liter or $3.00/gallon	$2.50/liter or $10.00/gallon	5¢	± 2½¢	± 5¢

S.1.6.5.2.   Money-Value Divisions, Digital. – A computing type device with digital indications shall comply with the requirements of paragraph G.S.5.5. Money-Values, Mathematical Agreement, and the total price computation shall be based on quantities not exceeding 0.05 L for devices indicating in metric units and 0.01 gal intervals for devices indicating in U.S. customary units.

(Added 1980)

S.1.6.5.3.   *Auxiliary Elements.* – *If a system is equipped with auxiliary indications, all indicated money-value divisions of the auxiliary element shall be identical with those of the primary element.*
[Nonretroactive and Enforceable as of January 1, 1985]

S.1.6.5.4.   Selection of Unit Price. – A system shall not permit a change to the unit price during delivery of product. When a product or grade is offered for sale at more than one unit price through a computing device, the following conditions shall be met:

(a) *Except for a system only capable of applying a post-delivery discount(s), the selection of the unit price shall be made prior to delivery through a deliberate action of the customer to select the unit price for the fuel delivery.*
[Nonretroactive as of January 1, 1991]

(b) For a system only capable of applying a post-delivery discount(s), the selection of the unit price shall be made through a deliberate action of the customer to select the unit price for the fuel delivery.

(Added 2012)

Note: When a product is offered at more than one unit price, selection of the unit price may be through the deliberate action of the customer: 1) using controls on the device; 2) through the customer's use of personal or vehicle-mounted electronic equipment communicating with the system; or 3) verbal instructions by the customer.
(Added 2012)

The provisions in (a) and (b) do not apply to dispensers used exclusively for fleet sales, other price contract sales, and truck refueling (e.g., truck stop dispensers used only to refuel trucks).
(Added 1989) (Amended 1991, 1992, 1993, 1996, and 2012)

S.1.6.5.5.   *Display of Quantity and Total Price.* – *Except for aviation refueling applications, when a delivery is completed, the total price and quantity for that transaction shall be displayed on the face of the dispenser for at least five minutes or until the next transaction is initiated by using controls on the device or other customer-activated controls.*
[Nonretroactive as of January 1, 1994]
(Added 1992) (Amended 1996 and 2007)

S.1.6.5.6.   *Display of Quantity and Total Price, Aviation Refueling Applications.*

(a) *The quantity shall be displayed throughout the transaction.*

(b) *The total price shall also be displayed under one of the following conditions:*

(1) The total price can appear on the face of the dispenser or through a controller adjacent to the device.

(2) If a device is designed to continuously compute and display the total price, then the total price shall be computed and displayed throughout the transaction for the quantity delivered.

(c) *The total price and quantity shall be displayed for at least five minutes or until the next transaction is initiated by using controls on the device or other customer-activated controls.*

(d) *A printed receipt shall be available and shall include, at a minimum, the total price, quantity, and unit price.*
[Nonretroactive as of January 1, 2008]
(Added 2007)

S.1.6.6.   Agreement Between Indications.

(a) When a quantity value indicated or recorded by an auxiliary element is a derived or computed value based on data received from a retail motor fuel dispenser, the value may differ from the quantity value displayed on the dispenser, provided the following conditions are met:

(1) all total money-values for an individual sale that are indicated or recorded by the system agree; and

(2) *within each element, the values indicated or recorded meet the formula (quantity × unit price = total sales price) to the closest cent.*
[Nonretroactive as of January 1, 1988]

(b) When a system applies a post-delivery discount(s) to a fuel's unit price through an auxiliary element, the following conditions shall apply for computed values:

(1) the total volume of the delivery shall be in agreement between all elements in the system.
(Added 2012)

(Added 1985) (Amended 1987, 1988, and 2012)

S.1.6.7.   Recorded Representations. – Except for fleet sales and other price contract sales and for transactions where a post-delivery discount is provided, a printed receipt providing the following information shall be available through a built-in or separate recording element for all transactions conducted with point-of-sale systems or devices activated by debit cards, credit cards, and/or cash:

(a) the total volume of the delivery;

(b) the unit price;

(c) the total computed price; and

(d) the product identity by name, symbol, abbreviation, or code number.
[Nonretroactive as of January 1, 1986]

(Added 1985) (Amended 1997, 2012, and 2014)

S.1.6.8.   Recorded Representations for Transactions Where a Post-Delivery Discount(s) is Provided. – Except for fleet sales and other price contract sales, a printed receipt providing the following information shall be available through a built-in or separate recording element that is part of the system for transactions involving a post-delivery discount:

(a) the product identity by name, symbol, abbreviation, or code number;

(b) transaction information as shown on the dispenser at the end of the delivery and prior to any post-delivery discount(s), including the:

(1) total volume of the delivery;

(2) unit price; and

(3) total computed price of the fuel sale.

(c) an itemization of the post-delivery discounts to the unit price; and

(d) the final total price of the fuel sale after all post-delivery discounts are applied.
(Added 2012) (Amended 2014)

S.1.6.9.   Lubricant Devices, Travel of Indicator. – The indicator shall move at least 2.5 cm (1 in) in relation to the graduations, if provided, for a delivery of 0.5 L (1 pt).

S.1.6.10.   Automatic Timeout – Pay-At-Pump Retail Motor-Fuel Devices. – *Once a device has been authorized, it must de-authorize within two minutes if not activated. Re-authorization of the device must be performed before any product can be dispensed. If the time limit to de-authorize the device is programmable, it shall not accept an entry greater than two minutes.*
[Nonretroactive as of January 1, 2017]
(Added 2016)

S.1.7.   Additional Operating Requirements, Wholesale Devices Only.

S.1.7.1.   Travel of Indicator. – A wholesale device shall be readily operable to deliver accurately any quantity from 200 L (50 gal) to the capacity of the device. If the most sensitive element of the indicating system utilizes an indicator and graduations, the relative movement of these parts corresponding to a delivery of 4 L (1 gal) shall be not less than 5 mm (0.20 in).
(Amended 1987)

S.1.7.2.   Money-Values – Mathematical Agreement. – Any digital money-value indication and any recorded money-value on a computing-type device shall be in mathematical agreement with its associated quantity indication or representation to within 1 cent of money-value.

S.2.   Measuring Elements.

S.2.1.   Vapor Elimination.

(a) A liquid-measuring device shall be equipped with a vapor or air eliminator or other automatic means to prevent the passage of vapor and air through the meter.

(b) Vent lines from the air or vapor eliminator shall be made of metal tubing or other rigid material.
(Amended 1975)

S.2.1.1.   Vapor Elimination on Loading Rack Metering Systems.

(a) A loading rack metering system shall be equipped with a vapor or air eliminator or other automatic means to prevent the passage of vapor and air through the meter unless the system is designed or operationally controlled by a method, approved by the weights and measures jurisdiction having control over the device, such that air and/or vapor cannot enter the system.

(b) Vent lines from the air or vapor eliminator (if present) shall be made of metal tubing or other rigid material.
(Added 1994)

S.2.2.   Provision for Sealing. – Adequate provision shall be made for an approved means of security (e.g., data change audit trail) or for physically applying a security seal in such a manner that requires the security seal to be broken before an adjustment or interchange can be made of:

(a)   any measuring or indicating element;

(b)   any adjustable element for controlling delivery rate when such rate tends to affect the accuracy of deliveries; and

(c)   any metrological parameter that will affect the metrological integrity of the device or system.

When applicable, the adjusting mechanism shall be readily accessible for purposes of affixing a security seal. *[Audit trails shall use the format set forth in Table S.2.2.]**
[*Nonretroactive and Enforceable as of January 1, 1995]

(Amended 1991, 1993, 1995, and 2006)

Table S.2.2. *Categories of Device and Methods of Sealing*	
Categories of Device	*Methods of Sealing*
***Category 1:** No remote configuration capability.*	*Seal by physical seal or two event counters: one for calibration parameters and one for configuration parameters.*
***Category 2:** Remote configuration capability, but access is controlled by physical hardware.* *The device shall clearly indicate that it is in the remote configuration mode and record such message if capable of printing in this mode or shall not operate while in this mode.*	*[The hardware enabling access for remote communication must be on-site. The hardware must be sealed using a physical seal or an event counter for calibration parameters and an event counter for configuration parameters. The event counters may be located either at the individual measuring device or at the system controller; however, an adequate number of counters must be provided to monitor the calibration and configuration parameters of the individual devices at a location. If the counters are located in the system controller rather than at the individual device, means must be provided to generate a hard copy of the information through an on-site device.]** *[*Nonretroactive as of January 1, 1996]*
***Category 3:** Remote configuration capability access may be unlimited or controlled through a software switch (e.g., password).* *[Nonretroactive as of January 1, 1995]* *The device shall clearly indicate that it is in the remote configuration mode and record such message if capable of printing in this mode or shall not operate while in this mode.* *[Nonretroactive as of January 1, 2001]*	*An event logger is required in the device; it must include an event counter (000 to 999), the parameter ID, the date and time of the change, and the new value of the parameter. A printed copy of the information must be available on demand through the device or through another on-site device. The information may also be available electronically. The event logger shall have a capacity to retain records equal to 10 times the number of sealable parameters in the device, but not more than 1000 records are required. (**Note:** Does not require 1000 changes to be stored for each parameter.)*

[Nonretroactive as of January 1, 1995]

(Table Added 1993) (Amended 1995, 1998, 1999, 2006, and 2015)

S.2.3.   Directional Flow Valves. – Valves intended to prevent reversal of flow shall be automatic in operation.

S.2.4.   Stop Mechanism.

S.2.4.1.   Indication. – The delivery for which the device is set shall be conspicuously indicated.
(Amended 1983)

S.2.4.2.   Stroke Limiting Elements. – Stops or other stroke limiting elements subject to direct pressure or impact shall be:

(a) made secure by positive, nonfrictional engagement of these elements; and

(b) adjustable to provide for deliveries within tolerances.
(Amended 1983)

S.2.4.3.   Setting. – If two or more stops or other elements may be selectively brought into operation to permit predetermined quantities of deliveries:

(a) the position for the proper setting of each such element shall be accurately defined; and

(b) any inadvertent displacement from the proper setting shall be obstructed.
(Amended 1983)

S.2.5.   Zero-Set-Back Interlock, Retail Motor-Fuel Devices. – A device shall be constructed so that:

(a) after a delivery cycle has been completed by moving the starting lever to any position that shuts off the device, an automatic interlock prevents a subsequent delivery until the indicating elements, and recording elements if the device is equipped and activated to record, have been returned to their zero positions;

(b) the discharge nozzle cannot be returned to its designed hanging position (that is, any position where the tip of the nozzle is placed in its designed receptacle and the lock can be inserted) until the starting lever is in its designed shut-off position and the zero-set-back interlock has been engaged; and

(c) in a system with more than one dispenser supplied by a single pump, an effective automatic control valve in each dispenser prevents product from being delivered until the indicating elements on that dispenser are in a correct zero position.
(Amended 1981 and 1985)

S.2.6.   Temperature Determination – Wholesale Devices. – For test purposes, means shall be provided (e.g., thermometer well) to determine the temperature of the liquid either:

(a) in the liquid chamber of the meter; or

(b) in the meter inlet or discharge line immediately adjacent to the meter.
[Nonretroactive as of January 1, 1985]
(Added 1984) (Amended 1986)

S.2.7.   Wholesale Devices Equipped with Automatic Temperature Compensators.

S.2.7.1.   Automatic Temperature Compensation. – A device may be equipped with an automatic means for adjusting the indication and registration of the measured volume of product to the volume at 15 °C (60 °F).

S.2.7.2.   Provision for Deactivating. – On a device equipped with an automatic temperature-compensating mechanism that will indicate or record only in terms of gallons compensated to 15 °C (60 °F),

provision shall be made for deactivating the automatic temperature-compensating mechanism so that the meter can indicate and record, if it is equipped to record, in terms of the uncompensated volume.
(Amended 1972)

S.2.7.3.   Provision for Sealing Automatic Temperature-Compensating Systems. – Provision shall be made for applying security seals in such a manner that an automatic temperature-compensating system cannot be disconnected and that no adjustment may be made to the system without breaking the seal.

S.2.7.4.   Temperature Determination with Automatic Temperature-Compensation. – For test purposes, means shall be provided (e.g., thermometer well) to determine the temperature of the liquid either:

(a) in the liquid chamber of the meter; or

(b) immediately adjacent to the meter in the meter inlet or discharge line.
(Amended 1987)

S.2.8.   Exhaustion of Supply, Lubricant Devices Other than Meter Types. – When the level of the supply of lubricant becomes so low as to compromise the accuracy of measurement, the device shall:

(a) automatically become inoperable; or

(b) give a conspicuous and distinct warning.

S.3.   Discharge Lines and Valves.

S.3.1.   Diversion of Measured Liquid. – No means shall be provided by which any measured liquid can be diverted from the measuring chamber of the meter or its discharge line. Two or more delivery outlets may be installed only if automatic means are provided to ensure that:

(a) liquid can flow from only one outlet at a time; and

(b) the direction of flow for which the mechanism may be set at any time is clearly and conspicuously indicated.

An outlet that may be opened for purging or draining the measuring system or for recirculating, if recirculation is required in order to maintain the product in a deliverable state, shall be permitted only when the system is measuring food products, agri-chemicals, biodiesel, or biodiesel blends. Effective automatic means shall be provided to prevent passage of liquid through any such outlet during normal operation of the measuring system and to inhibit meter indications (or advancement of indications) and recorded representations while the outlet is in operation.
(Amended 1991, 1995, 1996, and 2007)

S.3.2.   Exceptions. – The provisions of S.3.1. Diversion of Measured Liquid shall not apply to truck refueling devices when diversion of flow to other than the receiving vehicle cannot readily be accomplished and is readily apparent. Allowable deterrents include, but are not limited to, physical barriers to adjacent driveways, visible valves, or lighting systems that indicate which outlets are in operation, and explanatory signs.
(Amended 1982, 1990, 1991, and 2002)

S.3.3.   Pump-Discharge Unit. – A pump-discharge unit equipped with a flexible discharge hose shall be of the wet-hose type.

S.3.4.   Gravity-Discharge Unit. – On a gravity-discharge unit:

(a) the discharge hose or equivalent pipe shall be of the dry-hose type with no shutoff valve at its outlet end unless the hose or pipe drains to the same level under all conditions of use;

(b) the dry-hose shall be sufficiently stiff and only as long as necessary to facilitate drainage;

(c) an automatic vacuum breaker, or equivalent mechanism, shall be incorporated to prevent siphoning and to ensure rapid and complete drainage; and

(d) the inlet end of the hose or outlet pipe shall be high enough to ensure complete drainage.

S.3.5.   Discharge Hose, Reinforcement. – A discharge hose shall be reinforced so that the performance of the device is not affected by the expansion or contraction of the hose.

S.3.6.   Discharge Valve. – A discharge valve may be installed in the discharge line only if the device is of the wet-hose type.   Any other shutoff valve on the discharge side of the meter shall be of the automatic or semiautomatic predetermined-stop type or shall be operable only:

(a) by means of a tool (but not a pin) entirely separate from the device; or

(b) by mutilation of a security seal with which the valve is sealed open.

S.3.7.   Anti-drain Means. – In a wet-hose pressure-type device, means shall be incorporated to prevent the drainage of the discharge hose.
(Amended 1990)

S.4.   Marking Requirements.

S.4.1.   Limitation on Use. – The limitations on its use shall be clearly and permanently marked on any device intended to measure accurately only:

(a) products having particular properties;

(b) under specific installation or operating conditions; or

(c) when used in conjunction with specific accessory equipment.

S.4.2.   Air Pressure. – If a device is operated by air pressure, the air pressure gauge shall show by special graduations or other means the maximum and minimum working pressures recommended by the manufacturer.

S.4.3.   Wholesale Devices.

S.4.3.1.   Discharge Rates. – A wholesale device shall be marked to show its designed maximum and minimum discharge rates.   However, the minimum discharge rate shall not exceed 20 % of the maximum discharge rate.

S.4.3.2.   Temperature Compensation. – If a device is equipped with an automatic temperature compensation, the primary indicating elements, recording elements, and recorded representation shall be clearly and conspicuously marked to show that the volume delivered has been adjusted to the volume at 15 °C (60 °F).

S.4.4.   Retail Devices.

S.4.4.1.   Discharge Rates. – *On a retail device with a designed maximum discharge rate of 115 L (30 gal) per minute or greater, the maximum and minimum discharge rates shall be marked in accordance with*

S.4.4.2.	***Location of Marking Information; Retail Motor-Fuel Dispensers.*** *– The marked minimum discharge rate shall not exceed 20 % of the marked maximum discharge rate.*
[Nonretroactive as of January 1, 1985]

(Added 1984) (Amended 2003)

Example: With a marked maximum discharge rate of 230 L/min (60 gpm), the marked minimum discharge rate shall be 45 L/min (12 gpm) or less (e.g., 40 L/min [10 gpm] is acceptable). A marked minimum discharge rate greater than 45 L/min (12 gpm) (e.g., 60 L/min [15 gpm]) is not acceptable.

S.4.4.2.	***Location of Marking Information; Retail Motor-Fuel Dispensers.*** *– The marking information required in the General Code, paragraph G-S.1. Identification shall appear as follows:*

(a) *within 60 cm (24 in) to 150 cm (60 in) from the base of the dispenser;*

(b) *either internally and/or externally provided the information is permanent and easily read; and*

(c) *on a portion of the device that cannot be readily removed or interchanged (i.e., not on a service access panel).*

Note: *The use of a dispenser key or tool to access internal marking information is permitted for retail liquid-measuring devices.*
[Nonretroactive as of January 1, 2003]

(Added 2002) (Amended 2004)

S.5.	***Totalizers for Retail Motor-Fuel Dispensers.*** *– Retail motor-fuel dispensers shall be equipped with a non-resettable totalizer for the quantity delivered through the metering device.*
[Nonretroactive as of January 1, 1995]

(Added 1993) (Amended 1994)

N. Notes

N.1.	**Test Liquid.**

N.1.1.	**Type of Liquid.** – The liquid used for testing a liquid-measuring device shall be the type the device is used to measure, or another liquid with the same general physical characteristics.

N.1.2.	**Labeling.** – Following the completion of a successful examination of a wholesale device, the weights and measures official should attach a label or tag indicating the type of liquid used during the test.

N.2.	**Volume Change.** – Care shall be taken to minimize changes in volume of the test liquid due to temperature changes and evaporation losses.

N.3.	**Test Drafts.**

N.3.1.	**Retail Piston-Type and Visible-Type Devices.** – Test drafts shall include the full capacity delivery and each intermediate delivery for which the device is designed.

N.3.2.	**Slow-flow Meters.** – Test drafts shall be equal to at least four times the minimum volume that can be measured and indicated through either a visible indication or an audible signal.

N.3.3.	**Lubricant Devices.** – Test drafts shall be 1 L (1 qt). Additional test drafts may include 0.5 L (1 pt), 4 L (4 qt), and 6 L (6 qt).

N.3.4.   Other Retail Devices. – On devices with a designed maximum discharge rate of:

(a) less than 80 L (20 gal) per minute, tests shall include drafts of one or more amounts, including a draft of at least 19 L (5 gal).

(b) 80 L (20 gal) per minute or greater, tests shall include drafts of one or more amounts, including a draft of at least the amount delivered by the device in one minute at the maximum flow rate of the installation.

(Amended 1984)

N.3.5.   Wholesale Devices. – The delivered quantity should be equal to at least the amount delivered by the device in one minute at its maximum discharge rate, and shall in no case be less than 200 L (50 gal).

(Amended 1987 and 1996)

N.4.   Testing Procedures.

N.4.1.   Normal Tests. – The "normal" test of a device shall be made at the maximum discharge flow rate developed under the conditions of installation. Any additional tests conducted at flow rates down to and including one-half of the sum of the maximum discharge flow rate and the rated minimum discharge flow rate shall be considered normal tests.

(Amended 1991)

N.4.1.1.   Wholesale Devices Equipped with Automatic Temperature-Compensating Systems. – On wholesale devices equipped with automatic temperature-compensating systems, normal tests shall be conducted:

(a) by comparing the compensated volume indicated or recorded to the actual delivered volume corrected to 15 °C (60 °F); and

(b) with the temperature-compensating system deactivated, comparing the uncompensated volume indicated or recorded to the actual delivered volume.

The first test shall be performed with the automatic temperature-compensating system operating in the "as found" condition.

On devices that indicate or record both the compensated and uncompensated volume for each delivery, the tests in (a) and (b) may be performed as a single test.

(Amended 1987)

N.4.1.2.   Repeatability Tests. – Tests for repeatability should include a minimum of three consecutive test drafts of approximately the same size and be conducted under controlled conditions where variations in factors such as temperature, pressure, and flow rate are reduced to the extent that they will not affect the results obtained.

(Added 2001)

N.4.2.   Special Tests. – "Special" tests shall be made to develop the operating characteristics of a device and any special elements and accessories attached to or associated with the device. Any test except as set forth in N.4.1. Normal Tests shall be considered a special test.

N.4.2.1.   Slow-Flow Meters. – A "special" test shall be made at a flow rate:

(a) not larger than twice the actual minimum flow rate; and

(b) not smaller than the actual minimum flow rate of the installation.

N.4.2.2.   Retail Motor-Fuel Devices.

(a) Devices without a marked minimum flow-rate shall have a "special" test performed at the slower of the following rates:

    (1) 19 L (5 gal) per minute; or

    (2) the minimum discharge rate at which the device will deliver when equipped with an automatic discharge nozzle set at its slowest setting.

(b) Devices with a marked minimum flow-rate shall have a "special" test performed at or near the marked minimum flow rate.

(Added 1984) (Amended 2005)

N.4.2.3.   Other Retail Devices. – "Special" tests of other retail devices shall be made at the slower of the following rates:

(a) 50 % of the maximum discharge rate developed under the conditions of installation; or

(b) the minimum discharge rate marked on the device.

N.4.2.4.   Wholesale Devices. – "Special" tests shall be made to develop the operating characteristics of a measuring system and any special associated or attached elements and accessories. "Special" tests shall include a test at or slightly above the slower of the following rates:

(a) 20 % of the marked maximum discharge rate; or

(b) the minimum discharge rate marked on the device.

In no case shall the test be performed at a flow rate less than the minimum discharge rate marked on the device.

(Amended 2014)

N.4.3.   Money-Value Computation Tests.

N.4.3.1.   Laboratory Tests. – When testing the device in the laboratory:

(a) compliance with paragraph S.1.6.5. Money-Value Computations, shall be determined by using the cone gear as a reference for the total quantity delivered;

(b) the indicated quantity shall agree with the cone gear representation with the index of the indicator within the width of the graduation; and

(c) the maximum allowable variation of the indicated sales price shall be as shown in Table 1. Money-Value Divisions and Maximum Allowable Variations for Money-Value Computations on Mechanical Analog Computers.

(Amended 1984)

N.4.3.2.   Field Tests. – In the conduct of field tests to determine compliance with paragraph S.1.6.5. Money-Value Computations, the maximum allowable variation in the indicated sales price shall be as shown in Table 1. Money-Value Divisions and Maximum Allowable Variations for Money-Value Computations on Mechanical Analog Computers.

(Added 1982) (Amended 1984)

N.4.4. Pour and Drain Times.

N.4.4.1. Pour and Drain Times for Hand-held Test Measures. – Hand-held test measures require a 30-second (± 5 seconds) pour followed by a 10-second drain with the measure held at a 10-degree to 15-degree angle from vertical.

N.4.4.2. Drain Times for Bottom Drain Test Measures or Provers. – Bottom drain field standard provers require a 30-second drain time after main flow cessation.

(Added 2009)

N.4.5. Verification of Linearization Factors. – All enabled linearization factors shall be verified. The verification of enabled linearization factors shall be done through physical testing, or a combination of physical testing and empirical analysis at the discretion of the official with statutory authority.

(Added 2016)

N.5. Temperature Correction on Wholesale Devices. – Corrections shall be made for any changes in volume resulting from the differences in liquid temperatures between time of passage through the meter and time of volumetric determination in the prover. When adjustments are necessary, appropriate petroleum measurement tables should be used.

(Amended 1974)

T. Tolerances

T.1. Application to Underregistration and to Overregistration. – The tolerances hereinafter prescribed shall be applied to errors of underregistration and errors of overregistration, whether or not a device is equipped with an automatic temperature compensator.

T.2. Tolerance Values. – Maintenance, acceptance, and special test tolerances shall be as shown in Table T.2. Accuracy Classes and Tolerances for Liquid Measuring Devices Covered in NIST Handbook 44, Section 3.30.

Table T.2. Accuracy Classes and Tolerances for Liquid Measuring Devices Covered in NIST Handbook 44, Section 3.30.				
Accuracy Class	**Application**	**Acceptance Tolerance**	**Maintenance Tolerance**	**Special Test Tolerance[1]**
0.3	- Petroleum products delivered from large capacity (flow rates greater than 115 L/min or 30 gpm)** devices, including motor-fuel devices - Heated products (other than asphalt) at temperatures greater than 50 °C (122 °F) - Asphalt at temperatures equal to or below 50 °C (122 °F) - All other liquids not shown in the table where the typical delivery is over 200 L (50 gal)	0.2 %	0.3 %	0.5 %
0.3A	- Asphalt at temperatures greater than 50 °C (122 °F)	0.3 %	0.3 %	0.5 %
0.5*	- Petroleum products delivered from small capacity (at 4 L/min (1 gpm) through 115 L/min or 30 gpm)** motor-fuel devices - Agri-chemical liquids - All other applications not shown in the table where the typical delivery is ≤ 200 L (50 gal)	0.3 %	0.5 %	0.5 %
1.1	- Petroleum products and other normal liquids from devices with flow rates** less than 1 gpm. - Devices designed to deliver less than 1 gal	0.75 %	1.0 %	1.25 %

* For test drafts ≤ 40 L or 10 gal, the tolerances specified for Accuracy Class 0.5 in the table above do not apply. For these test drafts, the following applies:

    (a) Maintenance tolerances on normal and special tests shall be 20 mL plus 4 mL per indicated liter or 1 in^3 plus 1 in^3 per indicated gallon.

    (b) Acceptance tolerances on normal and special tests shall be one-half the maintenance tolerance values.

[1] Special test tolerances are not applicable to retail motor fuel dispensers.

** Flow rate refers to designed or marked maximum flow rate.

(Added 2002) (Amended 2006 and 2013)

T.3. Repeatability. – When multiple tests are conducted at approximately the same flow rate and draft size, the range of the test results for the flow rate shall not exceed 40 % of the absolute value of the maintenance tolerance and the results of each test shall be within the applicable tolerance. This tolerance does not apply to the test of the automatic temperature-compensating system. (Also see N.4.1.2. Repeatability Tests.)

(Added 1992) (Amended 2001 and 2002)

T.4. Automatic Temperature-Compensating Systems. *– The difference between the meter error (expressed as a percentage) for results determined with and without the automatic temperature-compensating system activated shall not exceed:*

    (a) 0.2 % for mechanical automatic temperature-compensating systems; and

    (b) 0.1 % for electronic automatic temperature-compensating systems.

The delivered quantities for each test shall be approximately the same size. The results of each test shall be within the applicable acceptance or maintenance tolerance.
[Nonretroactive as of January 1, 1988]
(Added 1987) (Amended 1992, 1996, and 2002)

UR.   User Requirements

UR.1.   Selection Requirements.

UR.1.1.   Discharge Hose.

UR.1.1.1.   Length. – The length of the discharge hose on a retail motor-fuel device:

(a) shall be measured from its housing or outlet of the discharge line to the inlet of the discharge nozzle;

(b) shall be measured with the hose fully extended if it is coiled or otherwise retained or connected inside a housing; and

(c) shall not exceed 5.5 m (18 ft) unless it can be demonstrated that a longer hose is essential to permit deliveries to be made to receiving vehicles or vessels.

An unnecessarily remote location of a device shall not be accepted as justification for an abnormally long hose.
(Amended 1972 and 1987)

UR.1.1.2.   Marinas and Airports.

UR.1.1.2.1.   Length. – The length of the discharge hose shall be as short as practicable, and shall not exceed 15 m (50 ft) unless it can be demonstrated that a longer hose is essential.

UR.1.1.2.2.   Protection. – Discharge hoses exceeding 8 m (26 ft) in length shall be adequately protected from weather and other environmental factors when not in use.
(Made retroactive 1974 and Amended 1984)

UR.2.   Installation Requirements.

UR.2.1.   Manufacturer's Instructions. – A device shall be installed in accordance with the manufacturer's instructions, and the installation shall be sufficiently secure and rigid to maintain this condition.
(Added 1987)

UR.2.2.   Discharge Rate. – A device shall be installed so that the actual maximum discharge rate will not exceed the rated maximum discharge rate. Automatic means for flow regulation shall be incorporated in the installation if necessary.

UR.2.3.   Suction Head. – A piston-type device shall be installed so that the total effective suction head will not be great enough to cause vaporization of the liquid being dispensed under the highest temperature and lowest barometric pressure likely to occur.

UR.2.4.   Diversion of Liquid Flow. – A motor-fuel device equipped with two delivery outlets used exclusively in the fueling of trucks shall be so installed that any diversion of flow to other than the receiving vehicle cannot be readily accomplished and is readily apparent. Allowable deterrents include, but are not limited to, physical barriers to adjacent driveways, visible valves, or lighting systems that indicate which outlets are in operation, and explanatory signs.
(Amended 1991)

UR.2.5.   Product Storage Identification.

(a) The fill connection for any petroleum product storage tank or vessel supplying motor-fuel devices shall be permanently, plainly, and visibly marked as to product contained.

(b) When the fill connection device is marked by means of a color code, the color code key shall be conspicuously displayed at the place of business.

(Added 1975) (Amended 1976)

UR.3.   Use of Device.

UR.3.1.   Return of Indicating and Recording Elements to Zero. – On any dispenser used in making retail deliveries, the primary indicating element, and recording element if so equipped, shall be returned to zero before each delivery.

Exceptions to this requirement are totalizers on key-lock-operated or other self-operated dispensers and the primary recording element if the device is equipped to record.

UR.3.2.   Unit Price and Product Identity.

(a) The following information shall be conspicuously displayed or posted on the face of a retail dispenser used in direct sale:

(1) except for unit prices resulting from any post-delivery discount and dispensers used exclusively for fleet sales, other price contract sales, and truck refueling (e.g., truck stop dispensers used only to refuel trucks), all of the unit prices at which the product is offered for sale; and

(2) in the case of a computing type or money-operated type, the unit price at which the dispenser is set to compute.

Provided that the dispenser complies with S.1.6.4.1. Display of Unit Price, it is not necessary that all the unit prices for all grades, brands, blends, or mixtures be simultaneously displayed or posted.

(b) The following information shall be conspicuously displayed or posted on each side of a retail dispenser used in direct sale:

(1) the identity of the product in descriptive commercial terms; and

(2) the identity of the grade, brand, blend, or mixture that a multi-product dispenser is set to deliver.

(Amended 1972, 1983, 1987, 1989, 1992, 1993, and 2012)

UR.3.3.   Computing Device. – Any computing device used in an application where a product or grade is offered for sale at one or more unit prices shall be used only for sales for which the device computes and displays the sales price for the selected transaction.

(Became retroactive 1999)

(Added 1989) (Amended 1992)

The following exceptions apply:

(a) Fleet sales and other price contract sales are exempt from this requirement.

(b) A truck stop dispenser used exclusively for refueling trucks is exempt from this requirement provided that:

   (1) all purchases of fuel are accompanied by a printed receipt of the transaction containing the applicable price per gallon, the total gallons delivered, and the total price of the sale; and
   (Added 1993)

   (2) unless a dispenser complies with S.1.6.4.1. Display of Unit Price, the price posted on the dispenser and the price at which the dispenser is set to compute shall be the highest price for any transaction which may be conducted.
   (Added 1993)

(c) A dispenser used in an application where a price per unit discount is offered following the delivery is exempt from this requirement, provided the following conditions are satisfied:

   (1) the unit price posted on the dispenser and the unit price at which the dispenser is set to compute prior to the application of any discount shall be the highest unit price for any transaction;
   (Amended 2014)

   (2) all purchases of fuel are accompanied by a receipt recorded by the system. The receipt shall contain:

      a. the product identity by name, symbol, abbreviation, or code number;

      b. transaction information as shown on the dispenser at the end of the delivery and prior to any post-delivery discount including the:

         1. total volume of the delivery;

         2. unit price; and

         3. total computed price of the fuel sale prior to post-delivery discounts being applied.

      c. an itemization of the post-delivery discounts to the unit price; and

      d. the final total price of the fuel sale.
   (Added 2012) (Amended 2014)
(Added 1989) (Amended 1992, 1993, 2012, and 2014)

UR.3.4.   Printed Ticket. – The total price, the total volume of the delivery, and the price per liter or gallon shall be shown, either printed by the device or in clear hand script, on any printed ticket issued by a device and containing any one of these values.
(Amended 2001)

UR.3.5.   Steps after Dispensing. – After delivery to a customer from a retail motor-fuel device:

(a) the starting lever shall be returned to its shutoff position and the zero-set-back interlock engaged; and

(b) the discharge nozzle shall be returned to its designed hanging position unless the primary indicating elements, and recording elements, if the device is equipped and activated to record, have been returned to a definite zero indication.

UR.3.6.   Temperature Compensation, Wholesale.

UR.3.6.1.   Automatic.

UR.3.6.1.1.   When to be Used. – If a device is equipped with a mechanical automatic temperature compensator, it shall be connected, operable, and in use at all times. An electronic or mechanical automatic temperature-compensating system may not be removed, nor may a compensated device be replaced with an uncompensated device, without the written approval of the responsible weights and measures jurisdiction.

Note: This requirement does not specify the method of sale for product measured through a meter.
(Amended 1989)

UR.3.6.1.2.   Invoices.

(a) A written invoice based on a reading of a device that is equipped with an automatic temperature compensator shall show that the volume delivered has been adjusted to the volume at 15 °C (60 °F).

(b) The invoice issued from an electronic wholesale device equipped with an automatic temperature-compensating system shall also indicate:

(1) the API gravity, specific gravity or coefficient of expansion for the product;

(2) product temperature; and

(3) gross reading.
(Amended 1987)

UR.3.6.2.   Nonautomatic.

UR.3.6.2.1.   Temperature Determination. – If the volume of the product delivered is adjusted to the volume at 15 °C (60 °F), the product temperature shall be taken during the delivery in:

(a) the liquid chamber of the meter; or

(b) the meter inlet or discharge line adjacent to the meter; or

(c) the compartment of the receiving vehicle at the time it is loaded.

UR.3.6.2.2.   Invoices. – The accompanying invoice shall indicate that the volume of the product has been adjusted for temperature variations to a volume at 15 °C (60 °F) and shall also state the product temperature used in making the adjustment.

UR.3.6.3.   Period of Use. – When fuel is bought or sold on an automatic or non-automatic temperature-compensated basis, it shall be bought or sold using this method over at least a consecutive 12-month period, unless otherwise agreed to by both the buyer and seller in writing.
(Added 2003)

U.R.4.   Maintenance Requirements.

U.R.4.1.   Use of Adjustments. – Whenever a device is adjusted, all enabled linearization factors shall be verified to determine that the errors are in tolerance and any adjustments which are made shall be made so as

to bring performance errors as close as practicable to zero value. The verification of enabled linearization factors shall be done through physical testing or a combination of testing and empirical analysis.
(Added 2016)

Liquid-Measuring Device Code Index

THIS PAGE INTENTIONALLY LEFT BLANK

Table of Contents

Section 3.31.   Vehicle-Tank Meters

A. Application

A.1.   General. – This code applies to meters mounted on vehicle tanks including those used for the measurement and delivery of petroleum products or agri-chemical liquids such as fertilizers, feeds, pesticides, defoliants, and bulk deliveries of water.
(Amended 1985 and 1995)

A.2.   Exceptions. – This code does not apply to the following devices:

(a) Devices used for dispensing liquefied petroleum gases, or other liquids that do not remain in a liquid state at atmospheric pressures and temperatures. (Also see Section 3.32. Code for Liquefied Petroleum Gas and Anhydrous Ammonia Liquid-Measuring Devices.)

(b) Devices used solely for dispensing a product in connection with operations in which the amount dispensed does not affect customer charges.

(c) Vehicle tanks used as measures. (Also see Section 4.40. Code for Vehicle Tanks Used as Measures.)

(d) Mass flow meters. (Also see Section 3.37. Code for Mass Flow Meters.)
(Added 1994)

(e) Devices used to measure cryogenic liquids. (Also see Section 3.34. Code for Cryogenic Liquid-Measuring Devices.)

(f) Devices used to measure carbon dioxide liquids. (Also see Section 3.38. Code for Carbon Dioxide Liquid-Measuring Devices.)

A.3.   Additional Code Requirements. – In addition to the requirements of this code, Vehicle-Tank Meters shall meet the requirements of 1.10. General Code requirements.

S. Specifications

S.1.   Design of Indicating and Recording Elements and of Recorded Representations.

S.1.1.   Primary Elements.

S.1.1.1.   General. – A meter shall be equipped with a primary indicating element and may also be equipped with a primary recording element.

Note: Except for systems used solely for the sale of aviation fuel into aircraft and for aircraft-related operations, vehicle-tank meters shall be equipped with a primary recording element as required by paragraph UR.2.2. Ticket Printer; Customer Ticket.
(Amended 1993)

S.1.1.2.   Units.

(a) A meter shall indicate, and record if the meter is equipped to record, its deliveries in terms of liters or gallons. Fractional parts of the liter or gallon shall be in terms of either decimal or binary subdivisions.

(b) When it is an industry practice to purchase and sell milk by weight based upon 1.03 kg/L (8.6 lb/gal), the primary indicating element may indicate in kilograms or pounds and decimal kilograms or pounds. The weight value division shall be a decimal multiple or submultiple of 1, 2, or 5. (Also see Section S.5.5. Conversion Factor.)

S.1.1.3.   Value of Smallest Unit. – The value of the smallest unit of indicated delivery, and recorded delivery if the meter is equipped to record, shall not exceed the equivalent of:

(a) 0.5 L (0.1 gal) or 0.5 kg (1 lb) on milk-metering systems;

(b) 0.5 L (0.1 gal) on meters with a rated maximum flow rate of 750 L/min (200 gal/min) or less;

(c) 5 L (1 gal) on meters with a rated maximum flow of 375 L/min (100 gal/min) or more used for jet fuel aviation refueling systems; or
(Added 2006)

(d) 5 L (1 gal) on other meters.
(Amended 1989, 1994 and 2006)

S.1.1.4.   Advancement of Indicating and Recording Elements. – Primary indicating and recording elements shall be susceptible to advancement only by the mechanical operation of the meter. However, a meter may be cleared by advancing its elements to zero, but only if:

(a) the advancing movement, once started, cannot be stopped until zero is reached; or

(b) in the case of indicating elements only, such elements are automatically obscured until the elements reach the correct zero position.

S.1.1.5.   Return to Zero. – Primary indicating elements shall be readily returnable to a definite zero indication. Means shall be provided to prevent the return of primary indicating elements, and of primary recording elements if these are returnable to zero, beyond their correct zero position. Primary indicating elements shall not be resettable to zero during a delivery.
(Amended 2016)

S.1.2. Graduations.

S.1.2.1.   Length. – Graduations shall be so varied in length that they may be conveniently read.

S.1.2.2.   Width. – In any series of graduations, the width of a graduation shall in no case be greater than the width of the minimum clear interval between graduations, and the width of main graduations shall be not more than 50 % greater than the width of subordinate graduations. Graduations shall in no case be less than 0.2 mm (0.008 in) wide.

S.1.2.3.   Clear Interval Between Graduations. – The clear interval shall be not less than 2.5 mm (0.10 in). If the graduations are not parallel, the measurement shall be made:

(a) along the line of relative movement between the graduations at the end of the indicator; or

(b) if the indicator is continuous, at the point of widest separation of the graduations.
(Amended 1986)

S.1.3. Indicators.

S.1.3.1.   Symmetry. – The index of an indicator shall be symmetrical with respect to the graduations at least throughout that portion of its length associated with the graduations.

S.1.3.2.   Length. – The index of an indicator shall reach to the finest graduations with which it is used, unless the indicator and the graduations are in the same plane, in which case the distance between the end of the indicator and the ends of the graduations, measured along the line of the graduations, shall be not more than 1.0 mm (0.04 in).

S.1.3.3.   Width. – The width of the index of an indicator in relation to the series of graduations with which it is used shall be not greater than:

(a) *the width of the narrowest graduation;* and*
     [*Nonretroactive as of January 1, 2002]
     (Amended 2001)

(b) the width of the minimum clear interval between graduations.

When the index of an indicator extends along the entire length of a graduation, that portion of the index of the indicator that may be brought into coincidence with the graduation shall be of the same width throughout the length of the index that coincides with the graduation.

S.1.3.4.   Clearance. – The clearance between the index of an indicator and the graduations shall in no case be more than 1.5 mm (0.06 in).

S.1.3.5.   Parallax. – Parallax effects shall be reduced to the practicable minimum.

S.1.3.6.   Travel of Indicator. – If the most sensitive element of the primary indicating element utilizes an indicator and graduations, the relative movement of these parts corresponding to the smallest indicated value shall not be less than 5 mm (0.20 in).

S.1.4.   Computing-Type Device.

S.1.4.1.   Display of Unit Price. – In a device of the computing type, means shall be provided for displaying, in a manner clear to the operator and an observer, the unit price at which the device is set to compute. The unit price is not required to be displayed continuously.
(Amended 1983 and 2005)

S.1.4.2.   Printed Ticket. – If a computing-type device issues a printed ticket which displays the total computed price, the ticket shall also have printed clearly thereon the total quantity of the delivery, the appropriate fraction of the quantity, and the price per unit of quantity.
(Amended 1989)

S.1.4.3.   Money-Value Computations. – Money-value computations shall be of the full-computing type in which the money-value at a single unit price, or at each of a series of unit prices, shall be computed for every delivery within either the range of measurement of the device or the range of the computing elements, whichever is less. Value graduations shall be supplied and shall be accurately positioned. The value of each graduated interval shall be one cent. On electronic devices with digital indications, the total price may be computed on the basis of the quantity indicated when the value of the smallest division indicated is equal to or less than 0.2 L (0.1 gal) or 0.2 kg (1 lb).
(Amended 1979 and 1989)

S.1.4.4.   Money-Values, Mathematical Agreement. – Any digital money-value indication and any recorded money-value on a computing-type device shall be in mathematical agreement with its associated quantity indication or representation to within one cent of money-value.

S.2.   Design of Measuring Elements.

S.2.1.   Vapor Elimination. – A metering system shall be equipped with an effective vapor or air eliminator or other automatic means to prevent the passage of vapor and air through the meter. Vent lines from the air or vapor eliminator shall be made of metal tubing or some other suitable rigid material.
(Amended 1993)

S.2.2.   Provision for Sealing. – Adequate provision shall be made for an approved means of security (e.g., data change audit trail) or for physically applying a security seal in such a manner that requires the security seal to be broken before a change or an adjustment or interchange may be made of:

(a) any measuring or indicating element;

(b) any adjustable element for controlling delivery rate when such rate tends to affect the accuracy of deliveries; and

(c) any metrological parameter that will affect the metrological integrity of the device or system.

When applicable, the adjusting mechanism shall be readily accessible for purposes of affixing a security seal. *[Audit trails shall use the format set forth in Table S.2.2. Categories of Device and Methods Sealing.]* *
[*Nonretroactive as of January 1, 1995]
(Amended 2006)

Table S.2.2. *Categories of Device and Methods of Sealing*	
Categories of Device	***Methods of Sealing***
Category 1: *No remote configuration capability.*	*Seal by physical seal or two event counters: one for calibration parameters and one for configuration parameters.*
Category 2: *Remote configuration capability, but access is controlled by physical hardware.* *The device shall clearly indicate that it is in the remote configuration mode and record such message if capable of printing in this mode or shall not operate while in this mode.*	*The hardware enabling access for remote communication must be on-site. The hardware must be sealed using a physical seal or an event counter for calibration parameters and an event counter for configuration parameters. The event counters may be located either at the individual measuring device or at the system controller; however, an adequate number of counters must be provided to monitor the calibration and configuration parameters of the individual devices at a location. If the counters are located in the system controller rather than at the individual device, means must be provided to generate a hard copy of the information through an on-site device.*
Category 3: *Remote configuration capability access may be unlimited or controlled through a software switch (e.g., password).* *The device shall clearly indicate that it is in the remote configuration mode and record such message if capable of printing in this mode or shall not operate while in this mode.*	*An event logger is required in the device; it must include an event counter (000 to 999), the parameter ID, the date and time of the change, and the new value of the parameter. A printed copy of the information must be available on demand through the device or through another on-site device. The information may also be available electronically. The event logger shall have a capacity to retain records equal to 10 times the number of sealable parameters in the device, but not more than 1000 records are required. (**Note:** Does not require 1000 changes to be stored for each parameter.)*

[Nonretroactive as of January 1, 1995]

(Table Added 2006) (Amended 2016)

S.2.3.   Directional Flow Valves. – Valves intended to prevent reversal of flow shall be automatic in operation. However, on equipment used exclusively for fueling aircraft, such valves may be manual in operation.

S.2.4.   Zero-Set-Back Interlock, Vehicle-Tank Meters, Electronic. – *Except for vehicle-mounted metering systems used solely for the delivery of aviation fuel, a device shall be so constructed that after an individual or multiple deliveries at one location have been completed, an automatic interlock system shall engage to prevent a subsequent delivery until the indicating and, if equipped, recording elements have been returned to their zero position. For individual deliveries, if there is no product flow for three minutes the transaction must be completed before additional product flow is allowed. The 3-minute timeout shall be a sealable feature on an indicator.*
[Nonretroactive as of January 1, 2006]

(Added 2005)

S.2.5.   Automatic Temperature Compensation for Refined Petroleum Products.

S.2.5.1.   Automatic Temperature Compensation for Refined Petroleum Products. – A device may be equipped with an automatic means for adjusting the indication and registration of the measured volume of product to the volume at 15 °C for liters or the volume at 60 °F for gallons and decimal subdivisions or fractional equivalents thereof where not prohibited by state law.

S.2.5.2.   Provision for Deactivating. – On a device equipped with an automatic temperature-compensating mechanism that will indicate or record only in terms of liters compensated to 15°C or gallons compensated to 60 °F, provision shall be made for deactivating the automatic temperature-compensating mechanism so the meter can indicate and record, if it is equipped to record, in terms of the uncompensated volume.

S.2.5.3.   Gross and Net Indications. – A device equipped with automatic temperature compensation shall indicate or record, if equipped to record, both the gross (uncompensated) and net (compensated) volume for testing purposes. It is not necessary that both net and gross volume be displayed simultaneously.

S.2.5.4.   Provision for Sealing Automatic Temperature-Compensating Systems. – Adequate provision shall be made for an approved means of security (e.g., data change audit trail) or physically applying security seals in such a manner that an automatic temperature-compensating system cannot be disconnected and no adjustment may be made to the system.

S.2.5.5.   Temperature Determination with Automatic Temperature Compensation. – For test purposes, means shall be provided (e.g., thermometer well) to determine the temperature of the liquid either:

(a)   in the liquid chamber of the meter; or

(b)   immediately adjacent to the meter in the meter inlet or discharge line.

(Added 2007)

S.2.6.   Thermometer Well, Temperature Determination. – *For test purposes, means shall be provided (e.g., thermometer well) to determine the temperature of the liquid either in the:*

(a)   liquid chamber of the meter; or

(b)   meter inlet or discharge line immediately adjacent to the meter.
[Nonretroactive as of January 1, 2012)

(Added 2011)

S.3.   Design of Discharge Lines and Discharge Line Valves.
(Not applicable to milk-metering systems.)

S.3.1.   Diversion of Measured Liquid. – Except on equipment used exclusively for fueling aircraft, no means shall be provided by which any measured liquid can be diverted from the measuring chamber of the meter or the discharge line thereof. However, two or more delivery outlets may be installed if means is provided to insure that:

(a)   liquid can flow from only one such outlet at one time; and

(b)   the direction of flow for which the mechanism may be set at any time is definitely and conspicuously indicated.

S.3.2.   Pump-Discharge Unit. – On a pump-discharge unit, the discharge hose shall be of the wet-hose type with a shutoff valve at its outlet end. However, a pump-discharge unit may be equipped also with a dry-hose without a shutoff valve at its outlet end, but only if:

(a)   the dry-hose is as short as practicable; and

(b)   there is incorporated in the discharge piping, immediately adjacent to the meter, effective means to insure that liquid can flow through only one of the discharge hoses at any one time and that the meter and the wet-hose remain full of liquid at all times.

S.3.3.   Gravity-Discharge Unit. – On a gravity-discharge unit, the discharge hose or equivalent pipe shall be of the dry-hose type with no shutoff valve at its outlet end. The dry-hose shall be of such stiffness and only of such length as to facilitate its drainage. The inlet end of the hose or of an equivalent outlet pipe shall be of such height as to provide for proper drainage of the hose or pipe. There shall be incorporated an automatic vacuum breaker or equivalent means to prevent siphoning and to ensure the rapid and complete drainage.

S.3.4.   Discharge Hose. – A discharge hose shall be adequately reinforced.

S.3.5.   Discharge Valve. – A discharge valve may be installed in the discharge line only if the device is of the wet-hose type, in which case such valve shall be at the discharge end of the line. Any other shutoff valve on the discharge side of the meter shall be of the automatic or semiautomatic predetermined-stop type or shall be operable only:

   (a)   by means of a tool (but not a pin) entirely separate from the device; or

   (b)   by mutilation of a security seal with which the valve is sealed open.

S.3.6.   Antidrain Valve. – In a wet-hose, pressure-type device, an effective antidrain valve shall be incorporated in the discharge valve or immediately adjacent thereto. The antidrain valve shall function so as to prevent the drainage of the discharge hose. However, a device used exclusively for fueling and defueling aircraft may be of the pressure type without an antidrain valve.

S.4.   Design of Intake Lines (for Milk-Metering Systems).

S.4.1.   Diversion of Liquid to be Measured. – No means shall be provided by which any liquid can be diverted from the supply tank to the receiving tank without being measured by the device.

S.4.2.   Intake Hose. – The intake hose shall be:

   (a)   of the dry-hose type;

   (b)   adequately reinforced;

   (c)   not more than 6 m (20 ft) in length, unless it can be demonstrated that a longer hose is essential to permit pickups from a supply tank; and

   (d)   connected to the pump at horizontal or above, to permit complete drainage of the hose.

S.5.   Marking Requirements.

S.5.1.   Limitation of Use. – If a meter is intended to measure accurately only liquids having particular properties, or to measure accurately only under specific installation or operating conditions, or to measure accurately only when used in conjunction with specific accessory equipment, these limitations shall be clearly and permanently stated on the meter.

S.5.2.   Discharge Rates. – A meter shall be marked to show its designed maximum and minimum discharge rates. However, the minimum discharge rate shall not exceed 20 % of the maximum discharge rate.

Note: Also see example in Section 3.30. Liquid-Measuring Devices Code, paragraph S.4.4.1. Discharge Rates.
(Added 2003)

S.5.3.   Measuring Components, Milk-Metering System. – All components that affect the measurement of milk that are disassembled for cleaning purposes shall be clearly and permanently identified with a common serial number.

S.5.4.   Flood Volume, Milk-Metering System. – When applicable, the volume of product necessary to flood the system when dry shall be clearly, conspicuously, and permanently marked on the air eliminator.

S.5.5.   Conversion Factor. – When the conversion factor of 1.03 kg/L (8.6 lb/gal) is used to convert the volume of milk to weight, the conversion factor shall be clearly marked on the primary indicating element and recorded on the delivery ticket.
(Added 1989)

S.5.6.   Temperature Compensation for Refined Petroleum Products. – If a device is equipped with an automatic temperature compensator, the primary indicating elements, recording elements, and recorded representations shall be clearly and conspicuously marked to show the volume delivered has been adjusted to the volume at 15 °C for liters or the volume at 60 °F for gallons and decimal subdivisions or fractional equivalents thereof.
(Added 2007)

S.5.7.   Meter Size. *– Except for milk meters, if the meter model identifier does not provide a link to the meter size (in terms of pipe diameter) on an NTEP Certificate of Conformance, the meter shall be marked to show meter size.*
[Nonretroactive as of January 1, 2009]
(Added 2008)

N. Notes

N.1.   Test Liquid.

(a) A measuring system shall be tested with the liquid to be commercially measured or with a liquid of the same general physical characteristics. Following a satisfactory examination, the weights and measures official should attach a seal or tag indicating the product used during the test.
(Amended 1975)

(b) A milk-measuring system shall be tested with the type of milk to be measured when the accuracy of the system is affected by the characteristics of milk (e.g., positive displacement meters).
(Added 1989)
(Amended 1975 and 1989)

N.2.   Evaporation and Volume Change. – Care shall be exercised to reduce to a minimum, evaporation losses and volume changes resulting from changes in temperature of the test liquid.

N.3.   Test Drafts. – Test drafts should be equal to at least the amount delivered by the device in 1 minute at its maximum discharge rate, and shall in no case be less than 180 L (50 gal) or 225 kg (500 lb).
(Amended 1989)

N.4.   Testing Procedures.

N.4.1.   Normal Tests. – The "normal" test of a measuring system shall be made at the maximum discharge rate that may be anticipated under the conditions of the installation. Any additional tests conducted at flow rates down to and including one-half of the sum of the maximum discharge flow rate and the rated minimum discharge flow rate shall be considered normal tests.
(Amended 1992)

N.4.1.1.   Milk Measuring System. – The "normal" test shall include a determination of the effectiveness of the air elimination system.

N.4.1.2.   Repeatability Tests. – Tests for repeatability should include a minimum of three consecutive test drafts of approximately the same size and be conducted under controlled conditions where variations in factors such as temperature, pressure, and flow rate are reduced to the extent that they will not affect the results obtained.

(Added 2001)

N.4.1.3.   Automatic Temperature-Compensating Systems for Refined Petroleum Products. – On devices equipped with automatic temperature-compensating systems, normal tests shall be conducted:

(a) by comparing the compensated volume indicated or recorded to the actual delivered volume corrected to 15 °C for liters or 60 °F for gallons and decimal subdivisions or fractional equivalents thereof; and

(b) with the temperature-compensating system deactivated, comparing the uncompensated volume indicated or recorded to the actual delivered volume.

The first test shall be performed with the automatic temperature-compensating system operating in the "as-found" condition. On devices that indicate or record both the compensated and uncompensated volume for each delivery, the tests in (a) and (b) may be performed as a single test.

(Added 2007)

N.4.2.   Special Tests (Except Milk-Measuring Systems). – "Special" tests shall be made to develop the operating characteristics of a measuring system and any special elements and accessories attached to or associated with the device. Any test except as set forth in N.4.1. Normal Tests and N.4.5. Product Depletion Test shall be considered a special test. Special tests of a measuring system shall be made at a minimum discharge rate of 20 % of the marked maximum discharge rate or at the minimum discharge rate marked on the device, whichever is less.

(Amended 1978 and 2005)

N.4.3.   Antidrain Valve Test. – The effectiveness of the antidrain valve shall be tested after the pump pressure in the measuring system has been released and a valve between the supply tank and the discharge valve is closed.

N.4.4.   System Capacity. – The test of a milk-measuring system shall include the verification of the volume of product necessary to flood the system as marked on the air eliminator.

N.4.5.   Product Depletion Test. – Except for vehicle-mounted metering systems used solely for the delivery of aviation fuel, the effectiveness of the vapor eliminator or vapor elimination means shall be tested by dispensing product at the normal flow rate until the product supply is depleted and continuing until the lack of fluid causes the meter indication to stop completely for at least 10 seconds. If the meter indication fails to stop completely for at least 10 seconds, continue to operate the system for 3 minutes. Finish the test by switching to another compartment with sufficient product to complete the test on a multi-compartment vehicle or by adding sufficient product to complete the test to a single compartment vehicle. When adding product to a single compartment vehicle, allow appropriate time for any entrapped vapor to disperse before continuing the test. Test drafts shall be of the same size and run at approximately the same flow rate.

(Added 2005)

N.4.6.   Verification of Linearization Factors. – All enabled linearization factors shall be verified. The verification of enabled linearization factors shall be done through physical testing or a combination of physical testing and empirical analysis at the discretion of the official with statutory authority.

(Added 2016)

N.5.   Temperature Correction for Refined Petroleum Products. – Corrections shall be made for any changes in volume resulting from the differences in liquid temperatures between the time of passage through the meter and the time of volumetric determination in the prover. When adjustments are necessary, appropriate petroleum measurement tables should be used.

(Added 2007)

T. Tolerances

T.1. Application.

T.1.1. To Underregistration and to Overregistration. – The tolerances hereinafter prescribed shall be applied to errors of underregistration and errors of overregistration.

T.2. Tolerance Values. – Tolerances shall be as shown in Table 1. Accuracy Classes and Tolerances for Vehicle-Tank Meters and Table 2. Tolerances for Vehicle-Mounted Milk Meters.

(Amended 1995)

Accuracy Class	Application		Acceptance Tolerance	Maintenance Tolerance	Special Test Tolerance
0.3	- Petroleum products delivered from large capacity (flow rates over 115 L/min or 30 gpm)** devices, including motor-fuel devices - Heated products (other than asphalt) at temperatures greater than 50 °C (122 °F) - Asphalt at temperatures equal to or below 50 °C (122 °F) - All other liquids not shown in the table where the typical delivery is greater than 200 L (50 gal)		0.15 %	0.3 %	0.45 %
0.3A	- Asphalt at temperatures greater than 50 °C (122 °F)		0.3 %	0.3 %	0.5 %
0.5*	- Petroleum products delivered from small capacity (at 4 L/min (1 gpm) through 115 L/min or 30 gpm)** motor-fuel devices - Agri-chemical liquids - All other applications not shown in the table where the typical delivery is ≤ 200 L (50 gal)		0.3 %	0.5 %	0.5 %
1.1	- Petroleum products and other normal liquids from devices with flow rates** less than 4 L/min (1 gpm) and - Devices designed to deliver less than 4 L (1 gal)		0.75 %	1.0 %	1.25 %
1.5	- Water	Overregistration	1.5 %	1.5 %	1.5 %
		Underregistration	1.5 %	1.5 %	5.0 %

Table 1.
Accuracy Classes and Tolerances for Vehicle-Tank Meters

* For 5 gal and 10 gal test drafts, the tolerances specified for Accuracy Class 0.5 in the table above do not apply. For these test drafts, the maintenance tolerances on normal and special tests for 5 gal and 10 gal test drafts are 6 in^3 and 11 in^3, respectively. Acceptance tolerances on normal and special tests are 3 in^3 and 5.5 in^3.
** Flow rate refers to designed or marked maximum flow rate.

(Added 2002) (Amended 2013)

Table 2. Tolerances for Vehicle-Mounted Milk Meters		
Indication (gallons)	Maintenance Tolerance (gallons)	Acceptance Tolerance (gallons)
100	0.5	0.3
200	0.7	0.4
300	0.9	0.5
400	1.1	0.6
500	1.3	0.7
Over 500	Add 0.002 gallon per indicated gallon over 500	Add 0.001 gallon per indicated gallon over 500

(Added 1989)

T.2.1.   Automatic Temperature-Compensating Systems. – The difference between the meter error (expressed as a percentage) for results determined with and without the automatic temperature-compensating system activated shall not exceed:

(a)   0.2 % for mechanical automatic temperature-compensating systems; and

(b)   0.1 % for electronic automatic temperature-compensating systems.

The delivered quantities for each test shall be approximately the same size. The results of each test shall be within the applicable acceptance or maintenance tolerance.

(Added 2007) (Amended 2010)

T.3.   Repeatability. – When multiple tests are conducted at approximately the same flow rate and draft size, the range of the test results for the flow rate shall not exceed 40 % of the absolute value of the maintenance tolerance and the results of each test shall be within the applicable tolerance. (Also see N.4.1.2. Repeatability Tests.)

(Added 1992) (Amended 2001 and 2002)

T.4.   Product Depletion Test. – The difference between the test result for any normal test and the product depletion test shall not exceed 0.5 % of the volume delivered in one minute at the maximum flow rate marked on the meter for meters rated higher than 380 Lpm (100 gpm) or 0.6 % of the volume delivered in one minute at the maximum flow rate marked on the meter for meters rated 380 Lpm (100 gpm) or lower. Test drafts shall be of the same size and run at approximately the same flow rate.

Note: The result of the product depletion test may fall outside of the applicable test tolerance as specified in Table 1. Accuracy Classes and Tolerances for Vehicle-Tank Meters.

(Amended 2013)

UR.   User Requirements

UR.1.   Installation Requirements.

UR.1.1.   Discharge Rate. – A meter shall be so installed that the actual maximum discharge rate will not exceed the rated maximum discharge rate. If necessary, means for flow regulation shall be incorporated in the installation, in which case this shall be fully effective and automatic in operation.

UR.1.2.   Unit Price. – There shall be displayed on the face of a device of the computing type the unit price at which the device is set to compute.

UR.1.3.   Intake Hose. – The intake hose in a milk-metering system shall be installed to permit complete drainage and ensure that all available product is measured following each pickup.

UR.1.4.   Liquid Measured. – A vehicle-tank meter shall continue to be used to measure the same liquid or one with the same general physical properties as that used for calibration and weights and measures approval unless the meter is recalibrated with a different product and tested by a registered service agency or a weights and measures official and approved by the weights and measures jurisdiction having statutory authority over the device.
(Added 2003)

UR.2.   Use Requirements.

UR.2.1.   Return of Indicating and Recording Elements to Zero. – The primary indicating elements (visual), and the primary recording elements, when these are returnable to zero, shall be returned to zero immediately before each delivery is begun and after the pump has been activated and the product to be measured has been supplied to the measuring system.
(Amended 1981)

UR.2.2.   Ticket Printer, Customer Ticket. – Vehicle-Mounted metering systems shall be equipped with a ticket printer which shall be used for all sales where product is delivered through the meter. A copy of the ticket issued by the device shall be left with the customer at the time of delivery or as otherwise specified by the customer.
(Added 1993) (Amended 1994)

> **UR.2.2.1.   Exceptions for the Sale of Aviation Fuel.** – The provisions of UR.2.2. Ticket Printer, Customer Ticket shall not apply to vehicle-mounted metering systems used solely for the delivery of aviation fuel into aircraft and for aircraft-related operations.
> (Added 1999)

UR.2.3.   Ticket in Printing Device. – A ticket shall not be inserted into a device equipped with a ticket printer until immediately before a delivery is begun, and in no case shall a ticket be in the device when the vehicle is in motion while on a public street, highway, or thoroughfare.

UR.2.4.   Credit for Flood Volume. – The volume of product necessary to flood the system as marked on the air eliminator shall be individually recorded on the pickup ticket of each seller affected.

UR.2.5.   Automatic Temperature Compensation for Refined Petroleum Products.

> **UR.2.5.1.   When to be Used.** – In a state that does not prohibit, by law or regulation, the sale of temperature-compensated product, a device equipped with an activated automatic-temperature compensator shall be connected, operable, and in use at all times.   An electronic or mechanical automatic temperature-compensating device or system may not be removed or deactivated, nor may a compensated device be replaced with an uncompensated device or system, without the written approval of the responsible weights and measures jurisdiction.
>
> **Note:** This requirement does not specify the method of sale for products measured through a meter.
> (Amended 2009)

> **UR.2.5.2.   Period of Use.** – When fuel is bought or sold on an automatic temperature compensation basis, it shall be bought or sold using this basis over at least a consecutive 12-month period unless otherwise agreed to by both the buyer and seller in writing.
> (Added 2009)

UR.2.5.3.   Invoices. – An invoice based on a reading of a device that is equipped with an automatic temperature compensator shall show that the volume delivered has been adjusted to the volume at 15 °C for liters or the volume at 60 °F for gallons and decimal subdivisions or fractional equivalents thereof.

(Added 2007)

U.R.3.   Maintenance Requirements.

UR.3.1. Use of Adjustments. – Whenever a device is adjusted, all enabled linearization factors shall be verified to determine that the errors are in tolerance and any adjustments which are made shall be made so as to bring performance errors as close as practicable to zero value. The verification of enabled linearization factors shall be done through physical testing or a combination of physical testing and empirical analysis.

(Added 2016)

THIS PAGE INTENTIONALLY LEFT BLANK

Table of Contents

Section 3.32.   Liquefied Petroleum Gas and Anhydrous Ammonia Liquid-Measuring Devices[1]

A. Application

A.1.   General. – This code applies to devices used for the measurement of liquefied petroleum gas and anhydrous ammonia in the liquid state, whether such devices are installed in a permanent location or mounted on a vehicle.

A.2.   Devices Used to Measure Other Liquid Products not Covered in Specific Codes. – Insofar as they are clearly appropriate, the requirements and provisions of the code may be applied to devices used for the measurement of other liquids that do not remain in a liquid state at atmospheric pressures and temperatures.

A.3.   Exceptions. – This code does not apply to mass flow meters. (Also see Section 3.37. Code for Mass Flow Meters.)

(Added 1994)

A.4.   Additional Code Requirements. – In addition to the requirements of this code, LPG and Anhydrous Ammonia Liquid-Measuring Devices shall meet the requirements of Section 1.10. General Code.

S. Specifications

S.1.   Design of Indicating and Recording Elements and of Recorded Representations.

S.1.1.   Primary Elements.

S.1.1.1.   General. – A device shall be equipped with a primary indicating element and may also be equipped with a primary recording element.

Note: Vehicle-mounted metering systems shall be equipped with a primary recording element as required by paragraph UR.2.6. Ticket Printer; Customer Ticket.

S.1.1.2.   Units. – A device shall indicate, and record if the device is equipped to record, its deliveries in terms of liters, gallons, quarts, pints, or binary-submultiple or decimal subdivisions of the liter or gallon.

(Amended 1987)

S.1.1.3.   Value of Smallest Unit. – The value of the smallest unit of indicated delivery, and recorded delivery if the device is equipped to record, shall not exceed the equivalent of:

(a) 0.5 L (1 pt) on retail devices; or

(b) 5 L (1 gal) on wholesale devices.

(Amended 1987)

S.1.1.4.   Advancement of Indicating and Recording Elements. – Primary indicating and recording elements shall be susceptible to advancement only by the mechanical operation of the device. However, a device may be cleared by advancing its elements to zero, but only if:

(a) the advancing movement, once started, cannot be stopped until zero is reached; or

---

[1]Title amended 1986.

(b) in the case of indicating elements only, such elements are automatically obscured until the elements reach the correct zero position.

S.1.1.5.   Money-Values, Mathematical Agreement. – Any digital money-value indication and any recorded money-value on a computing-type device shall be in mathematical agreement with its associated quantity indication or representation to within 1 cent of money-value; except that a stationary retail computing-type device must compute and indicate to the nearest 1 cent of money-value. (Also see Section 1.10. General Code, G-S.5.5. Money-Values, Mathematical Agreement.)
(Amended 1984 and 1988)

S.1.1.6.   Printed Ticket. – Any printed ticket issued by a device of the computing type on which there is printed the total computed price, shall have printed clearly thereon the total volume of the delivery in terms of liters or gallons, and the appropriate decimal fraction of the liter or gallon, and the corresponding price per liter or gallon.
(Added 1979) (Amended 1987)

S.1.2.   Graduations.

S.1.2.1.   Length. – Graduations shall be so varied in length that they may be conveniently read.

S.1.2.2.   Width. – In any series of graduations:

(a) the width of a graduation shall in no case be greater than the width of the minimum clear interval between graduations;

(b) the width of main graduations shall be not more than 50 % greater than the width of subordinate graduations; and

(c) graduations shall in no case be less than 0.2 mm (0.008 in) in width.

S.1.2.3.   Clear Interval between Graduations. – The clear interval shall be not less than 1.0 mm (0.04 in). If the graduations are not parallel, the measurement shall be made:

(a) along the line of relative movement between the graduations at the end of the indicator; or

(b) if the indicator is continuous, at the point of widest separation of the graduations.

S.1.3.   Indicators.

S.1.3.1.   Symmetry. – The index of an indicator shall be symmetrical with respect to the graduations, at least throughout that portion of its length associated with the graduations.

S.1.3.2.   Length. – The index of an indicator shall reach to the finest graduations with which it is used, unless the indicator and the graduations are in the same plane, in which case the distance between the end of the indicator and the ends of the graduations, measured along the line of graduations, shall be not more than 1.0 mm (0.04 in).

S.1.3.3.   Width. – The width of the index of an indicator in relation to the series of graduations with which it is used shall be not greater than:

(a) *the width of the narrowest graduation;* * and
[*Nonretroactive as of January 1, 2002]
(Amended 2001)

(b) the width of the minimum clear interval between graduations.

When the index of an indicator extends along the entire length of a graduation, that portion of the index of the indicator that may be brought into coincidence with the graduation shall be of the same width throughout the length of the index that coincides with the graduation.

S.1.3.4.   Clearance. – The clearance between the index of an indicator and the graduations shall in no case be more than 1.5 mm (0.06 in).

S.1.3.5.   Parallax. – Parallax effects shall be reduced to the practicable minimum.

S.1.4.   For Retail Devices Only.

S.1.4.1.   Indication of Delivery. – A retail device shall automatically show on its face the initial zero condition and the quantity delivered up to the nominal capacity of the device. However, the following requirements shall apply:

(a) For electric devices manufactured prior to January 1, 2006, the first 0.03 L (or 0.009 gal) of a delivery and its associated total sales price need not be indicated.

(b) *For electronic devices manufactured on or after January 1, 2006, the measurement, indication of delivered quantity, and the indication of total sales price shall be inhibited until the fueling position reaches conditions necessary to ensure that the delivery starts at zero.*
[Nonretroactive as of January 1, 2006]
(Amended 2016)

S.1.4.2.   Return to Zero.

(a) Primary indicating elements shall be readily returnable to a definite zero indication.

(b) Primary recording elements on a stationary retail device shall be readily returnable to a definite zero indication if the device is equipped to record.

(c) Means shall be provided to prevent the return of primary indicating elements and of primary recording elements if these are returnable to zero, beyond their correct zero position.

(d) Primary indicating elements shall not be resettable to zero during a delivery.
(Amended 1990 and 2016)

S.1.5.   For Stationary Retail Devices Only.

S.1.5.1.   Display of Unit Price and Product Identity. – A device of the computing type shall display on each face the unit price at which the device is set to compute or to deliver, and there shall be conspicuously displayed on each side of the device the identity of the product that is being dispensed.

Except for dispensers used exclusively for fleet sales and other price contract sales, all of the unit prices at which that product is offered for sales shall meet the following conditions:

(a) *For a system that applies a discount prior to the delivery, all unit prices shall be displayed or shall be capable of being displayed on the dispenser through a deliberate action of the purchaser prior to the delivery of the product. It is not necessary that all of the unit prices be simultaneously displayed prior to the delivery of the product.*
[Nonretroactive as of January 1, 2016]

(b) For a system that offers post-delivery discounts on fuel sales, display of pre-delivery unit price information is exempt from (a) above, provided the system complies with S.1.5.5. Recorded Representations for Transactions Where a Post-Delivery Discount(s) is Provided.

Note: When a product is offered at more than one unit price, display of the unit price information may be through the deliberate action of the customer: 1) using controls on the device; 2) through the customer's use of personal or vehicle-mounted electronic equipment communicating with the system; or 3) verbal instructions by the customer.

(Amended 2016)

S.1.5.2.   Money-Value Computations. – A computing device shall compute the total sales price at any single-purchase unit price (excluding fleet sales and other price contract sales) for which the product is offered for sale at any delivery possible within either the measurement range of the device or the range of the computing elements, whichever is less. The analog money-value indication shall not differ from the mathematically computed money-value (quantity × unit price = sales price), for any delivered quantity, by an amount greater than the values shown in Table 1. Money-Value Divisions and Maximum Allowable Variations for Money-Value Computations on Mechanical Analog Computers.

(Amended 1995)

Table 1. Money-Value Divisions and Maximum Allowable Variations for Money-Value Computations on Mechanical Analog Computers				
Unit Price		**Money-Value Division**	**Maximum Allowable Variation**	
From	**To and Including**		**Design Test**	**Field Test**
0	$0.25/liter or $1.00/gallon	1¢	± 1¢	± 1¢
$0.25/liter or $1.00/gallon	$0.75/liter or $3.00/gallon	1¢ or 2¢	± 1¢	± 2¢
$0.75/liter or $3.00/gallon	$2.50/liter or $10.00/gallon	1¢ or 2¢	± 1¢	± 2¢
$0.75/liter or $3.00/gallon	$2.50/liter or $10.00/gallon	5¢	± 2½¢	± 5¢

S.1.5.2.1.   Money-Value Divisions, Analog. – The value of the graduated intervals representing money-values on a computing-type device with analog indications shall be as follows:

(a) Not more than 1 cent at all unit prices up to and including $0.25 per liter or $1.00 per gallon.

(b) Not more than 2 cents at unit prices greater than $0.25 per liter or $1.00 per gallon up to and including $0.75 per liter or $3.00 per gallon.

(c) Not more than 5 cents at all unit prices greater than $0.75 per liter or $3.00 per gallon.

(Amended 1984)

S.1.5.2.2.   Money-Value Divisions, Digital. – A computing-type device with digital indications shall comply with the requirements of paragraph G.-S.5.5. Money-Values, Mathematical Agreement, and the total price computation shall be based on quantities not exceeding 0.01 gal intervals for devices indicating in U.S. customary units and 0.05 L for devices indicating in metric units.

S.1.5.2.3.   Money-Value Divisions, Auxiliary Indications. *– In a system equipped with auxiliary indications, all indicated money-value divisions shall be identical.*
[Nonretroactive as of January 1, 1985.]

S.1.5.3.   Agreement Between Indications.

(a) *When a quantity value indicated or recorded by an auxiliary element is a derived or computed value based on data received from a device, the value may differ from the quantity value displayed on the dispenser, provided that the following conditions are met:*

   (1) *All total values for an individual sale that are indicated or recorded by the system agree, and*

   (2) *Within each element, the values indicated or recorded meet the formula (quantity × unit price = total sale price) to the closest cent.*

(b) *When a system applies a post-delivery discount(s) to a fuel's unit price through an auxiliary element, the total volume of the delivery shall be in agreement between all elements in the system.*

[Nonretroactive as of January 1, 2016]

(Added 2016)

S.1.5.4.   Recorded Representations. – Except for fleet sales and other price contract sales and for transactions where a post-delivery discount is provided, a receipt providing the following information shall be available through a built-in or separate recording element for all transactions conducted with point-of-sale systems or devices activated by debit cards, credit cards, and/or cash:

(a) the total volume of the delivery;

(b) the unit price;

(c) the total computed price; and

(d) the product identity by name, symbol, abbreviation, or code number.

(Added 2014) (Amended 2016)

S.1.5.5.   Recorded Representations for Transactions Where a Post-Delivery Discount(s) is Provided. – Except for fleet sales and other price contract sales, a printed receipt providing the following information shall be available through a built-in or separate recording element that is part of the system for transactions involving a post-delivery discount:

(a) the product identity by name, symbol, abbreviation, or code number;

(b) transaction information as shown on the dispenser at the end of the delivery and prior to any post-delivery discount(s), including the:

(1) total volume of the delivery;

(2) unit price; and

(3) total computed price of the fuel sale.

(c) an itemization of the post-delivery discounts to the unit price; and

(d) the final total price of the fuel sale after all post-delivery discounts are applied.
(Added 2016)

S.1.5.6.   *Transaction Information, Power Loss.* – *In the event of a power loss, the information needed to complete any transaction in progress at the time of the power loss (such as the quantity and unit price, or sales price) shall be determinable for at least 15 minutes at the device or another on-site device accessible to the customer.*
[Nonretroactive as of January 1, 2017]
(Added 2016)

S.1.5.7.   *Totalizers for Retail Motor-Fuel Dispensers.* – *Retail motor-fuel dispensers shall be equipped with a nonresettable totalizer for the quantity delivered through the metering device.*
[Nonretroactive as of January 1, 2017]
(Added 2016)

S.1.6.   For Wholesale Devices Only.

S.1.6.1.   Travel of Indicator. – A wholesale device shall be readily operable to deliver accurately any quantity from 180 L (50 gal) to the capacity of the device. If the most sensitive element of the indicating system uses an indicator and graduations, the relative movement of these parts corresponding to a delivery of 5 L (1 gal) shall be not less than 5 mm (0.20 in).
(Amended 1987)

S.2.   Design of Measuring Elements.

S.2.1.   Vapor Elimination.

(a) A device shall be equipped with an effective automatic means to prevent the passage of vapor through the meter.

(b) Vent lines from the vapor eliminator shall be made of appropriate non-collapsible material.
(Amended 2016)

S.2.2.   Provision for Sealing. – Adequate provision shall be made for an approved means of security (e.g., data change audit trail) or for physically applying a security seal in such a manner that requires the security seal to be broken before an adjustment or interchange may be made of:

(a) any measuring or indicating element;

(b) any adjustable element for controlling delivery rate, when such rate tends to affect the accuracy of deliveries; and

(c) any metrological parameter that will affect the metrological integrity of the device or system.

When applicable, the adjusting mechanism shall be readily accessible for purposes of affixing a security seal. *[Audit trails shall use the format set forth in Table S.2.2. Categories of Device and Methods of Sealing.]* *[*Nonretroactive as of January 1, 1995]*

(Amended 2006)

Table S.2.2. Categories of Device and Methods of Sealing	
Categories of Device	*Methods of Sealing*
Category 1: *No remote configuration capability.*	*Seal by physical seal or two event counters: one for calibration parameters and one for configuration parameters.*
Category 2: *Remote configuration capability, but access is controlled by physical hardware.* *The device shall clearly indicate that it is in the remote configuration mode and record such message if capable of printing in this mode or shall not operate while in this mode.*	*The hardware enabling access for remote communication must be on-site. The hardware must be sealed using a physical seal or an event counter for calibration parameters and an event counter for configuration parameters. The event counters may be located either at the individual measuring device or at the system controller; however, an adequate number of counters must be provided to monitor the calibration and configuration parameters of the individual devices at a location. If the counters are located in the system controller rather than at the individual device, means must be provided to generate a hard copy of the information through an on-site device.*
Category 3: *Remote configuration capability access may be unlimited or controlled through a software switch (e.g., password).* *The device shall clearly indicate that it is in the remote configuration mode and record such message if capable of printing in this mode or shall not operate while in this mode.*	*An event logger is required in the device; it must include an event counter (000 to 999), the parameter ID, the date and time of the change, and the new value of the parameter. A printed copy of the information must be available on demand through the device or through another on-site device. The information may also be available electronically. The event logger shall have a capacity to retain records equal to 10 times the number of sealable parameters in the device, but not more than 1000 records are required. (**Note:** Does not require 1000 changes to be stored for each parameter.)*

[Nonretroactive as of January 1, 1995]

(Table Added 2006) (Amended 2016)

S.2.3.   Directional Flow Valves. – A measuring system shall be equipped with a valve or other effective means, automatic in operation and installed in or adjacent to the measuring element, to prevent reversal of flow of the product being measured.

(Amended 1982)

S.2.4.   Maintenance of Liquid State. – A device shall be so designed and installed that the product being measured will remain in a liquid state during the passage through the meter.

S.2.5.   *Zero-Set-Back Interlock for Stationary Retail Motor-Fuel Devices.* – A device shall be constructed so that:

(a) after a delivery cycle has been completed by moving the starting lever to any position that shuts off the device, an automatic interlock prevents a subsequent delivery until the indicating elements and recording elements, if the device is equipped and activated to record, have been returned to their zero positions;

(b) the discharge nozzle cannot be returned to its designed hanging position (that is, any position where the tip of the nozzle is placed in its designed receptacle and the lock can be inserted) until the starting lever is in its designed shut-off position and the zero-set-back interlock has been engaged; and

(c) in a system with more than one dispenser supplied by a single pump, an effective automatic control valve in each dispenser prevents product from being delivered until the indicating elements on that dispenser are in a correct zero position.

[Nonretroactive as of January 1, 2017]

(Added 2016)

S.2.6. Thermometer Well. – For test purposes, means shall be provided to determine the temperature of the liquid either:

(a) in the liquid chamber of the meter; or

(b) in the meter inlet or discharge line and immediately adjacent to the meter.

(Amended 1987)

S.2.7. Automatic Temperature Compensation. – A device may be equipped with an adjustable automatic means for adjusting the indication and registration of the measured volume of product to the volume at 15 °C (60 °F).

S.2.7.1. Provision for Deactivating. – On a device equipped with an automatic temperature-compensating mechanism that will indicate or record only in terms of liters or gallons adjusted to 15 °C (60 °F), provision shall be made to facilitate the deactivation of the automatic temperature-compensating mechanism so that the meter may indicate, and record if it is equipped to record, in terms of the uncompensated volume.

(Amended 1972)

S.2.7.2. Provision for Sealing. – Provision shall be made for applying security seals in such a manner that an automatic temperature-compensating system cannot be disconnected and that no adjustment may be made to the system.

S.3. Design of Discharge Lines and Discharge Line Valves.

S.3.1. Diversion of Measured Liquid. – No means shall be provided by which any measured liquid can be diverted from the measuring chamber of the meter or the discharge line therefrom. However, two or more delivery outlets may be permanently installed if means are provided to insure that:

(a) liquid can flow from only one such outlet at one time; and

(b) the direction of flow for which the mechanism may be set at any time is definitely and conspicuously indicated.

In addition, a manually controlled outlet that may be opened for the purpose of emptying a portion of the system to allow for repair and maintenance operations shall be permitted. Effective means shall be provided to prevent the passage of liquid through any such outlet during normal operation of the device and to indicate clearly and unmistakably when the valve controls are so set as to permit passage of liquid through such outlet.

(Amended 1975)

S.3.2.   **Delivery Hose.** – The delivery hose of a retail device shall be of the wet-hose type with a shutoff valve at its outlet end.

S.4.   **Marking Requirements.**

S.4.1.   **Limitation of Use.** – If a device is intended to measure accurately only products having particular properties, or to measure accurately only under specific installation or operating conditions, or to measure accurately only when used in conjunction with specific accessory equipment, these limitations shall be clearly and permanently stated on the device.

S.4.2.   **Discharge Rates.** – A device shall be marked to show its designed maximum and minimum discharge rates. The marked minimum discharge rate shall not exceed:

(a) 20 L (5 gal) per minute for stationary retail devices; or

(b) 20 % of the marked maximum discharge rate for other retail devices and for wholesale devices.
(Amended 1987)

Note: Also see example in Section 3.30. Liquid-Measuring Devices Code, paragraph S.4.4.1. Discharge Rates.
(Added 2003)

S.4.3.   ***Location of Marking Information; Retail Motor-Fuel Dispensers.*** – *The marking information required in General Code, paragraph G-S.1. Identification shall appear as follows:*

(a) within 60 cm (24 in) to 150 cm (60 in) from the base of the dispenser;

(b) either internally and/or externally provided the information is permanent and easily read; and

(c) on a portion of the device that cannot be readily removed or interchanged (i.e., not on a service access panel).

Note: The use of a dispenser key or tool to access internal marking information is permitted for retail motor-fuel dispensers. *[Nonretroactive as of January 1, 2003]*
(Added 2006)

S.4.4.   **Temperature Compensation.** – If a device is equipped with an automatic temperature compensator, the primary indicating elements, recording elements, and recorded representation shall be clearly and conspicuously marked to show that the volume delivered has been adjusted to the volume at 15 °C (60 °F).

N. Notes

N.1.   **Test Liquid.** – A device shall be tested with the liquid to be commercially measured or with a liquid of the same general physical characteristics.

N.2.   **Vaporization and Volume Change.** – Care shall be exercised to reduce to a minimum, vaporization and volume changes.

N.3.   **Test Drafts.** – Test drafts should be equal to at least the amount delivered by the device in one minute at its normal discharge rate.
(Amended 1982)

N.4.   Testing Procedures.

N.4.1.   Normal Tests. – The "normal" test of a device shall be made at the maximum discharge flow rate developed under the conditions of the installation. Any additional tests conducted at flow rates down to and including one-half the sum of the maximum discharge flow rate and the rated minimum discharge flow rate shall be considered normal tests.
(Amended 1998)

> **N.4.1.1.   Automatic Temperature Compensation.** – On devices equipped with automatic temperature-compensating systems, normal tests shall be conducted as follows:
>
> (a) by comparing the compensated volume indicated or recorded to the actual delivered volume adjusted to 15 °C (60 °F); and
>
> (b) with the temperature-compensating system deactivated, comparing the uncompensated volume indicated or recorded to the actual delivered volume.

The first test shall be performed with the automatic temperature-compensating system operating in the "as found" condition. On devices that indicate or record both the compensated and uncompensated volume for each delivery, the tests in (a) and (b) may be performed as a single test.
(Amended 1987)

> **N.4.1.2.   Repeatability Tests.** – Tests for repeatability should include a minimum of three consecutive test drafts of approximately the same size and be conducted under controlled conditions where variations in factors such as temperature, pressure, and flow rate are reduced to the extent that they will not affect the results obtained.
> (Added 2001)

N.4.2.   Special Tests. – "Special" tests shall be made to develop the operating characteristics of a device and any special elements and accessories attached to or associated with the device. Any test except as set forth in N.4.1. Normal Tests shall be considered a special test.

> **N.4.2.1.   For Motor-Fuel Devices.** – A motor-fuel device shall be so tested at a minimum discharge rate of:
>
> (a) 20 L (5 gal) per minute; or
>
> (b) the minimum discharge rate marked on the device, whichever is less.

> **N.4.2.2.   For Other Retail Devices.** – A retail device other than a motor-fuel device shall be tested at a minimum discharge rate of the:
>
> (a) minimum discharge rate that can be developed under the conditions of installation; or
>
> (b) minimum discharge rate marked on the device, whichever is greater.
> (Amended 1973)

N.4.2.3.   For Wholesale Devices. – A wholesale device shall be so tested at a minimum discharge rate of:

(a) 40 L (10 gal) per minute for a device with a rated maximum discharge less than 180 L (50 gal) per minute.

(b) 20 % of the marked maximum discharge rate for a device with a rated maximum discharge of 180 L (50 gal) per minute or more, or

(c) the minimum discharge rate marked on the device, whichever is least.

(Amended 1987)

N.4.3.   Money-Value Computation Tests.

N.4.3.1.   Laboratory Design Evaluation Tests. – In the conduct of laboratory design evaluation tests, compliance with paragraph S.1.5.2. Money-Value Computations shall be determined by using the cone gear as a reference for the total quantity delivered. The indicated delivered quantity shall agree with the cone gear representation with the index of the indicator within the width of the graduation. The maximum allowable variation of the indicated sales price shall be as shown in Table 1. Money-Value Divisions and Maximum Allowable Variations for Money-Value Computations on Mechanical Analog Computers.

N.4.3.2.   Field Tests. – In the conduct of field tests to determine compliance with paragraph S.1.5.2. Money-Value Computations the maximum allowable variation in the indicated sales price shall be as shown in Table 1. Money-Value Divisions and Maximum Allowable Variations for Money-Value Computations on Mechanical Analog Computers.

(Added 1984)

N.5.   Temperature Correction. – Adjustments shall be made for any changes in volume resulting from the differences in liquid temperatures between time of passage through the meter and time of volumetric determination in the prover. When adjustments are necessary, appropriate petroleum measurement tables should be used.

T. Tolerances

T.1.   Application.

T.1.1.   To Underregistration and to Overregistration. – The tolerances hereinafter prescribed shall be applied to errors of underregistration and errors of overregistration, whether or not a device is equipped with an automatic temperature compensator.

T.2.   Tolerance Values. – The maintenance and acceptance tolerances for normal and special tests shall be as shown in Table T.2. Accuracy Classes and Tolerances for LPG and Anhydrous Ammonia Liquid-Measuring Devices.

(Amended 2003)

Table T.2. Accuracy Classes and Tolerances for LPG and Anhydrous Ammonia Liquid-Measuring Devices				
Accuracy Class	Application	Acceptance Tolerance	Maintenance Tolerance	Special Test Tolerance
1.0	Anhydrous ammonia, LPG (including vehicle-mounted meters)	0.6 %	1.0 %	1.0 %

(Added 2003)

T.3.   Repeatability. – When multiple tests are conducted at approximately the same flow rate and draft size, the range of the test results for the flow rate shall not exceed 40 % of the absolute value of the maintenance tolerance and

the results of each test shall be within applicable tolerance. This tolerance does not apply to the test of the automatic temperature-compensating system. (Also see N.4.1.2. Repeatability Tests).
(Added 1992) (Amended 1997 and 2001)

T.4. Automatic Temperature-Compensating Systems. – The difference between the meter error (expressed as a percentage) for results determined with and without the automatic temperature-compensating system activated shall not exceed:

(a) 1.0 % for mechanical automatic temperature-compensating systems; and

(b) 0.5 % for electronic automatic temperature-compensating systems.

The delivered quantities for each test shall be approximately the same size. The results of each test shall be within the applicable acceptance or maintenance tolerance.
(Added 1991) (Amended 1992, 1996, and 1997)

UR.   User Requirements

UR.1.   Installation Requirements.

UR.1.1.   Discharge Rate. – A device shall be so installed that the actual maximum discharge rate will not exceed the rated maximum discharge rate. If necessary, means for flow regulation shall be incorporated in the installation, in which case this shall be fully effective and automatic in operation.

UR.1.2.   Length of Discharge Hose. – The length of the discharge hose on a stationary motor-fuel device shall not exceed 5.5 m (18 ft), measured from the outside of the housing of the device to the inlet end of the discharge nozzle, unless it can be demonstrated that a longer hose is essential to permit deliveries to be made to receiving vehicles or vessels. Unnecessarily remote location of a device shall not be accepted as justification for an abnormally long hose.
(Amended 1991)

UR.2.   Use Requirements.

UR.2.1.   Return of Indication and Recording Elements to Zero. – The primary indicating elements (visual), and the primary recording elements when these are returnable to zero, shall be returned to zero before each delivery.

UR.2.2.   Condition of Fill of Discharge Hose. – The discharge hose shall be completely filled with liquid before the "zero" condition is established prior to the start of a commercial delivery, whether this condition is established by resetting the primary indicating elements to zero indication or by recording the indications of the primary indicating elements. (Also see UR.2.1. Return of Indication and Recording Elements to Zero.)

UR.2.3.   Vapor-Return Line. – During any metered delivery of liquefied petroleum gas from a supplier's tank to a receiving container, a vapor-return line from the receiving container to the supplier's tank is prohibited except:

(a) in the case of any receiving container to which normal deliveries cannot be made without the use of such vapor-return line; or

(b) in the case of any top spray-fill receiving container when the ambient temperature is at or above 32 °C (90 °F).
(Amended 2016)

UR.2.4.   Temperature Compensation.

UR.2.4.1.   Use of Automatic Temperature Compensators. – If a device is equipped with an automatic temperature compensator, this shall be connected, operable, and in use at all times. Such automatic temperature compensator may not be removed, nor may a compensated device be replaced with an uncompensated device, without the written approval of the weights and measures authority having jurisdiction over the device.

UR.2.4.2.   Temperature Compensated Sale. – All sales of liquefied petroleum gas in a liquid state, when the quantity is determined by an approved measuring system equipped with a temperature-compensating mechanism, or by weight and converted to liters or gallons, or by a calibrated container, shall be in terms of liters or the U.S. gallon of 231 in^3 at 15 °C (60 °F).

(Added 1984)

UR.2.4.3.   Invoices. – Any invoice based on a reading of a device that is equipped with an automatic temperature compensator or based on a weight converted to gallons, or based on the volume of a calibrated container, shall have shown thereon that the volume delivered has been adjusted to the volume at 15 °C (60 °F).

(Amended 1984)

UR.2.4.4.   Automated Temperature-Compensating Systems. – Means for determining the temperature of measured liquid in an automatic temperature-compensating system shall be so designed and located that, in any "usual and customary" use of the system, the resulting indications and/or recorded representations are within applicable tolerances.

(Added 1987)

UR.2.5.   Ticket in Printing Device. – A ticket shall not be inserted into a device equipped with a ticket printer until immediately before a delivery is begun, and in no case shall a ticket be in the device when the vehicle is in motion while on a public street, highway, or thoroughfare.

UR.2.6.   Ticket Printer; Customer Ticket. – Vehicle-mounted metering systems shall be equipped with a ticket printer. The ticket printer shall be used for all sales; a copy of the ticket issued by the device shall be left with the customer at the time of delivery or as otherwise specified by the customer.

(Added 1992)

UR.2.7.   For Stationary Retail Computing-Type Systems Only, Installed After January 1, 2017.

UR.2.7.1.   Unit Price and Product Identity.

(a)   The following information shall be conspicuously displayed or posted on the face of a retail dispenser used in a direct sale:

(1)   except for unit prices resulting from any post-delivery discount and dispensers used exclusively for fleet sales, other price contract sales, and truck refueling (e.g., truck stop dispensers used only to refuel trucks), all of the unit prices at which the product is offered for sale; and

(2)   in the case of a computing-type device or money-operated type device, the unit price at which the dispenser is set to compute.

Provided that the dispenser complies with S.1.5.1. Display of Unit Price and Product Identity, it is not necessary that all the unit prices be simultaneously displayed or posted.

(b)   The following information shall be conspicuously displayed or posted on each side of a retail dispenser used in a direct sale:

(1) The identity of the product in descriptive commercial terms; and

(2) The identity of the grade, brand, blend, or mixture that a multi-product dispenser is set to deliver.

(Added 2016)

UR.2.7.2.   Computing Device. – Any computing device used in an application where a product or grade is offered for sale at one or more unit prices shall be used only for sales for which the device computes and displays the sales price for the selected transaction. The following exceptions apply:

(a) Fleet sales and other price contract sales are exempt from this requirement.

(b) A truck stop dispenser used exclusively for refueling trucks is exempt from this requirement provided that:

(1) all purchases of fuel are accompanied by a printed receipt of the transaction containing the applicable price per unit of measure, the total quantity delivered, and the total price of the sale; and

(2) unless a dispenser complies with S.1.5.1. Display of Unit Price, the price posted on the dispenser and the price at which the dispenser is set to compute shall be the highest price for any transaction which may be conducted.

(c) A dispenser used in an application where a price per unit discount is offered following the delivery is exempt from this requirement, provided the following conditions are satisfied:

(1) the unit price posted on the dispenser and the unit price at which the dispenser is set to compute shall be the highest unit price for any transaction;

(2) all purchases of fuel are accompanied by a receipt recorded by the system for the transaction containing:

a.   the product identity by name, symbol, abbreviation, or code number;

b.   transaction information as shown on the dispenser at the end of the delivery and prior to any post-delivery discount including the:

1.   total volume of the delivery;

2.   unit price; and

3.   total computed price of the fuel sale prior to post-delivery discounts being applied.

c.   an itemization of the post-delivery discounts to the unit price; and

d.   the final total price of the fuel sale after all post-delivery discounts are applied.

(Added 2016)

Table of Contents

THIS PAGE INTENTIONALLY LEFT BLANK

Section 3.33. Hydrocarbon Gas Vapor-Measuring Devices[1]

A. Application

A.1. General. – This code applies to devices used for the measurement of hydrocarbon gas in the vapor state, such as propane, propylene, butanes, butylenes, ethane, methane, natural gas, and any other hydrocarbon gas/air mix.
(Amended 1984, 1986, 1988, and 1991)

A.2. Exceptions. – This code does not apply to:

(a) Liquid-measuring devices used for dispensing liquefied petroleum gases in liquid form. (Also see Section 3.32. Code for Liquefied Petroleum Gas and Anhydrous Ammonia Liquid-Measuring Devices.)

(b) Natural, liquefied petroleum, and manufactured-gas-vapor meters when these are operated in a public utility system.

(c) Mass flow meters. (Also see Section 3.37. Code for Mass Flow Meters.)
(Added 1994)

A.3. Additional Code Requirements. – In addition to the requirements of this code, Hydrocarbon Gas Vapor-Measuring Devices shall meet the requirements of Section 1.10. General Code.

S. Specifications

S.1. Design of Indicating and Recording Elements and of Recorded Representations.

S.1.1. Primary Elements.

S.1.1.1. General. – A device shall be equipped with a primary indicating element and may also be equipped with a primary recording element.

S.1.1.2. Units. – A volume-measuring device shall indicate, and record if equipped to record, its deliveries in terms of cubic meters or cubic feet, or multiple or decimal subdivisions of cubic meters or cubic feet.
(Amended 1972 and 1991)

S.1.1.3. Value of Smallest Unit. – The value of the smallest unit of indicated delivery, and recorded delivery if the device is equipped to record, shall not exceed:

(a) 1 m^3 (1000 dm^3) (100 ft^3) when the maximum rated gas capacity is less than 280 m^3/h (10 000 ft^3/h);

(b) 10 m^3 (1000 ft^3) when the maximum rated gas capacity is 280 m^3/h (10 000 ft^3/h) up to, but not including, 1700 m^3/h (60 000 ft^3/h); and

(c) 100 m^3 (10 000 ft^3) when the maximum rated gas capacity is 1700 m^3/h (60 000 ft^3/h) or more.
(Amended 1972, 1988, and 1991)

S.1.1.4. Advancement of Indicating and Recording Elements. – Primary indicating and recording elements shall advance digitally or continuously and be susceptible to advancement only by the mechanical operation of the device.

---

[1]Title changed 1986.

S.1.1.5.   Proving Indicator. – Devices rated less than 280 m³/h (10 000 ft³/h) gas capacity shall be equipped with a proving indicator measuring 0.025, 0.05, 0.1, 0.2, or 0.25 m³ per revolution, (1, 2, 5, or 10 ft³ per revolution) for testing the meter. Devices with larger capacities shall be equipped as follows:

(a) Devices rated 280 m³ (10 000 ft³) up to but not including 1700 m³/h (60 000 ft³/h) gas capacity shall be equipped with a proving indicator measuring not greater than 1 m³ (100 ft³) per revolution.

(b) Devices rated 1700 m³/h (60 000 ft³/h) gas capacity or more shall be equipped with a proving indicator measuring not more than 10 m³ (1000 ft³) per revolution.

The test circle of the proving indicator shall be divided into ten equal parts. Additional subdivisions of one or more of such equal parts may be made.
(Amended 1973 and 1988)

S.1.2.   Graduations.

S.1.2.1.   Length. – Graduations shall be so varied in length that they may be conveniently read.

S.1.2.2.   Width. – In any series of graduations, the width of a graduation shall in no case be greater than the width of the minimum clear interval between graduations, and in no case should it exceed 1.0 mm (0.04 in) for indicating elements and 0.5 mm (0.02 in) for proving circles.

S.1.2.3.   Clear Interval Between Graduations. – The clear interval shall be not less than 1.0 mm (0.04 in). If the graduations are not parallel, the measurement shall be made:

(a) along the line of relative movement between the graduations at the end of the indicator; or

(b) if the indicator is continuous, at the point of widest separation of the graduations.

S.1.3.   Indicators.

S.1.3.1.   Symmetry. – The index of an indicator shall be symmetrical with respect to the graduations, at least throughout that portion of its length associated with the graduations.

S.1.3.2.   Length. – The index of an indicator shall reach to the finest graduations with which it is used.

S.1.3.3.   Width. – The width of the index of an indicator in relation to the series of graduations with which it is used shall be not greater than the:

(a) *width of the narrowest graduation;** and
    *[*Nonretroactive as of January 1, 2002]*
    (Amended 2001)

(b) width of the minimum clear interval between graduations.

When the index of an indicator extends along the entire length of a graduation, that portion of the index of the indicator that may be brought into coincidence with the graduation shall be of the same width throughout the length of the index that coincides with the graduation.

S.1.3.4.   Clearance. – The clearance between the index of an indicator and the graduations shall in no case be more than 1.5 mm (0.06 in).

S.1.3.5.   Parallax. – Parallax effects shall be reduced to the practicable minimum.

S.2.   Design of Measuring Elements.

S.2.1.   Pressure Regulation. – The vapor should be measured at a normal gauge pressure (psig) of:
(Amended 1991)

(a) 2740 Pa ± 685 Pa (11 in of water column (0.40 psig) ± 2.75 in of water column (0.10 psig)) for liquefied petroleum gas vapor; or

(b) 1744 Pa ± 436 Pa (7 in of water column (0.25 psig) ± 1.75 in of water column (0.06 psig)) for natural and manufactured gas.

When vapor is measured at a pressure other than what is specified above for the specific product, a volume multiplier shall be applied within the meter or to the billing invoice based on the following equation:

$$VPM = \frac{AAP + GP}{AAP + NGP}$$

Where:

VPM = Volume pressure multiplier
AAP = Assumed atmospheric pressure in Pa or psia
GP = Gauge pressure in Pa or psig
NGP = Normal gauge pressure in Pa or psig

The assumed atmospheric pressure is to be taken from Tables 2 and 2M.

When liquefied petroleum gas vapor is measured at a pressure of 6900 Pa (1 psig) or more, the delivery pressure shall be maintained within ± 1725 Pa (± 0.25 psig).

Pressure variations due to regulator lock off shall not increase the operating pressure by more than 25 %.
(Amended 1980, 1984, and 1991)

S.2.2.   Provision for Sealing. – Adequate provision shall be made for applying security seals in such a manner that no adjustment or interchange may be made of any measurement element.

S.2.3.   Maintenance of Vapor State. – A device shall be so designed and installed that the product being measured will remain in a vapor state during passage through the meter.

S.2.4.   Automatic Temperature Compensation. – A device may be equipped with an adjustable automatic means for adjusting the indication and registration of the measured volume of vapor product to the volume at 15 °C (60 °F).

S.3.   Design of Discharge Lines.

S.3.1.   Diversion of Measured Vapor. – No means shall be provided by which any measured vapor can be diverted from the measuring chamber of the meter or the discharge line therefrom.

S.4.   Marking Requirements.

S.4.1.   Limitations of Use. – If a device is intended to measure accurately only products having particular properties, or to measure accurately only under specific installation or operating conditions, or to measure accurately only when used in conjunction with specific accessory equipment, these limitations shall be clearly and permanently stated on the device.

S.4.2.   Discharge Rates. – A device shall be marked to show its rated gas capacity in cubic meters per hour or cubic feet per hour.

S.4.3.   Temperature Compensation. – If a device is equipped with an automatic temperature compensator, this shall be indicated on the badge or immediately adjacent to the badge of the device and on the register.

S.4.4.   Badge. – A badge affixed in a prominent position on the front of the device shall show the manufacturer's name, serial number and model number of the device, and capacity rate of the device for the particular products that it was designed to meter as recommended by the manufacturer.

N. Notes

N.1.   Test Medium. – The device shall be tested with air or the product to be measured.
(Amended 1991)

N.2.   Temperature and Volume Change. – Care should be exercised to reduce to a minimum any volume changes. The temperature of the air, bell-prover oil, and the meters under test should be within 1 °C (2 °F) of one another. The devices should remain in the proving room for at least 16 hours before starting any proving operations to allow the device temperature to approximate the temperature of the proving device.

N.3.   Test Drafts. – Except for low-flame tests, test drafts shall be at least equal to one complete revolution of the largest capacity proving indicator, and shall in no case be less than 0.05 m³ or 2 ft³. All flow rates shall be controlled by suitable outlet orifices.
(Amended 1973 and 1991)

Table 1. Capacity of Low-Flow Test Rate Orifices with Respect to Device Capacity			
Metric Units		U.S. Customary Units	
Rated Capacity	Low-Flow Test Rate	Rated Capacity	Low-Flow Test Rate
Up to and including 7 m³/h	0.007 m³/h	Up to and including 250 ft³/h	0.25 ft³/h
Over 7 m³/h up to and including 14 m³/h	0.014 m³/h	Over 250 ft³/h up to and including 500 ft³/h	0.50 ft³/h
Over 14 m³/h	0.1 % of capacity rate	Over 500 ft³/h	0.1 % of capacity rate

N.4.   Test Procedures. – If a device is equipped with an automatic temperature compensator, the proving device reading shall be corrected to 15 °C (60 °F), using an approved table.
(Amended 1972)

N.4.1.   Normal Tests. – The normal test of a device shall be made at a rate not to exceed the capacity rate given on the badge of the meter.
(Amended 1988)

N.4.1.1.   Automatic Temperature Compensation. – If a device is equipped with an automatic temperature compensator, the quantity of the test draft indication of the standard shall be corrected to 15 °C (60 °F).

N.4.1.2.   Repeatability Tests. – Tests for repeatability should include a minimum of three consecutive test drafts of approximately the same size and be conducted under controlled conditions where variations in factors such as temperature, pressure, and flow rate are reduced to the extent that they will not affect the results obtained.

(Added 2002)

N.4.2. Special Tests. – "Special" tests shall be made to develop the operating characteristics of a device and any special elements and accessories attached to or associated with the device. Any test except as set forth in N.4.1. Normal Tests shall be considered a special test.

    N.4.2.1. Slow Test. – The device shall be tested at a rate not less than 20 % of the marked capacity rate, or (at the check rate) not less than the minimum flow rate if marked on the device, whichever is less.

    (Amended 1988)

    N.4.2.2. Low-Flame Test. – The device shall be tested at an extremely low-flow rate as given in Table 1. The test shall consist of passing air at a pressure of 375 Pa (1.5 in water column) through the meter for not less than 60 minutes. The meter shall continue to advance at the conclusion of the test period.

    (Amended 1990 and 1991)

    N.4.2.3. Pressure Regulation Test. – On devices operating at a pressure of 6900 Pa (1 psig) or more, a pressure regulation test shall be made at both the minimum and maximum use load to determine the proper operation of the regulator and the proper sizing of the piping and dispensing equipment. These tests may include a test of 24 hours during which the pressure is recorded.

    (Added 1984)

N.5. Temperature Correction. – Corrections shall be made for any changes in volume resulting from the difference in air temperatures between time of passage through the device and time of volumetric determination in the proving device.

N.6. Frequency of Test. – A hydrocarbon gas vapor-measuring device shall be tested before installation and allowed to remain in service for 10 years from the time last tested without being retested, unless a test is requested by:

    (a) the purchaser of the product being metered;

    (b) the seller of the product being metered; or

    (c) the weights and measures official.

T. Tolerances

T.1. Tolerance Values on Normal Tests and on Special Tests Other Than Low-Flame Tests. – Maintenance and acceptance tolerances for normal and special tests for hydrocarbon gas vapor-measuring devices shall be as shown in Table T.1. Accuracy Classes and Tolerances for Hydrocarbon Gas Vapor-Measuring Devices.

(Amended 1981 and 2003)

Table T.1. Accuracy Classes and Tolerances for Hydrocarbon Gas Vapor-Measuring Devices				
Accuracy Class	**Application**		**Acceptance Tolerance**	**Maintenance Tolerance**
3.0	Gases at low pressure (for example, LPG vapor)	Overregistration	1.5 %	1.5 %
		Underregistration	3.0 %	3.0 %

(Added 2003)

T.2.   Repeatability. – When multiple tests are conducted at approximately the same flow rate and draft size, the range of the test results for the flow rate shall not exceed 0.9 % and the results of each test shall be within the applicable tolerance. (Also see N.4.1.2. Repeatability Test.)

(Added 2002)

UR.   User Requirements

UR.1.   Installation Requirements.

UR.1.1.   Capacity Rate. – A device shall be so installed that the actual maximum flow rate will not exceed the capacity rate except for short durations. If necessary, means for flow regulation shall be incorporated in the installation, in which case this shall be fully effective and automatic in operation.

UR.1.2.   Leakage. – The metering system shall be installed and maintained as a pressure-tight and leak-free system.

UR.2.   Use Requirements.

UR.2.1.   Automatic Temperature Compensation. – A compensated device may not be replaced with an uncompensated device without the written approval of the weights and measures authority having jurisdiction over the device.

UR.2.2.   Invoices. – A customer purchasing hydrocarbon gas measured by a vapor meter shall receive from the seller an invoice for each billing period. The invoice shall clearly and separately show the following:

(a)   The opening and closing meter readings and the dates of those readings.

(b)   The altitude correction factor.

(c)   The total cubic meters (cubic feet) billed, corrected for elevation.

(d)   The charge per cubic meter (cubic foot) after correction for elevation.

(e)   All periodic charges independent of the measured gas, such as meter charges, meter reading fees, service charges or a minimum charge for a minimum number of cubic meters (cubic feet).

(f)   The total charge for the billing period.

If the vapor meter is equipped with an automatic temperature compensator, or any other means are used to compensate for temperature, the invoice shall show that the volume has been adjusted to the volume at 15 °C (60 °F).

(Amended 1988 and 1991)

UR.2.3.   Correction for Elevation. – The metered volume of gas shall be corrected for changes in the atmospheric pressure with respect to elevation to the standard pressure of 101.56 kPa (14.73 psia).   The appropriate   altitude   correction   factor   from   Table 2M. Corrections   for   Altitude,   Metric   Units or Table 2. Corrections for Altitude, U.S. Customary Units shall be used. (The table is modified from NIST Handbook 117.)

(Amended 1988)

Elevation correction factors (ACF) were obtained by using the following equation:

$$ACF = \frac{GP \, of \, gas + AAP}{base \, pressure}$$

Where:

GP	= gauge pressure
AAP	= assumed atmospheric pressure
base pressure	= 101.560 kPa = 14.73 psia
2740 Pa	= 11 in of water column = 0.397 psig
1744 Pa	= 7 in of water column = 0.253 psig

(Added 1988)

UR.2.4.   *Valves and Test Tee.* – *All gas meter installations shall be provided with a shut-off valve located adjacent to and on the inlet side of the meter. In the case of a single meter installation utilizing a liquefied petroleum gas tank, the tank service valve may be used in lieu of the shut-off valve. All gas meter installations shall be provided with a test tee located adjacent to and on the outlet side of the meter.*
[Nonretroactive as of January 1, 1990]

(Added 1989)

UR.2.5.   Use of Auxiliary Heated Vaporizer Systems. – Automatic temperature compensation shall be used on hydrocarbon gas vapor meters equipped with an auxiliary heated vaporizer system unless there is sufficient length of underground piping to provide gas at a uniform temperature to the meter inlet. When required by weights and measures officials, a thermometer well (appropriately protected against freezing) shall be installed immediately up-stream of the meter.

(Added 1990)

			Altitude Correction Factor		Assumed Atmospheric Pressure	Assumed Atmospheric Pressure Plus Gauge Pressure	
Elevation (meters)			2.74 kPa Gauge Pressure	1.74 kPa Gauge Pressure	(kPa)	2.74 kPa Gauge Pressure	1.74 kPa Gauge Pressure
	− 50 to	120	1.02	1.01	100.85	103.59	102.58
above	120 to	300	1.00	0.99	98.82	101.56	100.54
above	300 to	470	0.98	0.97	96.79	99.53	98.51
above	470 to	650	0.96	0.95	94.76	97.50	96.48
above	650 to	830	0.94	0.93	92.73	95.47	94.45
above	830 to	1020	0.92	0.91	90.70	93.44	92.42
above	1020 to	1210	0.90	0.89	88.66	91.40	90.39
above	1210 to	1400	0.88	0.87	86.63	89.37	88.36
above	1400 to	1590	0.86	0.85	84.60	87.34	86.33
above	1590 to	1790	0.84	0.83	82.57	85.31	84.29
above	1790 to	2000	0.82	0.81	80.54	83.28	82.26
above	2000 to	2210	0.80	0.79	78.51	81.25	80.23
above	2210 to	2420	0.78	0.77	76.48	79.22	78.20
above	2420 to	2640	0.76	0.75	74.45	77.19	76.17
above	2640 to	2860	0.74	0.73	72.41	75.15	74.15
above	2860 to	3080	0.72	0.71	70.38	73.12	72.12
above	3080 to	3320	0.70	0.69	68.35	71.09	70.08
above	3320 to	3560	0.68	0.67	66.32	69.06	68.05
above	3560 to	3800	0.66	0.65	64.29	67.03	66.01
above	3800 to	4050	0.64	0.63	62.26	65.00	63.98
above	4050 to	4310	0.62	0.61	60.23	62.97	61.95
above	4310 to	4580	0.60	0.59	58.20	60.94	59.92

Table 2M.
Corrections for Altitude, Metric Units

Table 2. Corrections for Altitude, U.S. Customary Units						
Elevation (feet)		Altitude Correction Factor		Assumed Atmospheric Pressure	Assumed Atmospheric Pressure Plus Gauge Pressure	
		11 inch WC	7 inch WC	(psia)	11 inch WC (psia)	7 inch WC (psia)
	− 150 to 400	1.02	1.01	14.64	15.04	14.89
above	400 to 950	1.00	0.99	14.35	14.74	14.60
above	950 to 1 550	0.98	0.97	14.05	14.45	14.30
above	1 550 to 2 100	0.96	0.95	13.76	14.15	14.01
above	2 100 to 2 700	0.94	0.93	13.46	13.86	13.71
above	2 700 to 3 300	0.92	0.91	13.17	13.56	13.42
above	3 300 to 3 950	0.90	0.89	12.87	13.27	13.12
above	3 950 to 4 550	0.88	0.87	12.58	12.97	12.83
above	4 550 to 5 200	0.86	0.85	12.28	12.68	12.53
above	5 200 to 5 850	0.84	0.83	11.99	12.38	12.24
above	5 850 to 6 500	0.82	0.81	11.69	12.09	11.94
above	6 500 to 7 200	0.80	0.79	11.40	11.79	11.65
above	7 200 to 7 900	0.78	0.77	11.10	11.50	11.35
above	7 900 to 8 600	0.76	0.75	10.81	11.20	11.06
above	8 600 to 9 350	0.74	0.73	10.51	10.91	10.76
above	9 350 to 10 100	0.72	0.71	10.22	10.61	10.47
above	10 100 to 10 850	0.70	0.69	9.92	10.32	10.17
above	10 850 to 11 650	0.68	0.67	9.63	10.03	9.88
above	11 650 to 12 450	0.66	0.65	9.33	9.73	9.58
above	12 450 to 13 250	0.64	0.63	9.04	9.44	9.29
above	13 250 to 14 100	0.62	0.61	8.75	9.14	9.00
above	14 100 to 14 950	0.60	0.59	8.45	8.85	8.70

THIS PAGE INTENTIONALLY LEFT BLANK

Table of Contents

Section 3.34.   Cryogenic Liquid-Measuring Devices

A. Application

A.1.   General. – This code applies to devices used for the measurement of cryogenic liquids such as, but not limited to oxygen, nitrogen, hydrogen, and argon.
(Amended 1986 and 1995)

A.2.   Exceptions. – This code does not apply to the following:

(a) Devices used for dispensing liquefied petroleum gases (for which see Section 3.32. Code for Liquefied Petroleum Gas and Anhydrous Ammonia Liquid-Measuring Devices).

(b) Devices used solely for dispensing a product in connection with operations in which the amount dispensed does not affect customer charges.

(c) Devices used solely for dispensing liquefied natural gas.

(d) Mass flow meters. (Also see Section 3.37. Code for Mass Flow Meters.)
(Added 1994)

A.3.   Additional Code Requirements. – In addition to the requirements of this code, Cryogenic Liquid- Measuring Devices shall meet the requirements of Section 1.10. General Code.

S. Specifications

S.1.   Design of Indicating and Recording Elements and of Recorded Representations.

S.1.1.   Primary Elements.

S.1.1.1.   General. – A device shall be equipped with a primary indicating element and may also be equipped with a primary recording element.

S.1.1.2.   Units. – A device shall indicate and record, if equipped to record, its deliveries in terms of: kilograms or pounds; liters or gallons of liquid at the normal boiling point of the specific cryogenic product; cubic meters (cubic feet) of gas at a normal temperature of 21 °C (70 °F) and an absolute pressure of 101.325 kPa (14.696 psia); or decimal subdivisions or multiples of the measured units cited above.
(Amended 2002)

S.1.1.3.   Value of Smallest Unit. – The value of the smallest unit of indicated delivery, and recorded delivery, if the device is equipped to record, shall not exceed the equivalent of:

(a) for small delivery devices:

(1) 1 L;

(2) 0.1 gal;

(3) 1 kg;

(4) 1 lb;

    (5) 0.1 m³ of gas; or

    (6) 10 ft³ of gas.

  (b) for large delivery devices:

    (1) 10 L;

    (2) 1 gal;

    (3) 10 kg;

    (4) 10 lb;

    (5) 1 m³ of gas; or

    (6) 100 ft³ of gas.
(Amended 2002)

S.1.1.4. Advancement of Indicating and Recording Elements. – Primary indicating and recording elements shall be susceptible to advancement only by the normal operation of the device. However, a device may be cleared by advancing its elements to zero, but only if:

  (a) the advancing movement, once started, cannot be stopped until zero is reached; or

  (b) in the case of indicating elements only, such elements are automatically obscured until the elements reach the correct zero position.

S.1.1.5. Return to Zero. – Primary indicating and recording elements shall be readily returnable to a definite zero indication. Means shall be provided to prevent the return of primary indicating elements and of primary recording elements beyond their correct zero position.

S.1.2. Graduations.

S.1.2.1. Length. – Graduations shall be so varied in length that they may be conveniently read.

S.1.2.2. Width. – In any series of graduations, the width of a graduation shall in no case be greater than the width of the minimum clear interval between graduations, and the width of main graduations shall be not more than 50 % greater than the width of subordinate graduations. Graduations shall in no case be less than 0.2 mm (0.008 in) in width.

S.1.2.3. Clear Interval Between Graduations. – The clear interval shall be no less than 1.0 mm (0.04 in). If the graduations are not parallel, the measurement shall be made:

  (a) along the line of relative movement between the graduations at the end of the indicator; or

  (b) if the indicator is continuous, at the point of widest separation of the graduations.
(Also see S.1.3.6. Travel of Indicator.)

S.1.3. Indicators.

S.1.3.1. Symmetry. – The index of an indicator shall be symmetrical with respect to the graduations, at least throughout that portion of its length associated with the graduations.

S.1.3.2. Length. – The index of an indicator shall reach to the finest graduations with which it is used, unless the indicator and the graduations are in the same plane, in which case the distance between the end of the indicator and the ends of the graduations, measured along the line of the graduations, shall be not more than 1.0 mm (0.04 in).

S.1.3.3.   Width. – The width of the index of an indicator in relation to the series of graduations with which it is used shall be not greater than the:

(a) *width of the narrowest graduation;* * and
    [*Nonretroactive as of January 1, 2002]
    (Amended 2001)

(b) width of the minimum clear interval between graduations.

When the index of an indicator extends along the entire length of a graduation, that portion of the index of the indicator that may be brought into coincidence with the graduation shall be of the same width throughout the length of the index that coincides with the graduation.

S.1.3.4.   Clearance. – The clearance between the index of an indicator and the graduations shall in no case be more than 1.5 mm (0.06 in).

S.1.3.5.   Parallax. – Parallax effect shall be reduced to the practicable minimum.

S.1.3.6.   Travel of Indicator. – If the most sensitive element of the primary indicating element uses an indicator and graduations, the relative movement of these parts corresponding to the smallest indicated value shall be not less than 0.5 mm (0.20 in).

S.1.4.   Computing-Type Device.

S.1.4.1.   Printed Ticket. – Any printed ticket issued by a device of the computing type on which there is printed the total computed price shall have printed clearly thereon also the total quantity of the delivery and the price per unit.

S.1.4.2.   Money-Value Computations. – Money-value computations shall be of the full-computing type in which the money-value at a single unit price, or at each of a series of unit prices, shall be computed for every delivery within either the range of measurement of the device or the range of the computing elements, whichever is less. Value graduations shall be supplied and shall be accurately positioned. The total price shall be computed on the basis of the quantity indicated when the value of the smallest division indicated is equal to or less than the values specified in S.1.1.3. Value of Smallest Unit.

S.1.4.3.   Money-Values, Mathematical Agreement. – Any digital money-value indication and any recorded money-value on a computing type device shall be in mathematical agreement with its associated quantity indication or representation to within 1 cent of money-value.

S.2.   Design of Measuring Elements.

S.2.1.   Vapor Elimination. – A measuring system shall be equipped with an effective vapor eliminator or other effective means to prevent the measurement of vapor that will cause errors in excess of the applicable tolerances. (Also see Section T. Tolerances.)

S.2.2.   Directional Flow Valves. – A valve or valves or other effective means, automatic in operation, to prevent the reversal of flow shall be installed in or adjacent to the measuring device.
(Amended 1978)

S.2.3.   Maintenance of Liquid State. – A device shall be so designed that the product being measured will remain in a liquid state during passage through the device.

S.2.4.   Automatic Temperature or Density Compensation. – A device shall be equipped with automatic means for adjusting the indication and/or recorded representation of the measured quantity of the product, to indicate and/or record in terms of: kilograms or pounds; or liters or gallons of liquid at the normal boiling point of the specific cryogenic product; or the equivalent cubic meters (cubic feet) of gas at a normal temperature of 21 EC (70 EF) and an absolute pressure of 101.325 kPa (14.696 lb/in² absolute). *When a compensator system malfunctions, the indicating and recording elements may indicate and record in uncompensated volume if the mode of operation is clearly indicated, e.g., by a marked annunciator, recorded statement, or other obvious means.* *

*[*Nonretroactive as of January 1, 1992]*

(Amended 1991 and 2002)

S.2.5.   Provision for Sealing. – Adequate provision shall be made for an approved means of security (e.g., data change audit trail) or for physically applying a security seal in such a manner that requires the security seal to be broken before an adjustment or interchange may be made of:

(a)   any measuring or indicating element;

(b)   any adjustable element for controlling delivery rate when such rate tends to affect the accuracy of deliveries;

(c)   any automatic temperature or density compensating system; and

(d)   any metrological parameter that will affect the metrological integrity of the device or system.

When applicable, any adjusting mechanism shall be readily accessible for purposes of affixing a security seal.

[Audit trails shall use the format set forth in Table S.2.5. Categories of Device and Methods of Sealing] *
*[*Nonretroactive as of January 1, 1995]*

(Amended 2006)

Table S.2.5. Categories of Device and Methods of Sealing	
Categories of Device	*Methods of Sealing*
Category 1: *No remote configuration capability.*	*Seal by physical seal or two event counters: one for calibration parameters and one for configuration parameters.*
Category 2: *Remote configuration capability, but access is controlled by physical hardware.* *The device shall clearly indicate that it is in the remote configuration mode and record such message if capable of printing in this mode or shall not operate while in this mode.*	*The hardware enabling access for remote communication must be on-site. The hardware must be sealed using a physical seal or an event counter for calibration parameters and an event counter for configuration parameters. The event counters may be located either at the individual measuring device or at the system controller; however, an adequate number of counters must be provided to monitor the calibration and configuration parameters of the individual devices at a location. If the counters are located in the system controller rather than at the individual device, means must be provided to generate a hard copy of the information through an on-site device.*
Category 3: *Remote configuration capability access may be unlimited or controlled through a software switch (e.g., password).* *The device shall clearly indicate that it is in the remote configuration mode and record such message if capable of printing in this mode or shall not operate while in this mode.*	*An event logger is required in the device; it must include an event counter (000 to 999), the parameter ID, the date and time of the change, and the new value of the parameter. A printed copy of the information must be available on demand through the device or through another on-site device. The information may also be available electronically. The event logger shall have a capacity to retain records equal to 10 times the number of sealable parameters in the device, but not more than 1000 records are required. (**Note:** Does not require 1000 changes to be stored for each parameter.)*

[Nonretroactive as of January 1, 1995]

(Table Added 2006) (Amended 2016)

S.3.   Design of Discharge Lines and Discharge Line Valves.

S.3.1.   Diversion of Measured Liquid. – No means shall be provided by which any measured liquid can be diverted from the measuring chamber of the device or the discharge line therefrom, except that a manually controlled outlet that may be opened for purging or draining the measuring system shall be permitted. Effective means shall be provided to prevent the passage of liquid through any such outlet during normal operation of the device and to indicate clearly and unmistakably when the valve controls are so set as to permit passage of liquid through such outlet.

S.3.2.   Discharge Hose. – The discharge hose of a measuring system shall be of the completely draining dry-hose type.

S.4.   Marking Requirements.

S.4.1.   Limitation of Use. – If a measuring system is intended to measure accurately only liquids having particular properties, or to measure accurately only under specific installation or operating conditions, or to measure accurately only when used in conjunction with specific accessory equipment, these limitations shall be clearly and permanently marked on the device.

S.4.2.   Discharge Rates. – A meter shall be marked to show its designed maximum and minimum discharge rates.

S.4.3.   Temperature or Density Compensation. – Devices equipped with an automatic temperature or density compensator, shall be clearly and conspicuously marked on the primary indicating elements, recording elements, and recorded representations to show that the quantity delivered has been adjusted to the conditions specified in S.2.4. Automatic Temperature or Density Compensation.

N. Notes

N.1.   Test Liquid. – A meter shall be tested with the liquid to be commercially measured except that, in a type evaluation examination, nitrogen may be used.

N.2.   Vaporization and Volume Change. – Care shall be exercised to reduce to a minimum vaporization and volume changes. When testing by weight, the weigh tank and transfer systems shall be pre-cooled to liquid temperature prior to the start of the test to avoid the venting of vapor from the vessel being weighed.

N.3.   Test Drafts.

N.3.1.   Gravimetric Test. – Weight test drafts shall be equal to at least the amount delivered by the device in 2 minutes at its maximum discharge rate, and shall in no case be less than 907 kg (2000 lb).

N.3.2.   Transfer Standard Test. – When comparing a meter with a calibrated transfer standard, the test draft shall be equal to at least the amount delivered by the device in two minutes at its maximum discharge rate, and shall in no case be less than 180 L (50 gal) or equivalent thereof. When testing uncompensated volumetric meters in a continuous recycle mode, appropriate corrections shall be applied if product conditions are abnormally affected by this test mode.

(Amended 1976)

N.4.   Density. – Temperature and pressure of the metered test liquid shall be measured during the test for the determination of density or volume correction factors when applicable. For Liquid Density and Volume Correction Factors (with respect to temperature and pressure) the publications shown in Table N.4. Density or Volume Correction Factors shall apply.

(Amended 1986 and 2004)

Table N.4. Density or Volume Correction Factors	
Cryogenic Liquid	**Publication**
Argon	Tegeler, Ch., Span, R., Wagner, W. "A New Equation of State for Argon Covering the Fluid Region for Temperatures from the Melting Line to 700 K at Pressures up to 1000 Mpa." *J. Phys. Chem. Ref. Data*, 28(3):779-850, 1999.
Ethylene	Smukala, J., Span, R., Wagner, W. "New Equation of State for Ethylene Covering the Fluid Region for Temperatures from the Melting Line to 450 k at Pressures up to 300 Mpa." *J. Phys. Chem. Ref. Data*, 29(5):1053-1122, 2000.
Nitrogen	Span, R., Lemmon, E.W., Jacobsen, R.T, Wagner, W., and Yokozeki, A. "A Reference Thermodynamic Property Formulation for Nitrogen." *J. Phys. Chem. Ref. Data*, Volume 29, Number 6, pp. 1361-1433, 2000.
Oxygen	Schmidt, R., Wagner, W. "A New Form of the Equation of State for Pure Substances and its Application to Oxygen." *Fluid Phase Equilib.*, 19:175-200, 1985
Hydrogen	Leachman, J. W., Jacobsen, R. T., Lemmon, E.W., and Penoncello, S.G. "Fundamental Equations of State for Parahydrogen, Normal Hydrogen, and Orthohydrogen" *J. Phys. Chem. Ref. Data*, Volume 38, Number 3, pp. 565, 2009.

Note: A complete database program containing all of the most recent equations for calculating density for various cryogenic liquids is available at www.nist.gov/srd/nist23.cfm. There is a fee for download of this database.

(Added 2004)

N.5.   Testing Procedures.

N.5.1.   Normal Tests. – The "normal" tests of a device shall be made over a range of discharge rates that may be anticipated under the conditions of installation.

N.5.1.1.   Repeatability Tests. – Tests for repeatability should include a minimum of three consecutive test drafts of approximately the same size and be conducted under controlled conditions where variations in factors such as temperature, pressure, and flow rate are reduced to the extent that they will not affect the results obtained.

(Added 2001)

N.5.2.   Special Tests. – Any test except as set forth in N.5.1. Normal Tests shall be considered a "special" test. Tests shall be conducted, if possible, to evaluate any special elements or accessories attached to or associated with the device. A device shall be tested at a minimum discharge rate of:

(a)   50 % of the maximum discharge rate developed under the conditions of installation, or the minimum discharge rate marked on the device, whichever is less; or

(b)   the lowest discharge rate practicable under conditions of installation.

Special tests may be conducted to develop any characteristics of the device that are not normally anticipated under the conditions of installation.

N.6. Temperature Correction. – Corrections shall be made for any changes in volume resulting from the differences in liquid temperature between time of passage through the meter and time of volumetric determination of test draft.

N.7. Automatic Temperature or Density Compensation. – When a device is equipped with an automatic temperature or density compensator, the compensator shall be tested by comparing the quantity indicated or recorded by the device (with the compensator connected and operating) with the actual delivered quantity corrected to the normal boiling point of the cryogenic product being measured or to the normal temperature and pressure as applicable.

T. Tolerances

T.1. Application.

T.1.1. To Underregistration and to Overregistration. – The tolerances hereinafter prescribed shall be applied to errors of underregistration and errors of overregistration.

T.2. Tolerance Values. – The maintenance and acceptance tolerances for normal and special tests shall be as shown in Table T.2. Accuracy Classes and Tolerances for Cryogenic Liquid-Measuring Devices.
(Amended 2003)

Table T.2. Accuracy Classes and Tolerances for Cryogenic Liquid-Measuring Devices				
Accuracy Class	Application	Acceptance Tolerance	Maintenance Tolerance	Special Test Tolerance
2.5	Cryogenic products; liquefied compressed gases other than liquid carbon dioxide	1.5 %	2.5 %	2.5 %

(Added 2003)

T.3. On Tests Using Transfer Standards. – To the basic tolerance values that would otherwise be applied, there shall be added an amount equal to two times the standard deviation of the applicable transfer standard when compared to a basic reference standard.
(Added 1976)

T.4. Repeatability. – When multiple tests are conducted at approximately the same flow rate and draft size, the range of the test results for the flow rate shall not exceed 40 % of the absolute value of the maintenance tolerance and the results of each test shall be within the applicable tolerance. Also see N.5.1.1. Repeatability Tests.
(Added 2001)

UR. User Requirements

UR.1. Installation Requirements.

UR.1.1. Discharge Rate. – A device shall be so installed that the actual maximum discharge rate will not exceed the rated maximum discharge rate. If necessary, means for flow regulation shall be incorporated in the installation.

UR.1.2. Length of Discharge Hose. – The discharge hose shall be of such a length and design as to keep vaporization of the liquid to a minimum.

UR.1.3.   Maintenance of Liquid State. – A device shall be so installed and operated that the product being measured shall remain in the liquid state during passage through the meter.

UR.2.   Use Requirements.

UR.2.1.   Return of Indicating and Recording Elements to Zero. – The primary indicating elements (visual) and the primary recording elements shall be returned to zero immediately before each delivery.

UR.2.2.   Condition of Discharge System. – The discharge system, up to the measuring element, shall be precooled to liquid temperatures before a "zero" condition is established prior to the start of a commercial delivery.

UR.2.3.   Vapor Return Line. – A vapor return line shall not be used during a metered delivery.
(Amended 1976)

UR.2.4.   Drainage of Discharge Line. – On a dry-hose system, upon completion of a delivery, the vendor shall leave the discharge line connected to the receiving container with the valve adjacent to the meter in the closed position and the valve at the discharge line outlet in the open position for a period of at least:

(a) 1 minute for small delivery devices; and

(b) 3 minutes for large delivery devices

to allow vaporization of some product in the discharge line to force the remainder of the product in the line to flow into the receiving container.
(Amended 1976)

UR.2.5.   Conversion Factors. – Established conversion values (Also see references in Table N.4. Density or Volume Correction Factors.) shall be used whenever metered liquids are to be billed in terms of:

(a) kilograms or pounds based on a meter indication of liters, gallons, cubic meters of gas, or cubic feet of gas;

(b) cubic meters or cubic feet of gas based on a meter indication of liters or gallons, kilograms, or pounds; or

(c) liters or gallons based on a meter indication of kilograms or pounds, cubic meters of gas or cubic feet of gas.

All sales of cryogenics shall be based on either kilograms or pounds, liters or gallons of liquid at NBP,[1] cubic meters of gas or cubic feet of gas at NTP[1].
(Amended 1986)

UR.2.6.   Temperature or Density Compensation.

UR.2.6.1.   Use of Automatic Temperature or Density Compensators. – If a device is equipped with an automatic temperature or density compensator, this shall be connected, operable, and in use at all times. Such automatic temperature or density compensator may not be removed, nor may a compensated device be replaced with an uncompensated device, without the written approval of the weights and measures authority having jurisdiction over the device.

UR.2.6.2.   Tickets or Invoices. – Any written invoice or printed ticket based on a reading of a device that is equipped with an automatic temperature or density compensator shall have shown thereon that the quantity

----

[1] See Appendix D, Definitions.

delivered has been adjusted to the quantity at the NBP of the specific cryogenic product or the equivalent volume of gas at NTP.

UR.2.6.3.   Printed Ticket. – Any printed ticket issued by a device of the computing type on which there is printed the total computed price, the total quantity of the delivery, or the price per unit, shall also show the other two values (either printed or in clear hand script).

UR.2.6.4.   Ticket in Printing Device. – A ticket shall not be inserted into a device equipped with a ticket printer until immediately before a delivery is begun, and in no case shall a ticket be in the device when the vehicle is in motion while on a public street, highway, or thoroughfare.

UR.2.7.   Pressure of Tanks with Volumetric Metering Systems without Temperature Compensation. – When the saturation pressure of the product in the vendor's tank exceeds 240 kPa (35 psia), a correction shall be applied to the written invoice or printed ticket using the appropriate tables as listed in Table N.4. Density or Volume Correction Factors; or the saturation pressure shall be reduced to 207 kPa (30 psia) (if this can be safely accomplished) prior to making a delivery.

(Added 1976)

Table of Contents

Section 3.35.   Milk Meters

A. Application

A.1.   General. – This code applies to devices used for the measurement of milk; generally applicable to, but not limited to, meters used in dairies, milk processing plants, and cheese factories, to measure incoming bulk milk.

A.2.   Exceptions. – This code does not apply to mass flow meters. (Also see Section 3.37. Code for Mass Flow Meters.)

(Added 1994)

A.3.   Additional Code Requirements. – In addition to the requirements of this code, Milk Meters shall meet the requirements of Section 1.10. General Code.

S. Specifications

S.1.   Design of Indicating and Recording Elements and of Recorded Representations.

S.1.1.   Primary Elements.

S.1.1.1.   General. – A meter shall be equipped with a primary indicating element and may also be equipped with a primary recording element.

S.1.1.2.   Units.

(a) A meter shall indicate, and record if the meter is equipped to record, its deliveries in terms of liters or gallons. Fractional parts of the liter shall be in terms of decimal subdivisions. Fractional parts of the gallon shall be in terms of either decimal or binary subdivisions.

(b) When it is an industry practice to purchase and sell milk by weight based upon 1.03 kg/L (8.6 lb/gal), the primary indicating element may indicate in kilograms or pounds. The weight value division shall be a decimal multiple or submultiple of 1, 2, or 5. Fractional parts of the kilogram or pound shall be in decimal subdivisions. (Also see S.4.5. Conversion Factor.)

S.1.1.3.   Value of Smallest Unit. – The value of the smallest unit of indicated quantity and recorded quantity, if the meter is equipped to record, shall not exceed the equivalent of:

(a) 0.5 L or 0.5 kg (1 pt or 1 lb) when measuring quantities less than or equal to 4000 L or 4000 kg (1000 gal or 8600 lb); or

(b) 5 L or 5 kg (1 gal or 10 lb) when measuring quantities in excess of 4000 L or 4000 kg (1000 gal or 8600 lb).

(Amended 1989)

S.1.1.4.   Advancement of Indicating and Recording Elements. – Primary indicating and recording elements shall be susceptible to advancement only by the mechanical operation of the meter. However, a meter may be cleared by advancing its elements to zero, but only if:

(a) the advancing movement, once started, cannot be stopped until zero is reached; or

(b) in the case of indicating elements only, such elements are automatically obscured until the elements reach the correct zero position.

S.1.1.5.   Return to Zero. – Primary indicating elements and primary recording elements, if the device is equipped to record, shall be readily returnable to a definite zero indication. Means shall be provided to prevent the return of the primary indicating elements and the primary recording elements, if the device is so equipped, beyond their correct zero position.

S.1.1.6.   Indication of Measurement. – A meter shall be constructed to show automatically its initial zero condition and the volume measured up to the nominal capacity of the device.

S.1.2.   Graduations.

S.1.2.1.   Length. – Graduations shall be so varied in length that they may be conveniently read.

S.1.2.2.   Width. – In any series of graduations, the width of a graduation shall in no case be greater than the width of the minimum clear interval between graduations, and the width of main graduations shall be not more than 50 % greater than the width of subordinate graduations. Graduations shall in no case be less than 0.2 mm (0.008 in) in width.

S.1.2.3.   Clear Interval between Graduations. – The clear interval shall be not less than 1.0 mm (0.04 in). If the graduations are not parallel, the measurement shall be made:

(a) along the line of relative movement between the graduations at the end of the indicator; or

(b) if the indicator is continuous, at the point of widest separation of the graduations.

S.1.3.   Indicators.

S.1.3.1.   Symmetry. – The index of an indicator shall be symmetrical with respect to the graduations, at least throughout that portion of its length associated with the graduations.

S.1.3.2.   Length. – The index of an indicator shall reach to the finest graduations with which it is used, unless the indicator and the graduations are in the same plane, in which case the distance between the end of the indicator and the ends of the graduations, measured along the line of graduations, shall be not more than 1.0 mm (0.04 in).

S.1.3.3.   Width. – The width of the index of an indicator in relation to the series of graduations with which it is used shall be not greater than:

(a) *the width of the narrowest graduation;* * and
     *[*Nonretroactive as of January 1, 2002]*
     (Amended 2001)

(b) the width of the minimum clear interval between graduations.

When the index of an indicator extends along the entire length of a graduation, that portion of the index of the indicator that may be brought into coincidence with the graduation shall be of the same width throughout the length of the index that coincides with the graduation.

S.1.3.4.   Clearance. – The clearance between the index of an indicator and the graduations shall in no case be more than 1.5 mm (0.06 in).

S.1.3.5.   Parallax. – Parallax effects shall be reduced to the practicable minimum.

S.1.3.6.   Travel of Indicator. – If the most sensitive element of the primary indicating element utilizes an indicator and graduations, the relative movement of these parts corresponding to the smallest indicated value shall be not less than 5 mm (0.20 in).

S.1.4. Computing-Type Devices.

S.1.4.1.   Display of Unit Price. – In a device of the computing type, means shall be provided for displaying on the outside of the device, and in close proximity to the display of the total computed price, the price per unit at which the device is set to compute.

S.1.4.2.   Printed Ticket. – If a computing-type device issues a printed ticket which displays the total computed price, the ticket also shall have printed clearly thereon the total quantity of the delivery, the appropriate fraction of the quantity, and the price per unit of quantity.
(Amended 1989)

S.1.4.3.   Money-Value Computations. – Money-value computations shall be of the full-computing type in which the money-value at a single unit price, or at each of a series of unit prices, shall be computed for every delivery within either the range of measurement of the device or the range of the computing elements, whichever is less. Value graduations shall be supplied and shall be accurately positioned. The value of each graduated interval shall be 1 cent.

S.1.4.4.   Money-Values, Mathematical Agreement. – Any digital money-value indication and any recorded money-value on a computing-type device shall be in mathematical agreement with its associated quantity indicating or representation to within 1 cent of money-value.

S.2.   Design of Measuring Elements.

S.2.1.   Vapor Elimination. – A metering system shall be equipped with an effective vapor eliminator or other effective means automatic in operation to prevent the passage of vapor and air through the meter. Vent lines from the air (or vapor) eliminator shall be made of metal tubing or some other suitably rigid material.

S.2.2.   Maintaining Flooded Condition. – The vent on the vapor eliminator shall be positioned or installed in such a manner that the vapor eliminator cannot easily be emptied between uses.

S.2.3.   Provision for Sealing. – Adequate provision shall be made for an approved means of security (e.g., data change audit trail) or for physically applying a security seal in such a manner that requires the security seal to be broken before an adjustment or interchange may be made of any:

(a) measuring element or indicating element;

(b) adjustable element for controlling delivery rate, when such rate tends to affect the accuracy of deliveries; and

(c) metrological parameter that will affect the metrological integrity of the device or system.

When applicable, the adjusting mechanism shall be readily accessible for purposes of affixing a security seal.

[Audit trails shall use the format set forth in Table S.2.3. Categories of Device and Methods of Sealing]*
[*Nonretroactive as of January 1, 1995]
(Amended 2006)

Table S.2.3. *Categories of Device and Methods of Sealing*	
Categories of Device	*Methods of Sealing*
Category 1: *No remote configuration capability.*	*Seal by physical seal or two event counters: one for calibration parameters and one for configuration parameters.*
Category 2: *Remote configuration capability, but access is controlled by physical hardware.* *The device shall clearly indicate that it is in the remote configuration mode and record such message if capable of printing in this mode or shall not operate while in this mode.*	*The hardware enabling access for remote communication must be on-site. The hardware must be sealed using a physical seal or an event counter for calibration parameters and an event counter for configuration parameters. The event counters may be located either at the individual measuring device or at the system controller; however, an adequate number of counters must be provided to monitor the calibration and configuration parameters of the individual devices at a location. If the counters are located in the system controller rather than at the individual device, means must be provided to generate a hard copy of the information through an on-site device.*
Category 3: *Remote configuration capability access may be unlimited or controlled through a software switch (e.g., password).* *The device shall clearly indicate that it is in the remote configuration mode and record such message if capable of printing in this mode or shall not operate while in this mode.*	*An event logger is required in the device; it must include an event counter (000 to 999), the parameter ID, the date and time of the change, and the new value of the parameter. A printed copy of the information must be available on demand through the device or through another on-site device. The information may also be available electronically. The event logger shall have a capacity to retain records equal to 10 times the number of sealable parameters in the device, but not more than 1000 records are required. (**Note:** Does not require 1000 changes to be stored for each parameter.)*

[Nonretroactive as of January 1, 1995]
(Table Added 2006) (Amended 2016)

S.2.4. **Directional Flow Valves.** – Valves intended to prevent reversal of flow shall be automatic in operation.

S.3. **Design of Intake Lines.**

S.3.1. **Diversion of Liquid to be Measured.** – No means shall be provided by which any liquid can be diverted from the supply tank to the receiving tank without being measured by the device. A manually controlled outlet that may be opened for purging or draining the measuring system shall be permitted. Effective means shall be provided to prevent passage of liquid through any such outlet during normal operation of the measuring system.
(Amended 1994)

S.3.2. **Intake Hose.** – The intake hose shall be:

(a) of the dry-hose type;

(b) adequately reinforced;

(c) not more than 6 m (20 ft) in length unless it can be demonstrated that a longer hose is essential to permit transfer from a supply tank; and

(d) connected to the pump at horizontal or above to permit complete drainage of the hose.

(Amended 1991)

S.4. Marking Requirements.

S.4.1.   Limitation of Use. – If a meter is intended to measure accurately only liquids having particular properties, or to measure accurately only under specific installation or operating conditions, or to measure accurately only when used in conjunction with specific accessory equipment, these limitations shall be clearly and permanently stated on the meter.

S.4.2.   Discharge Rates. – A meter shall be marked to show its designed maximum and minimum discharge rates. The marked minimum discharge rate shall not exceed 20 % of the marked maximum discharge rate.

Note: Also see example in Section 3.30. Liquid-Measuring Devices Code, paragraph S.4.4.1. Discharge Rates.

(Added 2003)

S.4.3.   Measuring Components. – All components that affect the measurement of milk that are disassembled for cleaning purposes shall be clearly and permanently identified with a common serial number.

S.4.4.   Flood Volume. – When applicable, the volume of product (to the nearest minimum division of the meter) necessary to flood the system when dry shall be clearly, conspicuously, and permanently marked on the air eliminator.

S.4.5.   Conversion Factor. – When the conversion factor of 1.03 kg/L (8.6 lb/gal) is used to convert the volume of milk to weight, the conversion factor shall be clearly marked on the primary indicating element and recorded on the delivery ticket.

N. Notes

N.1. Test Liquid.

(a) A meter shall be tested with the liquid to be commercially measured or with a liquid of the same general physical characteristics. Following a satisfactory examination, the weights and measures official should attach a seal or tag indicating the product used during the test.

(Amended 1989)

(b) A milk measuring system shall be tested with the type of milk to be measured when the accuracy of the system is affected by the characteristics of milk (e.g., positive displacement meters).

(Added 1989)

N.2.   Evaporation and Volume Change. – Care shall be exercised to reduce to a minimum, evaporation losses and volume changes resulting from changes in temperature of the test liquid.

N.2.1.   Temperature Correction. – Corrections shall be made for any changes in volume resulting from the differences in liquid temperatures between time of passage through the meter and time of volumetric determination in the test measure. When adjustments are necessary, appropriate tables should be used.

N.3.   Test Drafts. – Test drafts should be equal to at least the amount delivered by the device in one minute at its maximum discharge rate, and shall in no case be less than 400 L or 400 kg (100 gal or 1 000 lb).

(Amended 1989)

N.4. Testing Procedures.

N.4.1. Normal Tests. – The "normal" test of a meter shall be made at the maximum discharge rate that may be anticipated under the conditions of the installation. The "normal" test shall include a determination of the effectiveness of the air elimination system.

N.4.1.1. Repeatability Tests. – Tests for repeatability should include a minimum of three consecutive test drafts of approximately the same size and be conducted under controlled conditions where variations in factors such as temperature, pressure, and flow rate are reduced to the extent that they will not affect the results obtained.

(Added 2002)

N.4.2. Special Tests. – "Special" tests shall be made to develop the operating characteristics of a device and any special elements and accessories attached to or associated with the device. Any test except as set forth in N.4.1. Normal Tests shall be considered a special test.

N.4.3. System Capacity. – The test of a milk-metering system shall include the verification of the volume of product necessary to flood the system as marked on the air eliminator.

T. Tolerances

T.1. Application.

T.1.1. To Underregistration and to Overregistration. – The tolerances hereinafter prescribed shall be applied to errors of underregistration and errors of overregistration.

T.2. Tolerance Values. – Maintenance and acceptance tolerances shall be as shown in Table 1. Tolerances for Milk Meters.

(Amended 1989)

Table 1. Tolerances for Milk Meters		
Indication	**Maintenance**	**Acceptance**
gallons	gallons	gallons
100	0.5	0.3
200	0.7	0.4
300	0.9	0.5
400	1.1	0.6
500	1.3	0.7
Over 500	Add 0.002 gallon per indicated gallon over 500	Add 0.001 gallon per indicated gallon over 500

(Added 1989)

T.3. Repeatability. – When multiple tests are conducted at approximately the same flow rate and draft size, the range of the test results for the flow rate shall not exceed 40 % of the absolute value of the maintenance tolerance and the results of each test shall be within the applicable tolerance. (Also see N.4.1.1. Repeatability Tests.)

(Added 2002)

UR.   User Requirements

UR.1.   Installation Requirements.

UR.1.1.   Plumb and Level Condition. – A device installed in a fixed location shall be installed plumb and level, and the installation shall be sufficiently strong and rigid to maintain this condition.

UR.1.2.   Discharge Rate. – A meter shall be so installed that the actual maximum discharge rate will not exceed the rated maximum discharge rate.   If necessary, means for flow regulation shall be incorporated in the installation, in which case this shall be fully effective and automatic in operation.

UR.1.3.   Unit Price. – There shall be displayed on the face of a device of the computing type the unit price at which the device is set to compute.

UR.1.4.   Intake Hose. – The intake hose shall be so installed as to permit complete drainage and that all available product is measured following each transfer.

UR.2.   Use Requirements.

UR.2.1.   Return of Indicating and Recording Elements to Zero. – The primary indicating elements (visual), and the primary recording elements when these are returnable to zero, shall be returned to zero before each transfer.

UR.2.2.   Printed Ticket. – Any printed ticket issued by a device of the computing type on which there is printed the total computed price, the total quantity, or the price per unit of quantity, shall also show the other two values (either printed or in clear script).
(Amended 1989)

UR.2.3.   Ticket in Printing Device. – A ticket shall not be inserted into a device equipped with a ticket printer until immediately before a transfer is begun.   If the meter is mounted on a vehicle, in no case shall a ticket be in the device when the vehicle is in motion while on a public street, highway, or thoroughfare.

UR.2.4.   Credit for Flood Volume. – The volume of product necessary to flood the system as marked on the air eliminator shall be individually recorded on the ticket of each transfer affected.

THIS PAGE INTENTIONALLY LEFT BLANK

Table of Contents

THIS PAGE INTENTIONALLY LEFT BLANK

Section 3.36. Water Meters

A. Application

A.1. General. – This code applies to devices used for the measurement of water; generally applicable to, but not limited to, utilities type meters installed in residences or business establishments and meters installed in batching systems.

(Amended 2002)

A.2. Exceptions. – This code does not apply to:

(a) water meters mounted on vehicle tanks (for which see Section 3.31. Vehicle-Tank Meters); or

(b) mass flow meters. (Also see Section 3.37. Mass Flow Meters.)

(Added 1994)

A.3. Additional Code Requirements. – In addition to the requirements of this code, Water Meters shall meet the requirements of Section 1.10. General Code.

S. Specifications

S.1. Design of Indicating and Recording Elements and of Recorded Representations.

S.1.1. Primary Elements.

S.1.1.1. General. – A water meter shall be equipped with a primary indicating element and may also be equipped with a primary recording element. Such elements shall be visible at the point of measurement or be stored in non-volatile and non-resettable memory. The display may be remotely located provided it is readily accessible to the customer.

(Amended 2002)

S.1.1.2. Units. – A water meter shall indicate and record, if the device is equipped to record, its deliveries in terms of liters, gallons or cubic feet or binary or decimal subdivisions thereof except batch plant meters, which shall indicate deliveries in terms of liters, gallons or decimal subdivisions of the liter or gallon only.

S.1.1.3. Value of Smallest Unit. – The value of the smallest unit of indicated delivery and recorded delivery, if the device is equipped to record, shall not exceed the equivalent of:

(a) 50 L (10 gal or 1 ft^3) on utility type meters, sizes 1 in and smaller; or

(b) 500 L (100 gal or 10 ft^3) on utility-type meters, sizes 1½ in and 2 in; or

(c) 0.2 L ($^1/_{10}$ gal or $^1/_{100}$ ft^3) on batching meters delivering less than 375 L/min (100 gal/min or 13 ft^3/min); or

(d) 5 L (1 gal or $^1/_{10}$ ft^3) on batching meters delivering 375 L/min (100 gal/min or 13 ft^3/min) or more.

(Amended 2009)

S.1.1.4. Advancement of Indicating and Recording Elements. – Primary indicating and recording elements shall be susceptible to advancement only by the mechanical operation of the device.

S.1.1.5. **Return to Zero.** – If the meter is so designed that the primary indicating elements are readily returnable to a definite zero indication, means shall be provided to prevent the return of these elements beyond their correct zero position.

S.1.1.6. **Proving indicator.** – Utility-type meters shall be equipped with a proving indicator. The individual graduations on a mechanical (analog) proving indicator shall indicate volumes no larger than $1/100$ of the value of the smallest unit of indicated delivery required in S.1.1.3. Value of Smallest Unit. For electronic (digital) proving indications, the smallest unit of volume displayed shall be no larger than $1/1000$ of the value of the smallest unit of indicated delivery required in S.1.1.3.

(Added 2009)

S.1.2. Graduations.

S.1.2.1. **Length.** – Graduations shall be so varied in length that they may be conveniently read.

S.1.2.2. **Width.** – In any series of graduations, the width of a graduation shall in no case be greater than the width of the minimum clear interval between graduations, and the width of main graduations shall be not more than 50 % greater than the width of subordinate graduations. Graduations shall in no case be less than 0.2 mm (0.008 in) in width.

S.1.2.3. **Clear Interval Between Graduations.** – The clear interval shall not be less than 1.0 mm (0.04 in). If the graduations are not parallel, the measurement shall be made:

(a) along the line of relative movement between the graduations at the end of the indicator; or

(b) if the indicator is continuous, at the point of widest separation of the graduations.

S.1.3. Indicators.

S.1.3.1. **Symmetry.** – The index of an indicator shall be symmetrical with respect to the graduations, at least throughout that portion of its length associated with the graduations.

S.1.3.2. **Length.** – The index of an indicator shall reach to the finest graduations with which it is used, the width of the minimum clear interval between graduations, and the width of main graduations shall be not more than 50 % greater than the width of subordinate graduations. Graduations shall in no case be less than 0.2 mm (0.008 in) in width.

S.1.3.3. **Width.** – The width of the index of an indicator in relation to the series of graduations with which it is used shall not be greater than:

(a) *the width of the narrowest graduation;* * and
   *[*Nonretroactive as of January 1, 2002]*
   (Amended 2001)

(b) the width of the minimum clear interval between graduations.

When the index of an indicator extends along the entire length of a graduation, that portion of the index of the indicator that may be brought into coincidence with the graduation shall be of the same width throughout the length of the index that coincides with the graduation.

S.1.3.4. **Clearance.** – The clearance between the index of an indicator and the graduations shall in no case be more than 1.5 mm (0.06 in).

S.1.3.5. **Parallax.** – Parallax effects shall be reduced to the practicable minimum.

S.2.   Design of Measuring Elements.

S.2.1.   Provision for Sealing. – Adequate provision shall be made for applying security seals in such a manner that no adjustment or interchange may be made of:

(a) any measurement elements; and

(b) any adjustable element for controlling delivery rate when such rate tends to affect the accuracy of deliveries.

The adjusting mechanism shall be readily accessible for purposes of affixing a security seal.

S.2.2.   Batching Meters Only.

S.2.2.1.   Air Elimination. – Batching meters shall be equipped with an effective air eliminator.

S.2.2.2.   Directional Flow Valves. – Valves intended to prevent reversal of flow shall be automatic in operation.

S.2.3.   Multi-Jet Meter Identification. – Multi-Jet water meters shall be clearly and permanently marked as such on the device or identified on the Certificate of Conformance.
(Added 2003)

S.3. Markings

S.3.1.   Location of Marking Information; Utility Type Meters. *– All required markings, including those required by G-S.1. Identification, shall be either on the meter body or primary indicator.*
[Nonretroactive as of January 1, 2013]
(Added 2012)

N. Notes

N.1.   Test Liquid. – A meter shall be tested with water.

N.2.   Evaporation and Volume Change. – Care shall be exercised to reduce to a minimum, evaporation losses and volume changes resulting from changes to temperature of the test liquid.

N.3.   Test Drafts. – Test drafts should be equal to at least the amount delivered by the device in two minutes and in no case less than the amount delivered by the device in one minute at the actual maximum flow rate developed by the installation. The test draft sizes shown in Table N.4.1. Flow Rate and Draft Size for Water Meters Normal Tests, shall be followed as closely as possible.
(Amended 2003)

N.4.   Testing Procedures.

N.4.1.   Normal Tests. – The normal test of a meter shall be made at the maximum discharge rate developed by the installation. Meters with maximum gallon per minute ratings higher than the values specified in Table N.4.1. Flow Rate and Draft Size for Water Meters Normal Tests may be tested up to the meter rating, with meter indications no less than those shown.
(Amended 1990, 2002, and 2003)

Table N.4.1. Flow Rate and Draft Size for Water Meters Normal Tests			
Meter Size (inches)	Rate of Flow (gal/min)	Maximum Rate	
		Meter Indication/Test Draft	
		gal	ft³
Less than ⅝	8	50	5
⅝	15	50	5
¾	25	50	5
1	40	100	10
1½	80	300	40
2	120	500	40
3	250	500	50
4	350	1000	100
6	700	1000	100

(Table Added 2003)

N.4.1.1.   Repeatability Tests. – Tests for repeatability should include a minimum of three consecutive test drafts of approximately the same size and be conducted under controlled conditions where variations in factors such as temperature, pressure, and flow rate are reduced to the extent that they will not affect the results obtained.

(Added 2002)

N.4.2.   Special Tests. – Special tests to develop the operating characteristics of meters may be made according to the rates and quantities shown in Table N.4.2.a. Flow Rate and Draft Size for Water Meters Special Tests and Table N.4.2.b. Flow Rate and Draft Size for Utility Type Water Meters Special Tests.

(Amended 2003 and 2010)

Table N.4.2.a. Flow Rate and Draft Size for Batching Water Meters Special Tests						
Meter Size (inches)	Intermediate Rate			Minimum Rate		
	Rate of Flow (gal/min)	Meter Indication/Test Draft		Rate of Flow (gal/min)	Meter Indication/Test Draft	
		gal	ft³		gal	ft³
Less than or equal to ⅝	2	10	1	¼	5	1
¾	3	10	1	½	5	1
1	4	10	1	¾	5	1
1½	8	50	5	1½	10	1
2	15	50	5	2	10	1
3	20	50	5	4	10	1
4	40	100	10	7	50	5
6	60	100	10	12	50	5

(Table Added 2003) (Table Amended 2010)

Table N.4.2.b. Flow Rate and Draft Size for Utility Type Water Meters Special Tests						
Meter Size (inches)	Intermediate Rate			Minimum Rate		
	Rate of Flow (gal/min)	Meter Indication/Test Draft		Rate of Flow (gal/min)	Meter Indication/Test Draft	
		gal	ft³		gal	ft³
Less than ⁵⁄₈	2	10	1	¼	5	1
⁵⁄₈	2	10	1	¼	5	1
⁵⁄₈ x ¾	2	10	1	¼	5	1
¾	3	10	1	½	5	1
1	4	10	1	¾	5	1
1½	8	100	10	1½	100	10
2	15	100	10	2	100	10

(Table Added 2010)

N.4.3. Batching Meter Tests. – Tests on batching meters should be conducted at the maximum and intermediate rates only.

T. Tolerances

T.1. Tolerance Values. – Maintenance and acceptance tolerances shall be as shown in Table T.1. Accuracy Classes and Tolerances for Water Meters.

(Amended 2003)

Table T.1. Accuracy Classes and Tolerances for Water Meters					
Accuracy Class	Application		Acceptance Tolerance	Maintenance Tolerance	Tolerance for Special Tests Conducted at the Minimum Flow Rate
1.5	Water, Other Than Multi-Jet Water Meters	Overregistration	1.5 %	1.5 %	1.5 %
		Underregistration	1.5 %	1.5 %	5.0 %
1.5	Water, Multi-Jet Water Meters	Overregistration	1.5 %	1.5 %	3.0 %
		Underregistration	1.5 %	1.5 %	3.0 %

(Table Added 2003)

T.1.1.   Repeatability. – When multiple tests are conducted at approximately the same flow rate, each test shall be within the applicable tolerances and the range of test results shall not exceed the values shown in Table T.1.1. Repeatability.

(Added 2002) (Amended 2010)

Table T.1.1. Repeatability		
	Batching Meters	**Utility-Type Meters**
Normal Flow Rates	0.6 %	0.6 %
Intermediate Flow Rates	0.6 %	2.0 %
Minimum Flow Rate	1.3 %	4.0 %

(Table Added 2010)

UR.   User Requirements

UR.1.   Batching Meters Only.

UR.1.1.   Strainer. – A filter or strainer shall be provided if it is determined that the water contains excessive amounts of foreign material.

UR.1.2.   Siphon Breaker. – An automatic siphon breaker or other effective means shall be installed in the discharge piping at the highest point of outlet, in no case below the top of the meter, to prevent siphoning of the meter and permit rapid drainage of the pipe or hose.

UR.1.3.   Provision for Testing. – Acceptable provisions for testing shall be incorporated into all meter systems. Such provisions shall include a two-way valve, or manifold valving, and a pipe or hose installed in the discharge line accessible to the proper positioning of the test measure.

UR.2.   Accessibility of Customer Indication. – An unobstructed standing space of at least 76 cm (30 in) wide, 91 cm (36 in) deep, and 198 cm (78 in) high shall be maintained in front of an indication intended for use by the customer to allow for reading the indicator. The customer indication shall be readily observable to a person located within the standing space without necessity of a separate tool or device.

(Added 2008)

Table of Contents

Section 3.37.   Mass Flow Meters

A. Application

A.1.   Liquids. – This code applies to devices that are designed to dynamically measure the mass, or the mass and density of liquids. It also specifies the relevant examinations and tests that are to be conducted.
(Amended 1997)

A.2.   Vapor (Gases). – This code applies to devices that are designed to dynamically measure the mass of hydrocarbon gas in the vapor state. Examples of these products are propane, propylene, butanes, butylenes, ethane, methane, natural gas and any other hydrocarbon gas/air mix.

A.3.   Additional Code Requirements. – In addition to the requirements of this code, Mass Flow Meters shall meet the requirements of Section 1.10. General Code.

S. Specifications

S.1.   Indicating and Recording Elements.

S.1.1.   Indicating Elements. – A measuring assembly shall include an indicating element. Indications shall be clear, definite, accurate, and easily read under normal conditions of operation of the instrument.

S.1.2.   Compressed Natural Gas and Liquefied Natural Gas Dispensers. – Except for fleet sales and other price contract sales, a compressed or liquefied natural gas dispenser used to refuel vehicles shall be of the computing type and shall indicate the quantity, the unit price, and the total price of each delivery. The dispenser shall display the mass measured for each transaction either continuously on an external or internal display accessible during the inspection and test of the dispenser, or display the quantity in mass units by using controls on the device.
(Added 1994) (Amended 2016)

S.1.3.   Units.

S.1.3.1.   Units of Measurement. – Deliveries shall be indicated and recorded in grams, kilograms, metric tons, pounds, tons, and/or liters, gallons, quarts, pints and decimal subdivisions thereof. The indication of a delivery shall be on the basis of apparent mass versus a density of 8.0 g/cm^3. The volume indication shall be based on the mass measurement and an automatic means to determine and correct for changes in product density.
(Amended 1993 and 1997)

S.1.3.1.1.   Compressed Natural Gas Used as an Engine Fuel. – When compressed natural gas is dispensed as an engine fuel, the delivered quantity shall be indicated in "gasoline gallon equivalent units (GGE)" or "diesel gallon equivalent units (DGE)," or in mass. (Also see Appendix D. Definitions.)
(Added 1994) (Amended 2016)

S.1.3.1.2.   Liquefied Natural Gas Used as an Engine Fuel. – When liquefied natural gas is dispensed as an engine fuel, the delivered quantity shall be indicated in diesel gallon equivalent units (DGE) or in mass. (Also see Appendix D. Definitions.)
(Added 2016)

S.1.3.2.   Numerical Value of Quantity-Value Divisions. – The value of a scale interval shall be equal to:

(a) 1, 2, or 5; or

(b) a decimal multiple or submultiple of 1, 2, or 5.

S.1.3.3.   Maximum Value of Quantity-Value Divisions.

(a) The maximum value of the quantity-value division for liquids shall not be greater than 0.2 % of the minimum measured quantity.

(b) For dispensers of compressed natural gas used to refuel vehicles, the value of the division for the gasoline liter equivalent shall not exceed 0.01 GLE; the division for gasoline gallon equivalent (GGE) shall not exceed 0.001 GGE. The maximum value of the mass division shall not exceed 0.001 kg or 0.001 lb.

(Amended 1994)

S.1.3.4.   Values Defined. – Indicated values shall be adequately defined by a sufficient number of figures, words, symbols, or combinations thereof. A display of "zero" shall be a zero digit for all displayed digits to the right of the decimal mark and at least one to the left.

S.2.   Operating Requirements.

S.2.1.   Return to Zero. – Except for measuring assemblies in a pipeline:

(a) One indicator and the primary recording elements, if the device is equipped to record, shall be provided with a means for readily returning the indication to zero either automatically or manually.

(b) It shall not be possible to return primary indicating elements, or primary recording elements, beyond the correct zero position.

(Amended 1993)

S.2.2.   Indicator Reset Mechanism. – The reset mechanism for the indicating element shall not be operable during a delivery. Once the zeroing operation has begun, it shall not be possible to indicate a value other than the latest measurement, or "zeros" when the zeroing operation has been completed.

S.2.3.   Non-resettable Indicator. – An instrument may also be equipped with a non-resettable indicator if the indicated values cannot be construed to be the indicated values of the resettable indicator for a delivered quantity.

S.2.4.   Provisions for Power Loss.

S.2.4.1.   Transaction Information. – In the event of a power loss, the information needed to complete any transaction in progress at the time of the power loss (such as the quantity and unit price, or sales price) shall be determinable for at least 15 minutes at the dispenser or at the console if the console is accessible to the customer.

(Added 1993)

S.2.4.2.   User Information. – The device memory shall retain information on the quantity of fuel dispensed and the sales price totals during power loss.

(Added 1993)

S.2.5.   Display of Unit Price and Product Identity.

S.2.5.1.   Unit Price. – A computing or money-operated device shall be able to display on each face the unit price at which the device is set to compute or to dispense.
(Added 1993)

S.2.5.2.   Product Identity. – A device shall be able to conspicuously display on each side the identity of the product being dispensed.
(Added 1993)

S.2.5.3.   Selection of Unit Price. – Except for dispensers used exclusively for fleet sales, other price contract sales, and truck refueling (e.g., truck stop dispensers used only to refuel trucks), when a product or grade is offered for sale at more than one unit price through a computing device, the selection of the unit price shall be made prior to delivery using controls on the device or other customer-activated controls. A system shall not permit a change to the unit price during delivery of a product.
[Nonretroactive as of January 1, 1998]
(Added 1997)

S.2.5.4.   Agreement Between Indications. – When a quantity value indicated or recorded by an auxiliary element is a derived or computed value based on data received from a retail motor-fuel dispenser, the value may differ from the quantity value displayed on the dispenser, provided the following conditions are met:

    (a) all total money-values for an individual sale that are indicated or recorded by the system agree; and

    (b) within each element the values indicated or recorded meet the formula (quantity x unit price = total sales price) to the closest cent.
[Nonretroactive as of January 1, 1998]
(Added 1997)

S.2.6.   Money-Value Computations. – A computing device shall compute the total sales price at any single-purchase unit price (i.e., excluding fleet sales, other price contract sales, and truck stop dispensers used only to refuel trucks) for which the product being measured is offered for sale at any delivery possible within either the measurement range of the device or the range of the computing elements, whichever is less.
(Added 1993)

S.2.6.1.   Auxiliary Elements. – If a system is equipped with auxiliary indications, all indicated money-value and quantity divisions of the auxiliary element shall be identical with those of the primary element.
(Added 1993)

S.2.6.2.   Display of Quantity and Total Price. – When a delivery is completed, the total price and quantity for that transaction shall be displayed on the face of the dispenser for at least 5 minutes or until the next transaction is initiated by using controls on the device or other user-activated controls.
(Added 1993)

S.2.7.   Recorded Representations, Point-of-Sale Systems. – The sales information recorded by cash registers when interfaced with a retail motor-fuel dispenser shall contain the following information for products delivered by the dispenser:

    (a) the total volume of the delivery;

    (b) the unit price;

(c) the total computed price; and

(d) the product identity by name, symbol, abbreviation, or code number.
[Nonretroactive as of January 1, 1986]

(Added 1993)

S.2.8.   Indication of Delivery. – *The device shall automatically show on its face the initial zero condition and the quantity delivered (up to the nominal capacity). However, the first 0.03 L (0.009 gal) of a delivery and its associated total sales price need not be indicated.*
[Nonretroactive as of January 1, 1998]

(Added 1997)

S.3.   Measuring Elements and Measuring Systems.

S.3.1.   Maximum and Minimum Flow-Rates.

(a) The ratio of the maximum to minimum flow-rates specified by the manufacturer for devices measuring liquefied gases shall be 5:1 or greater.

(b) The ratio of the maximum to minimum flow-rates specified by the manufacturer for devices measuring other than liquefied gases shall be 10:1 or greater.

S.3.2.   Adjustment Means. – An assembly shall be provided with the means to change the ratio between the indicated quantity and the quantity of liquid measured by the assembly. A bypass on the measuring assembly shall not be used for these means.

S.3.2.1.   Discontinuous Adjusting Means. – When the adjusting means changes the ratio between the indicated quantity and the quantity of measured liquid in a discontinuous manner, the consecutive values of the ratio shall not differ by more than 0.1 %.

S.3.3.   Vapor Elimination. – A liquid-measuring instrument or measuring system shall be equipped with an effective vapor or air eliminator or other effective means, automatic in operation, to prevent the measurement of vapor and air. Vent lines from the air or vapor eliminator shall be made of metal tubing or some other suitable rigid material.

(Amended 1999)

S.3.3.1.   Vapor Elimination on Loading Rack Liquid Metering Systems.

(a) A loading rack liquid metering system shall be equipped with a vapor or air eliminator or other automatic means to prevent the passage of vapor and air through the meter unless the system is designed or operationally controlled by a method, approved by the weights and measures jurisdiction having statutory authority over the device, such that neither air nor vapor can enter the system.

(b) Vent lines from the air or vapor eliminator (if present) shall be made of metal tubing or other rigid material.

(Added 1995)

S.3.4.   Maintenance of Liquid State. – A liquid-measuring device shall be installed so that the measured product remains in a liquid state during passage through the instrument.

S.3.5.   Provision for Sealing. – Adequate provision shall be made for an approved means of security (e.g., data change audit trail) or physically applying security seals in such a manner that no adjustment or interchange may be made of:

(a) any measuring or indicating element;

(b) any adjustable element for controlling delivery rate when such rate tends to affect the accuracy of deliveries;

(c) the zero adjustment mechanism; and

(d) any metrological parameter that will affect the metrological integrity of the device or system.

When applicable, the adjusting mechanism shall be readily accessible for purposes of affixing a security seal.

[Audit trails shall use the format set forth in Table S.3.5. Categories of Device and Methods of Sealing] *
[*Nonretroactive as of January 1, 1995]

(Amended 1992, 1995, and 2006)

Table S.3.5. Categories of Device and Methods of Sealing	
Categories of Device	**Methods of Sealing**
Category 1: *No remote configuration capability.*	*Seal by physical seal or two event counters: one for calibration parameters and one for configuration parameters.*
Category 2: *Remote configuration capability, but access is controlled by physical hardware.* *The device shall clearly indicate that it is in the remote configuration mode and record such message if capable of printing in this mode or shall not operate while in this mode.*	*[The hardware enabling access for remote communication must be on-site. The hardware must be sealed using a physical seal or an event counter for calibration parameters and an event counter for configuration parameters. The event counters may be located either at the individual measuring device or at the system controller; however, an adequate number of counters must be provided to monitor the calibration and configuration parameters of the individual devices at a location. If the counters are located in the system controller rather than at the individual device, means must be provided to generate a hard copy of the information through an on-site device.] ** *[*Nonretroactive as of January 1, 1996]*
Category 3: *Remote configuration capability access may be unlimited or controlled through a software switch (e.g., password).* *[Nonretroactive as of January 1, 1995]* *The device shall clearly indicate that it is in the remote configuration mode and record such message if capable of printing in this mode or shall not operate while in this mode.* *[Nonretroactive as of January 1, 2001]*	*An event logger is required in the device; it must include an event counter (000 to 999), the parameter ID, the date and time of the change, and the new value of the parameter. A printed copy of the information must be available on demand through the device or through another on-site device. The information may also be available electronically. The event logger shall have a capacity to retain records equal to 10 times the number of sealable parameters in the device, but not more than 1000 records are required. (**Note:** Does not require 1000 changes to be stored for each parameter.)*

[Nonretroactive as of January 1, 1995]

(Table Added 1995) (Amended 1995, 1998, 1999, 2006, and 2016)

S.3.6.   Automatic Density Correction.

(a)   An automatic means to determine and correct for changes in product density shall be incorporated in any mass flow metering system that is affected by changes in the density of the product being measured.

(b)   Volume-measuring devices with automatic temperature compensation used to measure natural gas as a motor vehicle engine fuel shall be equipped with an automatic means to determine and correct for changes in product density due to changes in the temperature, pressure, and composition of the product.

(Amended 1994 and 1997)

S.3.7.   Pressurizing the Discharge Hose. – The discharge hose for compressed natural gas shall automatically pressurize prior to the device beginning to register the delivery.

(Added 1993)

S.3.8.   Zero-Set-Back Interlock, Retail Motor-Fuel Devices. – A device shall be constructed so that:

(a) after a delivery cycle has been completed by moving the starting lever to any position that shuts off the device, an automatic interlock prevents a subsequent delivery until the indicating elements, and recording elements if the device is equipped and activated to record, have been returned to their zero positions;

(b) the discharge nozzle cannot be returned to its designed hanging position (that is, any position where the tip of the nozzle is placed in its designed receptacle and the lock can be inserted) until the starting lever is in its designed shut-off position and the zero-set-back interlock has been engaged; and

(c) in a system with more than one dispenser supplied by a single pump, an effective automatic control valve in each dispenser prevents product from being delivered until the indicating elements on that dispenser are in a correct zero position.

(Added 1993)

S.4.   Discharge Lines and Valves.

S.4.1.   Diversion of Measured Product. – No means shall be provided by which any measured product can be diverted from the measuring instrument. However, two or more delivery outlets may be permanently installed and operated simultaneously, provided that any diversion of flow to other than the intended receiving receptacle cannot be readily accomplished or is readily apparent. Such means include physical barriers, visible valves, or indications that make it clear which outlets are in operation, and explanatory signs if deemed necessary.

An outlet that may be opened for purging or draining the measuring system, or for recirculating product if recirculation is required in order to maintain the product in a deliverable state shall be permitted. Effective automatic means shall be provided to prevent the passage of liquid through any such outlet during normal operation of the measuring system and to inhibit meter indications (or advancement of indications) and recorded representations while the outlet is in operation.

(Amended 2002 and 2006)

S.4.2.   Pump-Discharge Unit. – A pump-discharge unit for liquids equipped with a flexible discharge hose shall be of the wet-hose type.

(Added 1993)

S.4.3.   Directional Flow Valves. – If a reversal of flow could result in errors that exceed the tolerance for the minimum measured quantity, a valve or valves or other effective means, automatic in operation (and equipped with a pressure limiting device, if necessary) to prevent the reversal of flow shall be properly installed in the system. (Also see N.1. Minimum Measured Quantity.)

S.4.4.   Discharge Valves. – A discharge valve may be installed on a discharge line only if the system is a wet-hose type. Any other shutoff valve on the discharge side of the instrument shall be of the automatic or semiautomatic predetermined-stop type or shall be operable only:

(a) by means of a tool (but not a pin) entirely separate from the device; or

(b) by means of a security seal with which the valve is sealed open.

S.4.5.   Antidrain Means. – In a wet-hose type device, effective means shall be provided to prevent the drainage of the hose between transactions.

S.4.6.   Other Valves. – Check valves and closing mechanisms that are not used to define the measured quantity shall have relief valves (if necessary) to dissipate any abnormally high pressure that may arise in the measuring assembly.

S.5.   Markings. – A measuring system shall be legibly and indelibly marked with the following information:

(a)   pattern approval mark (i.e., type approval number);

(b)   name and address of the manufacturer or his trademark and, if required by the weights and measures authority, the manufacturer's identification mark in addition to the trademark;

(c)   model identifier or product name selected by the manufacturer;

(d)   nonrepetitive serial number;

(e)   *the accuracy class of the meter as specified by the manufacturer consistent with Table T.2. Accuracy Classes for Mass Flow Meter Applications Covered in NIST Handbook 44, Section 3.37 Mass Flow Meters;* [*Nonretroactive as of January 1, 1995]*
(Added 1994)

(f)   maximum and minimum flow rates in pounds per unit of time;

(g)   maximum working pressure;

(h)   applicable range of temperature if other than − 10 °C to + 50 °C;

(i)   minimum measured quantity; and

(j)   product limitations, if applicable.

S.5.1.   *Location of Marking Information; Retail Motor-Fuel Dispensers.* – *The marking information required in General Code, paragraph G-S.1. Identification shall appear as follows:*

(a)   *within 60 cm (24 in) to 150 cm (60 in) from the base of the dispenser;*

(b)   *either internally and/or externally provided the information is permanent and easily read; and*

(c)   *on a portion of the device that cannot be readily removed or interchanged (i.e., not on a service access panel).*

Note: *The use of a dispenser key or tool to access internal marking information is permitted for retail liquid-measuring devices. [Nonretroactive as of January 1, 2003]*
(Added 2006)

S.5.2.   Marking of Equivalent Conversion Factors for Compressed Natural Gas. – A device dispensing compressed natural gas shall have either the statement "1 Gasoline Gallon Equivalent (GGE) means 5.660 lb of Compressed Natural Gas" or "1 Diesel Gallon Equivalent (DGE) means 6.384 lb of Compressed Natural Gas" permanently and conspicuously marked on the face of the dispenser according to the method of sale used.
(Added 1994) (Amended 2016)

S.5.3.   Marking of Equivalent Conversion Factor for Liquefield Natural Gas. – A device dispensing liquefied natural gas shall have the statement "1 Diesel Gallon Equivalent (DGE) means 6.059 lb of Liquefied Natural Gas" permanently and conspicuously marked on the face of the dispenser according to the method of sale used.
(Added 2016)

S.6.   Printer. – When an assembly is equipped with means for printing the measured quantity, the following conditions apply:

(a) the scale interval shall be the same as that of the indicator;

(b) the value of the printed quantity shall be the same value as the indicated quantity;

(c) *the printed quantity shall also include the mass value if the mass is not the indicated quantity;*
[Nonretroactive as of January 1, 2021]

(d) a quantity for a delivery (other than an initial reference value) cannot be recorded until the measurement and delivery has been completed;

(e) the printer is returned to zero when the resettable indicator is returned to zero; and

(f) the printed values shall meet the requirements applicable to the indicated values.

(Amended 2016)

    S.6.1.   Printed Receipt. – Any delivered, printed quantity shall include an identification number, the time and date, and the name of the seller. This information may be printed by the device or pre-printed on the ticket.

S.7.   *Totalizers for Retail Motor-Fuel Devices.* – *Retail motor-fuel dispensers shall be equipped with a nonresettable totalizer for the quantity delivered through the metering device.*
[Nonretroactive as of January 1, 1998]
(Added 1997)

N. Notes

N.1.   Minimum Measured Quantity. – The minimum measured quantity shall be specified by the manufacturer.

N.2.   Test Medium.

    N.2.1.   Liquid-Measuring Devices. – The device shall be tested with the liquid that the device is intended to measure or another liquid with the same general physical characteristics.

    N.2.2.   Vapor-Measuring Devices. – The device shall be tested with air or the product to be measured.

N.3.   Test Drafts. – The minimum test shall be one test draft at the maximum flow rate of the installation and one test draft at the minimum flow rate. More tests may be performed at these or other flow rates. (Also see T.3. Repeatability.)

N.4.   Minimum Measured Quantity. – The device shall be tested for a delivery equal to the declared minimum measured quantity when the device is likely to be used to make deliveries on the order of the minimum measured quantity.

N.5.   Motor-Fuel Dispenser. – When a device is intended for use as a liquid motor-fuel dispenser, the type evaluation test shall include a test for accuracy using five starts and stops during a delivery to simulate the operation of the automatic shut-off nozzle. This test may be conducted as part of the normal inspection and test of the meter.

N.6.   Testing Procedures.

    N.6.1.   Normal Tests. – The normal test of a meter shall be made at the maximum discharge rate developed by the installation. Any additional tests conducted at flow rates down to and including the rated minimum discharge flow rate shall be considered normal tests.

    (Added 1999)

N.6.1.1.   Repeatability Tests. – Tests for repeatability should include a minimum of three consecutive test drafts of approximately the same size and be conducted under controlled conditions where variations in factors such as temperature, pressure, and flow rate are reduced to the extent that they will not affect the results obtained.

(Added 2001)

N.6.2.   Special Tests. – "Special" tests shall be made to develop the operating characteristics of a device and any special elements and accessories attached to or associated with the device. Any test except as set forth in N.6.1. Normal Tests shall be considered a special test. Special tests of a measuring system shall be made to develop operating characteristics of the measuring systems during a split compartment delivery. (Also see Table T.2. Accuracy Classes and Tolerances for Mass Flow Meters.)

(Added 1999)

T. Tolerances

T.1.   Tolerances, General.

(a) The tolerances apply equally to errors of underregistration and errors of overregistration.

(b) The tolerances apply to all products at all temperatures measured at any flow rate within the rated measuring range of the meter.

(Amended 1999)

T.2.   Tolerances. – The tolerances for mass flow meters for specific liquids, gases, and applications are listed in Table T.2. Accuracy Classes and Tolerances for Mass Flow Meters.

(Amended 1994 and 1999)

Table T.2. Accuracy Classes and Tolerances for Mass Flow Meters				
Accuracy Class	Application or Commodity Being Measured	Acceptance Tolerance	Maintenance Tolerance	Special Tolerance
0.3	- Large capacity motor-fuel dispensers (maximum discharge flow rates greater than 100 L/min or 25 gal/min) - Heated products (other than asphalt) at temperatures greater than 50 °C (122 °F) - Asphalt at temperatures equal to or below 50 °C (122 °F) - Loading rack meters - Vehicle-tank meters - Home heating oil - Milk and other food products - All other liquid applications not shown in the table where the minimum delivery is at least 700 kg (1500 lb)	0.2 %	0.3 %	0.5 %
0.3A	- Asphalt at temperatures greater than 50 °C (122 °F)	0.3 %	0.3 %	0.5 %
0.5	- Small capacity (retail) motor-fuel dispensers - Agri-chemical liquids - All other liquid applications not shown in the table where the minimum delivery is less than 700 kg or 1500 lb	0.3 %	0.5 %	0.5 %
1.0	- Anhydrous ammonia - LP Gas (including vehicle-tank meters)	0.6 %	1.0 %	1.0 %
2.0	- Compressed natural gas as a motor-fuel	1.5 %	2.0 %	2.0 %
2.5	- Cryogenic liquid meters - Liquefied compressed gases other than LP Gas	1.5 %	2.5 %	2.5 %

(Added 1994) (Amended 1999, 2001, and 2013)

T.3. Repeatability. – When multiple tests are conducted at approximately the same flow rate and draft size, the range of the test results for the flow rate shall not exceed 40 % of the absolute value of the maintenance tolerance and the results of each test shall be within the applicable tolerance. (Also see N.6.1.1. Repeatability Tests.)
(Amended 1992, 1994, and 2001)

T.4. Type Evaluation Examinations for Liquid-Measuring Devices. – For type evaluation examinations, the tolerance values shall apply under the following conditions:

(a) with any one liquid within the range of liquids;

(b) at any one liquid temperature and pressure within the operating range of the meter; and

(c) at all flow rates within the range of flow rates.

(Added 1993) (Amended 1994)

UR.   User Requirements

UR.1.   Selection Requirements.

UR.1.1.   Discharge Hose-Length. – *The length of the discharge hose on a retail motor-fuel device shall not exceed 4.6 m (15 ft) unless it can be demonstrated that a longer hose is essential to permit deliveries to be made to receiving vehicles or vessels.*
[Nonretroactive as of January 1, 1998]
(Added 1997)

UR.1.2.   Minimum Measured Quantity.

(a)   The minimum measured quantity shall be specified by the manufacturer.

(b)   The minimum measured quantity appropriate for a transaction may be specified by the weights and measures authority. A device may have a minimum measured quantity smaller than that specified by the weights and measures authority; however, the device must perform within the performance requirements for the declared minimum measured quantity.

UR.2.   Installation Requirements.

UR.2.1.   Manufacturer's Instructions. – A device shall be installed in accordance with the manufacturer's instructions, and the installation shall be sufficiently secure and rigid to maintain this condition.
(Added 1997)

UR.2.2.   Discharge Rate. – A device shall be installed so that the actual maximum discharge rate will not exceed the rated maximum discharge rate. Automatic means of flow regulation shall be incorporated in the installation if necessary.
(Added 1997)

UR.2.3.   Low-Flow Cut-Off Valve. – If a metering system is equipped with a programmable or adjustable "low-flow cut-off" feature:

(a)   the low-flow cut-off value shall not be set at flow rates lower than the minimum operating flow rate specified by the manufacturer on the meter; and

(b)   the system shall be equipped with flow control valves which prevent the flow of product and stop the indicator from registering product flow whenever the product flow rate is less than the low-flow cut-off value.

(Added 1992)

UR.3.   Use of Device.

UR.3.1.   Unit Price and Product Identity for Retail Dispensers. – The following information shall be conspicuously displayed or posted on the face of a retail dispenser used in direct sale:

(a)   except for dispensers used exclusively for fleet sales, other price contract sales, and truck refueling (e.g., truck stop dispensers used only to refuel trucks), all of the unit prices at which the product is offered for sale; and

(b)   in the case of a computing type or money-operated type, the unit price at which the dispenser is set to compute.

(Added 1993)

UR.3.1.1.   Marking of Equivalent Conversion Factors for Compressed Natural Gas. – A device dispensing compressed natural gas shall have either the statement "1 Gasoline Gallon Equivalent (GGE) means 5.660 lb of Compressed Natural Gas" or "1 Diesel Gallon Equvalient (DGE) means 6.384 lb of Compressed Natural Gas" permanently and conspicuously marked on the face of the dispenser according to the method of sale used.

(Added 2016)

U.R.3.1.2.   Marking of Equivalent Conversion Factor for Liquefied Natural Gas. – A device dispensing liquefied natural gas shall have the statement "1 Diesel Gallon Equivalent (DGE) means 6.059 lb of Liquefied Natural Gas" permanently and conspicuously marked on the face of the dispenser according to the method of sale used.

(Added 2016)

UR.3.2.   Vapor-Return Line. – During any metered delivery of liquefied petroleum gas and other liquids from a supplier's tank to a receiving container, there shall be no vapor-return line from the receiving container to the supplier's tank:

(a) in the case of any receiving container to which normal deliveries can be made without the use of such vapor-return line; or

(b) in the case of any new receiving container when the ambient temperature is below 90 °F.

(Added 1993)

UR.3.3.   Ticket Printer; Customer Ticket. – Vehicle-mounted metering systems shall be equipped with a ticket printer which shall be used for all sales where product is delivered through the meter. A copy of the ticket issued by the device shall be left with the customer at the time of delivery or as otherwise specified by the customer.

(Added 1994)

UR.3.4.   Printed Ticket. – The total price, the total quantity of the delivery, and the price per unit shall be printed on any ticket issued by a device of the computing type and containing any one of these values.

(Added 1993)

UR.3.5.   Ticket in Printing Device. – A ticket shall not be inserted into a device equipped with a ticket printer until immediately before a delivery is begun, and in no case shall a ticket be in the device when the vehicle is in motion while on a public street, highway, or thoroughfare.

(Added 1993)

UR.3.6.   Steps After Dispensing. – After delivery to a customer from a retail motor-fuel device:

(a) the starting lever shall be returned to its shutoff position and the zero-set-back interlock engaged; and

(b) the discharge nozzle shall be returned to its designed hanging position unless the primary indicating elements, and recording elements, if the device is equipped and activated to record, have been returned to a definite zero indication.

(Added 1993)

UR.3.7.   Return of Indicating and Recording Elements to Zero. – The primary indicating elements (visual), and the primary recording elements when these are returnable to zero, shall be returned to zero immediately before each delivery. Exceptions to this requirement are totalizers on key-lock-operated or other self-operated dispensers and the primary recording element if the device is equipped to record.

(Added 1995) (Amended 1997)

UR.3.8.   Return of Product to Storage, Retail Compressed and Liquefied Natural Gas Dispensers. – Provisions at the site shall be made for returning product to storage or disposing of the product in a safe and timely manner during or following testing operations. Such provisions may include return lines, or cylinders adequate in size and number to permit this procedure.

(Added 1998) (Amended 2016)

Table of Contents

Section 3.38.   Carbon Dioxide Liquid-Measuring Devices

A. Application

A.1.   General. – This code applies to liquid-measuring devices used for the measurement of liquid carbon dioxide.

A.2.   Exceptions. – This code does not apply to devices used solely for dispensing a product in connection with operations in which the amount dispensed does not affect customer charges.

A.3.   Additional Code Requirements. – In addition to the requirements of this code, Carbon Dioxide Liquid-Measuring Devices shall meet the requirements of Section 1.10. General Code.

A.4.   Type Evaluation. – The National Type Evaluation Program will accept for type evaluation only those devices that comply with all requirements of this code.
(Added 1998)

S. Specifications

S.1.   Design of Indicating and Recording Elements and of Recorded Representations.

S.1.1.   Primary Elements.

S.1.1.1.   General. – A device shall be equipped with a primary indicating element and may also be equipped with a primary recording element.

S.1.1.2.   Units. – A device shall indicate and record, if equipped to record, its deliveries in terms of pounds or kilograms or decimal subdivisions or multiples thereof.

S.1.1.3.   Value of Smallest Unit. – The value of the smallest unit of indicated delivery, and recorded delivery, if the device is equipped to record, shall not exceed the equivalent of:

(a) for small delivery devices:

(1) 1 kilogram; or

(2) 1 pound

(b) for large delivery devices:

(1) 10 kilograms; or

(2) 10 pounds

S.1.1.4.   Advancement of Indicating and Recording Elements. – Primary indicating and recording elements shall be susceptible to advancement only by the normal operation of the device. However, a device may be cleared by advancing its elements to zero, but only if:

(a) the advancing movement, once started, cannot be stopped until zero is reached; or

(b) in the case of indicating elements only, such elements are automatically obscured until the elements reach the correct zero position.

S.1.1.5.   Return to Zero. – Primary indicating and recording elements shall be readily returnable to a definite zero indication. Means shall be provided to prevent the return of primary indicating elements and of primary recording elements beyond their correct zero position.

S.1.2.   Graduations.

S.1.2.1.   Length. – Graduations shall be so varied in length that they may be conveniently read.

S.1.2.2.   Width. – In any series of graduations, the width of a graduation shall in no case be greater than the width of the minimum clear interval between graduations. The width of main graduations shall be not more than 50 % greater than the width of subordinate graduations. Graduations shall in no case be less than 0.2 mm (0.008 in) in width.

S.1.2.3.   Clear Interval Between Graduations. – The clear interval shall be not less than 1.0 mm (0.04 in). If the graduations are not parallel, the measurement shall be made:

(a) along the line of relative movement between the graduations at the end of the indicator; or

(b) if the indicator is continuous, at the point of widest separation of the graduations.
(Also see S.1.3.6. Travel of Indicator.)

S.1.3.   Indicators.

S.1.3.1.   Symmetry. – The index of an indicator shall be of the same shape as the graduations at least throughout that portion of its length associated with the graduations.

S.1.3.2.   Length. – The index of an indicator shall reach to the finest graduations with which it is used, unless the indicator and the graduations are in the same plane, in which case the distance between the end of the indicator and the ends of the graduations, measured along the line of the graduations, shall be not more than 1.0 mm (0.04 in).

S.1.3.3.   Width. – The width of the index of the indicator in relation to the series of graduations with which it is used shall be not greater than:

(a) *the width of the narrowest graduation;* * and
      [*Nonretroactive as of January 1, 2002]
      (Amended 2001)

(b) the width of the minimum clear interval between graduations.

When the index of an indicator extends along the entire length of a graduation, that portion of the index of the indicator that may be brought into coincidence with the graduation shall be of the same width throughout the length of the index that coincides with the graduation.

S.1.3.4.   Clearance. – The clearance between the index of an indicator and the graduations shall in no case be more than 1.5 mm (0.06 in).

S.1.3.5.   Parallax. – Parallax effects shall be reduced to the practicable minimum.

S.1.3.6.   Travel of Indicator. – If the most sensitive element of the primary indicating element utilizes an indicator and graduations, the relative movement of these parts corresponding to the smallest indicated value shall be no less than 5 mm (0.20 in).

S.1.4.   Computing-Type Devices.

S.1.4.1.   Printed Ticket. – Any printed ticket issued by a device of the computing type on which there is printed the total computed price shall have printed clearly thereon also the total quantity of the delivery and the price per unit.

S.1.4.2.   Money-Value Computations. – Money-value computations shall be of the full-computing type in which the money-value at a single unit price, or at each of a series of unit prices, shall be computed for every delivery within either the range of measurement of the device or the range of the computing elements, whichever is less.

The total price shall be computed on the basis of the quantity indicated when the value of the smallest division indicated is equal to or less than the value specified in S.1.1.3. Value of Smallest Unit.

S.1.4.3.   Money-Values, Mathematical Agreement. – Any digital money-value indication and any recorded money-value on a computing-type device shall be in mathematical agreement with its associated quantity indication or representation to within 1 cent of money-value.

S.2.   Design of Measuring Elements.

S.2.1.   Vapor Elimination.

(a)   A device shall be equipped with an effective automatic means to prevent the passage of vapor through the meter.

(b)   Vent lines from the vapor eliminator shall be made of appropriate non-collapsible material.
(Amended 2016)

S.2.2.   Reverse Flow Measurement. – Effective means, automatic in operation, shall be installed to prevent reverse flow measurement.

S.2.3.   Maintenance of Liquid State. – A device shall be so designed that the product being measured will remain in a liquid state during passage through the device.

S.2.4.   Automatic Temperature or Density Compensation. – A volumetric device shall be equipped with automatic means for adjusting the indication and recorded representation of the measured quantity of the product to indicate or record the quantity of the product measured in terms of pounds.

S.2.5.   Provision for Sealing. – Adequate provision shall be made for an approved means of security (e.g., data change audit trail) or for physically applying a security seal in such a manner that requires the security seal to be broken before an adjustment or interchange may be made of:

(a)   any measuring or indicating element;

(b)   any adjustable element for controlling delivery rate when such rate tends to affect the accuracy of deliveries;

(c)   any automatic temperature or density compensating system; and

(d)   any metrological parameter that will affect the metrological integrity of the device or system.

When applicable any adjusting mechanism shall be readily accessible for purposes of affixing a security seal.
[Audit trails shall use the format set forth in Table S.2.5. Provision for Sealing] *
*[*Nonretroactive as of January 1, 1995]*
(Amended 2006)

Table S.2.5. Categories of Device and Methods of Sealing	
Categories of Device	**Methods of Sealing**
Category 1: *No remote configuration capability.*	*Seal by physical seal or two event counters: one for calibration parameters and one for configuration parameters.*
Category 2: *Remote configuration capability, but access is controlled by physical hardware.* *The device shall clearly indicate that it is in the remote configuration mode and record such message if capable of printing in this mode or shall not operate while in this mode.*	*The hardware enabling access for remote communication must be on-site. The hardware must be sealed using a physical seal or an event counter for calibration parameters and an event counter for configuration parameters. The event counters may be located either at the individual measuring device or at the system controller; however, an adequate number of counters must be provided to monitor the calibration and configuration parameters of the individual devices at a location. If the counters are located in the system controller rather than at the individual device, means must be provided to generate a hard copy of the information through an on-site device.*
Category 3: *Remote configuration capability access may be unlimited or controlled through a software switch (e.g., password).* *The device shall clearly indicate that it is in the remote configuration mode and record such message if capable of printing in this mode or shall not operate while in this mode.*	*An event logger is required in the device; it must include an event counter (000 to 999), the parameter ID, the date and time of the change, and the new value of the parameter. A printed copy of the information must be available on demand through the device or through another on-site device. The information may also be available electronically. The event logger shall have a capacity to retain records equal to 10 times the number of sealable parameters in the device, but not more than 1000 records are required. (**Note:** Does not require 1000 changes to be stored for each parameter.)*

[*Nonretroactive as of January 1, 1995*]

(Table Added 2006) (Amended 2016)

S.3.   Design of Discharge Lines and Discharge Line Valves.

S.3.1.   Diversion of Measured Liquid. – No means shall be provided by which any measured liquid can be diverted from the measuring chamber of the device or the discharge line therefrom, except that a manually controlled outlet that may be opened for purging or draining the measuring system shall be permitted. Effective means shall be provided to prevent the passage of liquid through any such outlet during normal operation of the device and to indicate clearly and unmistakably when the valve controls are so set as to permit passage of liquid through such outlet.

S.3.2.   Discharge Hose. – The discharge hose of a measuring system shall be of a wet hose type with a shutoff valve at its outlet end.

S.4.   Marking Requirements.

S.4.1.   Limitation of Use. – If a measuring system is intended to measure accurately only liquids having particular properties, or to measure accurately only under specific installation or operating conditions, or to measure accurately only when used in conjunction with specific accessory equipment, these limitations shall be clearly and permanently marked on the device.

S.4.2.   Discharge Rates. – A meter shall be marked to show its designed maximum and minimum discharge rates. The marked minimum discharge rate shall not exceed 20 % of the marked maximum discharge rate.

Note: Also see example in Section 3.30. Liquid-Measuring Devices Code, paragraph S.4.4.1. Discharge Rates.
(Note Added 2003)

N. Notes

N.1.   Test Liquid. – The test liquid shall be carbon dioxide in a compressed liquid state.

N.2.   Vaporization and Volume Change. – Care shall be exercised to reduce vaporization and volume changes to a minimum. When testing by weight, the weigh tank and transfer systems shall be pre-cooled to liquid temperature prior to the start of the test to avoid the venting of vapor from the vessel being weighed.

N.3.   Test Drafts.

N.3.1.   Gravimetric Test. – Weight test drafts shall be equal to at least the amount delivered by the device in 2 minutes at its maximum discharge rate.

N.3.2.   Transfer Standard Test. – When comparing a meter with a calibrated transfer standard, the test draft shall be equal to at least the amount delivered by the device in two minutes at its maximum discharge rate.

N.3.3.   Volumetric Prover Test Drafts. – Test drafts shall be equal to at least the amount delivered in one minute at its normal discharge rate.

N.4.   Testing Procedures.

N.4.1.   Normal Tests. – The "normal" test of a device shall be made at the maximum discharge flow rate developed under the conditions of installation. Any additional tests conducted at flow rates down to and including one-half of the sum of the maximum discharge flow rate and the rated minimum discharge flow rate shall be considered normal tests.

N.4.1.1.   Repeatability Tests. – Tests for repeatability should include a minimum of three consecutive test drafts of approximately the same size and be conducted under controlled conditions where variations in factors such as temperature, pressure, and flow rate are reduced to the extent that they will not affect the results obtained.

(Added 2002)

N.4.2.   Special Tests. – Any test except as set forth in N.4.1. Normal Tests shall be considered a special test. Tests shall be conducted, if possible, to evaluate any special elements or accessories attached to or associated with the device. A device shall be tested at a minimum discharge rate of:

(a) not less than the marked minimum discharge rate or 20 % of the maximum rated discharge rate of the device, whichever is less; or

(b) the lowest discharge rate practicable under the conditions of installation.

"Special" tests may be conducted to develop any characteristics of the device anticipated under the conditions of installation.

N.4.3.   Density. – Temperature and pressure of the metered test liquid shall be measured during the test for the determination of density or volume correction when applicable. The appropriate correction values shall apply as specified in Table N.4.4.

N.4.4.   Automatic Temperature or Density Compensation. – If a device is equipped with an automatic temperature or density compensator, the compensator shall be tested by comparing the quantity indicated or recorded by the device (with the compensator connected and operating) with the actual delivered quantity. The appropriate correction values shall apply as specified in Table N.4.4.

Temp °F	Pressure		Liquid Density		Vapor Density		Vapor Displacement %
	PSIA	PSIG	lb/gal	(lb-oz)/gal	lb/cu ft	lb/gal	
− 30.00	177.89	163.19	9.127	9 - 2.0	1.989	0.266	2.9
− 29.75	178.75	164.05	9.122	9 - 2.0	1.999	0.267	2.9
− 29.50	179.62	164.92	9.117	9 - 1.9	2.008	0.268	2.9
− 29.25	180.49	165.79	9.113	9 - 1.8	2.018	0.270	3.0
− 29.00	181.36	166.67	9.108	9 - 1.7	2.028	0.271	3.0
− 28.75	182.24	167.54	9.103	9 - 1.7	2.038	0.272	3.0
− 28.50	183.12	168.42	9.098	9 - 1.6	2.048	0.274	3.0
− 28.25	184.00	169.31	9.094	9 - 1.5	2.058	0.275	3.0
− 28.00	184.89	170.19	9.089	9 - 1.4	2.067	0.276	3.0
− 27.75	185.78	171.08	9.084	9 - 1.3	2.077	0.278	3.1
− 27.50	186.67	171.98	9.080	9 - 1.3	2.087	0.279	3.1
− 27.25	187.57	172.87	9.075	9 - 1.2	2.098	0.280	3.1
− 27.00	188.47	173.77	9.070	9 - 1.1	2.108	0.282	3.1
− 26.75	189.37	174.67	9.065	9 - 1.0	2.118	0.283	3.1
− 26.50	190.28	175.58	9.061	9 - 1.0	2.128	0.284	3.1
− 26.25	191.18	176.49	9.056	9 - 0.9	2.138	0.286	3.2
− 26.00	192.10	177.40	9.051	9 - 0.8	2.148	0.287	3.2
− 25.75	193.01	178.32	9.046	9 - 0.7	2.159	0.289	3.2
− 25.50	193.93	179.23	9.041	9 - 0.7	2.169	0.290	3.2
− 25.25	194.85	180.16	9.037	9 - 0.6	2.179	0.291	3.2
− 25.00	195.78	181.08	9.032	9 - 0.5	2.190	0.293	3.2
− 24.75	196.70	182.01	9.027	9 - 0.4	2.200	0.294	3.3
− 24.50	197.64	182.94	9.022	9 - 0.4	2.211	0.296	3.3
− 24.25	198.57	183.87	9.017	9 - 0.3	2.221	0.297	3.3
− 24.00	199.51	184.81	9.013	9 - 0.2	2.232	0.298	3.3
− 23.75	200.45	185.75	9.008	9 - 0.1	2.243	0.300	3.3
− 23.50	201.39	186.70	9.003	9 - 0.0	2.253	0.301	3.3
− 23.25	202.34	187.64	8.998	9 - 0.0	2.264	0.303	3.4
− 23.00	203.29	188.60	8.993	8 - 15.9	2.275	0.304	3.4

Table N.4.4.
Automatic Temperature or Density Compensation

Table N.4.4.
Automatic Temperature or Density Compensation

Temp °F	Pressure		Liquid Density		Vapor Density		Vapor Displacement %
	PSIA	PSIG	lb/gal	(lb-oz)/gal	lb/cu ft	lb/gal	
– 22.75	204.25	189.55	8.989	8 - 15.8	2.286	0.306	3.4
– 22.50	205.20	190.51	8.984	8 - 15.7	2.296	0.307	3.4
– 22.25	206.16	191.47	8.979	8 - 15.7	2.307	0.308	3.4
– 22.00	207.13	192.43	8.974	8 - 15.6	2.318	0.310	3.5
– 21.75	208.09	193.40	8.969	8 - 15.5	2.329	0.311	3.5
– 21.50	209.06	194.37	8.964	8 - 15.4	2.340	0.313	3.5
– 21.25	210.04	195.34	8.959	8 - 15.4	2.351	0.314	3.5
– 21.00	211.02	196.32	8.955	8 - 15.3	2.362	0.316	3.5
– 20.75	212.00	197.30	8.950	8 - 15.2	2.374	0.317	3.5
– 20.50	212.98	198.28	8.945	8 - 15.1	2.385	0.319	3.6
– 20.25	213.97	199.27	8.940	8 - 15.0	2.396	0.320	3.6
– 20.00	214.96	200.26	8.935	8 - 15.0	2.407	0.322	3.6
– 19.75	215.95	201.26	8.930	8 - 14.9	2.419	0.323	3.6
– 19.50	216.95	202.25	8.925	8 - 14.8	2.430	0.325	3.6
– 19.25	217.95	203.25	8.920	8 - 14.7	2.441	0.326	3.7
– 19.00	218.95	204.26	8.915	8 - 14.6	2.453	0.328	3.7
– 18.75	219.96	205.27	8.911	8 - 14.6	2.464	0.329	3.7
– 18.50	220.97	206.28	8.906	8 - 14.5	2.476	0.331	3.7
– 18.25	221.99	207.29	8.901	8 - 14.4	2.488	0.333	3.7
– 18.00	223.01	208.31	8.896	8 - 14.3	2.499	0.334	3.8
– 17.75	224.03	209.33	8.891	8 - 14.3	2.511	0.336	3.8
– 17.50	225.05	210.36	8.886	8 - 14.2	2.523	0.337	3.8
– 17.25	226.08	211.38	8.881	8 - 14.1	2.534	0.339	3.8
– 17.00	227.11	212.42	8.876	8 - 14.0	2.546	0.340	3.8
– 16.75	228.15	213.45	8.871	8 - 13.9	2.558	0.342	3.9
– 16.50	229.18	214.49	8.866	8 - 13.9	2.570	0.344	3.9
– 16.25	230.23	215.53	8.861	8 - 13.8	2.582	0.345	3.9
– 16.00	231.27	216.58	8.856	8 - 13.7	2.594	0.347	3.9
– 15.75	232.32	217.62	8.851	8 - 13.6	2.606	0.348	3.9
– 15.50	233.37	218.68	8.846	8 - 13.5	2.618	0.350	4.0
– 15.25	234.43	219.73	8.841	8 - 13.5	2.630	0.352	4.0
– 15.00	235.49	220.79	8.836	8 - 13.4	2.643	0.353	4.0
– 14.75	236.55	221.86	8.831	8 - 13.3	2.655	0.355	4.0
– 14.50	237.62	222.92	8.826	8 - 13.2	2.667	0.357	4.0
– 14.25	238.69	223.99	8.821	8 - 13.1	2.680	0.358	4.1

Table N.4.4. Automatic Temperature or Density Compensation							
Temp °F	**Pressure**		**Liquid Density**		**Vapor Density**		**Vapor Displacement %**
	PSIA	**PSIG**	**lb/gal**	**(lb-oz)/gal**	**lb/cu ft**	**lb/gal**	
− 14.00	239.76	225.07	8.816	8 - 13.1	2.692	0.360	4.1
− 13.75	240.84	226.14	8.811	8 - 13.0	2.704	0.362	4.1
− 13.50	241.92	227.22	8.806	8 - 12.9	2.717	0.363	4.1
− 13.25	243.00	228.31	8.801	8 - 12.8	2.729	0.365	4.1
− 13.00	244.09	229.39	8.796	8 - 12.7	2.742	0.367	4.2
− 12.75	245.18	230.49	8.791	8 - 12.7	2.755	0.368	4.2
− 12.50	246.28	231.58	8.786	8 - 12.6	2.767	0.370	4.2
− 12.25	247.37	232.68	8.781	8 - 12.5	2.780	0.372	4.2
− 12.00	248.48	233.78	8.776	8 - 12.4	2.793	0.373	4.3
− 11.75	249.58	234.89	8.771	8 - 12.3	2.806	0.375	4.3
− 11.50	250.69	236.00	8.765	8 - 12.2	2.819	0.377	4.3
− 11.25	251.80	237.11	8.760	8 - 12.2	2.832	0.379	4.3
− 11.00	252.92	238.22	8.755	8 - 12.1	2.845	0.380	4.3
− 10.75	254.04	239.34	8.750	8 - 12.0	2.858	0.382	4.4
− 10.50	255.16	240.47	8.745	8 - 11.9	2.871	0.384	4.4
− 10.25	256.29	241.60	8.740	8 - 11.8	2.884	0.386	4.4
− 10.00	257.42	242.73	8.735	8 - 11.8	2.897	0.387	4.4
− 9.75	258.56	243.86	8.730	8 - 11.7	2.911	0.389	4.5
− 9.50	259.70	245.00	8.725	8 - 11.6	2.924	0.391	4.5
− 9.25	260.84	246.14	8.719	8 - 11.5	2.937	0.393	4.5
− 9.00	261.98	247.29	8.714	8 - 11.4	2.951	0.394	4.5
− 8.75	263.13	248.44	8.709	8 - 11.3	2.964	0.396	4.5
− 8.50	264.29	249.59	8.704	8 - 11.3	2.978	0.398	4.6
− 8.25	265.44	250.75	8.699	8 - 11.2	2.991	0.400	4.6
− 8.00	266.60	251.91	8.694	8 - 11.1	3.005	0.402	4.6
− 7.75	267.77	253.07	8.688	8 - 11.0	3.019	0.404	4.6
− 7.50	268.93	254.24	8.683	8 - 10.9	3.032	0.405	4.7
− 7.25	270.11	255.41	8.678	8 - 10.8	3.046	0.407	4.7
− 7.00	271.28	256.59	8.673	8 - 10.8	3.060	0.409	4.7
− 6.75	272.46	257.76	8.668	8 - 10.7	3.074	0.411	4.7
− 6.50	273.64	258.95	8.662	8 - 10.6	3.088	0.413	4.8
− 6.25	274.83	260.13	8.657	8 - 10.5	3.102	0.415	4.8
− 6.00	276.02	261.32	8.652	8 - 10.4	3.116	0.417	4.8
− 5.75	277.21	262.52	8.647	8 - 10.3	3.130	0.418	4.8
− 5.50	278.41	263.72	8.641	8 - 10.3	3.144	0.420	4.9

Temp °F	Pressure		Liquid Density		Vapor Density		Vapor Displacement %
	PSIA	PSIG	lb/gal	(lb-oz)/gal	lb/cu ft	lb/gal	
− 5.25	279.61	264.92	8.636	8 - 10.2	3.159	0.422	4.9
− 5.00	280.82	266.12	8.631	8 - 10.1	3.173	0.424	4.9
− 4.75	282.03	267.33	8.626	8 - 10.0	3.187	0.426	4.9
− 4.50	283.24	268.55	8.620	8 - 9.9	3.202	0.428	5.0
− 4.25	284.46	269.76	8.615	8 - 9.8	3.216	0.430	5.0
− 4.00	285.68	270.98	8.610	8 - 9.8	3.231	0.432	5.0
− 3.75	286.90	272.21	8.604	8 - 9.7	3.245	0.434	5.0
− 3.50	288.13	273.44	8.599	8 - 9.6	3.260	0.436	5.1
− 3.25	289.37	274.67	8.594	8 - 9.5	3.275	0.438	5.1
− 3.00	290.60	275.91	8.589	8 - 9.4	3.289	0.440	5.1
− 2.75	291.84	277.15	8.583	8 - 9.3	3.304	0.442	5.1
− 2.50	293.09	278.39	8.578	8 - 9.2	3.319	0.444	5.2
− 2.25	294.33	279.64	8.573	8 - 9.2	3.334	0.446	5.2
− 2.00	295.58	280.89	8.567	8 - 9.1	3.349	0.448	5.2
− 1.75	296.84	282.14	8.562	8 - 9.0	3.364	0.450	5.3
− 1.50	298.10	283.40	8.556	8 - 8.9	3.379	0.452	5.3
− 1.25	299.36	284.67	8.551	8 - 8.8	3.395	0.454	5.3
− 1.00	300.63	285.93	8.546	8 - 8.7	3.410	0.456	5.3
− 0.75	301.90	287.21	8.540	8 - 8.6	3.425	0.458	5.4
− 0.50	303.18	288.48	8.535	8 - 8.6	3.440	0.460	5.4
− 0.25	304.46	289.76	8.530	8 - 8.5	3.456	0.462	5.4
0.00	305.74	291.74	8.524	8 - 8.4	3.471	0.464	5.4
0.25	307.03	292.33	8.519	8 - 8.3	3.487	0.466	5.5
0.50	308.32	293.62	8.513	8 - 8.2	3.503	0.468	5.5
0.75	309.61	294.92	8.508	8 - 8.1	3.518	0.470	5.5
1.00	310.91	296.21	8.502	8 - 8.0	3.534	0.472	5.6
1.25	312.21	297.52	8.497	8 - 8.0	3.550	0.475	5.6
1.50	313.52	298.82	8.491	8 - 7.9	3.566	0.477	5.6
1.75	314.83	300.13	8.486	8 - 7.8	3.582	0.479	5.6
2.00	316.15	301.45	8.480	8 - 7.7	3.598	0.481	5.7
2.25	317.46	302.77	8.475	8 - 7.6	3.614	0.483	5.7
2.50	318.79	304.09	8.469	8 - 7.5	3.630	0.485	5.7
2.75	320.11	305.42	8.464	8 - 7.4	3.646	0.487	5.8
3.00	321.45	306.75	8.458	8 - 7.3	3.662	0.490	5.8
3.25	322.78	308.08	8.453	8 - 7.2	3.679	0.492	5.8

Table N.4.4.
Automatic Temperature or Density Compensation

Temp °F	Pressure		Liquid Density		Vapor Density		Vapor Displacement %
	PSIA	**PSIG**	**lb/gal**	**(lb-oz)/gal**	**lb/cu ft**	**lb/gal**	

Table N.4.4.
Automatic Temperature or Density Compensation

Temp °F	PSIA	PSIG	lb/gal	(lb-oz)/gal	lb/cu ft	lb/gal	Vapor Displacement %
3.50	324.12	309.42	8.447	8 - 7.2	3.695	0.494	5.8
3.75	325.46	310.77	8.442	8 - 7.1	3.712	0.496	5.9
4.00	326.81	312.11	8.436	8 - 7.0	3.728	0.498	5.9
4.25	328.16	313.46	8.431	8 - 6.9	3.745	0.501	5.9
4.50	329.52	314.82	8.425	8 - 6.8	3.761	0.503	6.0
4.75	330.88	316.18	8.420	8 - 6.7	3.778	0.505	6.0
5.00	332.24	317.54	8.414	8 - 6.6	3.795	0.507	6.0
5.25	333.61	318.91	8.408	8 - 6.5	3.812	0.510	6.1
5.50	334.98	320.28	8.403	8 - 6.4	3.829	0.512	6.1
5.75	336.35	321.66	8.397	8 - 6.4	3.846	0.514	6.1
6.00	337.73	323.04	8.392	8 - 6.3	3.863	0.516	6.2
6.25	339.12	324.42	8.386	8 - 6.2	3.880	0.519	6.2
6.50	340.51	325.81	8.380	8 - 6.1	3.897	0.521	6.2
6.75	341.90	327.20	8.375	8 - 6.0	3.915	0.523	6.3
7.00	343.30	328.60	8.369	8 - 5.9	3.932	0.526	6.3
7.25	344.70	330.00	8.363	8 - 5.8	3.949	0.528	6.3
7.50	346.10	331.41	8.358	8 - 5.7	3.967	0.530	6.3
7.75	347.51	332.82	8.352	8 - 5.6	3.984	0.533	6.4
8.00	348.92	334.23	8.346	8 - 5.5	4.002	0.535	6.4
8.25	350.34	335.65	8.341	8 - 5.4	4.020	0.537	6.4
8.50	351.76	337.07	8.335	8 - 5.4	4.038	0.540	6.5
8.75	353.19	338.49	8.335	8 - 5.4	4.038	0.540	6.5
9.00	354.62	339.92	8.323	8 - 5.2	4.073	0.545	6.5
9.25	356.06	341.36	8.318	8 - 5.1	4.091	0.547	6.6
9.50	357.49	342.80	8.312	8 - 5.0	4.110	0.549	6.6
9.75	358.94	344.24	8.306	8 - 4.9	4.128	0.552	6.6
10.00	360.38	345.69	8.300	8 - 4.8	4.146	0.554	6.7
10.25	361.84	347.14	8.295	8 - 4.7	4.164	0.557	6.7
10.50	363.29	348.60	8.289	8 - 4.6	4.183	0.559	6.7
10.75	364.75	350.06	8.283	8 - 4.5	4.201	0.562	6.8
11.00	366.22	351.52	8.277	8 - 4.4	4.220	0.564	6.8
11.25	367.68	352.99	8.271	8 - 4.3	4.238	0.567	6.8
11.50	369.16	354.46	8.266	8 - 4.2	4.257	0.569	6.9
11.75	370.64	355.94	8.260	8 - 4.2	4.276	0.572	6.9
12.00	372.12	357.42	8.254	8 - 4.1	4.295	0.574	7.0

Temp °F	Pressure		Liquid Density		Vapor Density		Vapor Displacement %
	PSIA	PSIG	lb/gal	(lb-oz)/gal	lb/cu ft	lb/gal	
12.25	373.60	358.91	8.248	8 - 4.0	4.314	0.577	7.0
12.50	375.09	360.40	8.242	8 - 3.9	4.333	0.579	7.0
12.75	376.59	361.89	8.236	8 - 3.8	4.352	0.582	7.1
13.00	378.09	363.39	8.230	8 - 3.7	4.371	0.584	7.1
13.25	379.59	364.89	8.224	8 - 3.6	4.390	0.587	7.1
13.50	381.10	366.40	8.219	8 - 3.5	4.410	0.589	7.2
13.75	382.61	367.91	8.213	8 - 3.4	4.429	0.592	7.2
14.00	384.13	369.43	8.207	8 - 3.3	4.449	0.595	7.2
14.25	385.65	370.95	8.201	8 - 3.2	4.468	0.597	7.3
14.50	387.17	372.48	8.195	8 - 3.1	4.488	0.600	7.3
14.75	388.70	374.01	8.189	8 - 3.0	4.508	0.603	7.4
15.00	390.24	375.54	8.183	8 - 2.9	4.527	0.605	7.4
15.25	391.78	377.08	8.177	8 - 2.8	4.547	0.608	7.4
15.50	393.32	378.62	8.171	8 - 2.7	4.567	0.611	7.5
15.75	394.87	380.17	8.165	8 - 2.6	4.587	0.613	7.5
16.00	396.42	381.72	8.159	8 - 2.5	4.608	0.616	7.5
16.25	397.98	383.28	8.153	8 - 2.4	4.628	0.619	7.6
16.50	399.54	384.84	8.147	8 - 2.3	4.648	0.621	7.6
16.75	401.10	386.41	8.141	8 - 2.2	4.669	0.624	7.7
17.00	402.67	387.98	8.134	8 - 2.2	4.689	0.627	7.7
17.25	404.25	389.55	8.128	8 - 2.1	4.710	0.630	7.7
17.50	405.82	391.13	8.122	8 - 2.0	4.731	0.632	7.8
17.75	407.41	392.71	8.116	8 - 1.9	4.751	0.635	7.8
18.00	409.00	394.30	8.110	8 - 1.8	4.772	0.638	7.9
18.25	410.59	395.89	8.104	8 - 1.7	4.793	0.641	7.9
18.50	412.19	397.49	8.098	8 - 1.6	4.814	0.644	7.9
18.75	413.79	399.09	8.092	8 - 1.5	4.835	0.646	8.0
19.00	415.39	400.70	8.085	8 - 1.4	4.857	0.649	8.0
19.25	417.00	402.31	8.079	8 - 1.3	4.878	0.652	8.1
19.50	418.62	403.92	8.073	8 - 1.2	4.900	0.655	8.1
19.75	420.24	405.54	8.067	8 - 1.1	4.921	0.658	8.2
20.00	421.86	407.17	8.061	8 - 1.0	4.943	0.661	8.2

Table N.4.4. Automatic Temperature or Density Compensation

T. Tolerances

T.1. Application.

T.1.1. To Underregistration and to Overregistration. – The tolerances hereinafter prescribed shall be applied to errors of underregistration and errors of overregistration.

T.2. Tolerance Values. – The maintenance and acceptance tolerances for normal and special tests shall be as shown in Table T.2. Accuracy Classes and Tolerances for Carbon Dioxide Liquid-Measuring Devices.

Table T.2. Accuracy Classes and Tolerances for Carbon Dioxide Liquid-Measuring Devices				
Accuracy Class	Application	Acceptance Tolerance	Maintenance Tolerance	Special Test Tolerance
2.5	Liquid carbon dioxide	1.5 %	2.5 %	2.5 %

(Table Added 2003) (Amended 2003)

T.2.1. Repeatability. – When multiple tests are conducted at approximately the same flow rate and draft size, the range of the test results for the flow rate shall not exceed 40 % of the absolute value of the maintenance tolerance and the results of each test shall be within the applicable tolerance. (Also see N.4.1.1. Repeatability Tests.)

(Added 2002)

T.3. On Tests Using Transfer Standards. – To the basic tolerance values that would otherwise be applied, there shall be added an amount equal to two times the standard deviation of the applicable transfer standard when compared to a basic reference standard.

UR. User Requirements

UR.1. Installation Requirements.

UR.1.1. Discharge Rate. – A device shall be so installed that the actual maximum discharge rate will not exceed the rated maximum discharge rate. If necessary, means for flow regulation shall be incorporated in the installation.

UR.1.2. Length of Discharge Hose. – The discharge hose shall be of such a length and design as to keep vaporization of the liquid to a minimum.

UR.1.3. Maintenance of Liquid State. – A device shall be so installed and operated that the product being measured shall remain in the liquid state during passage through the meter.

UR.2. Use Requirements.

UR.2.1. Return of Indicating and Recording Elements to Zero. – The primary indicating elements (visual) and the primary recording elements shall be returned to zero immediately before each delivery.

UR.2.2. Condition of Discharge System. – The discharge hose, up to the valve at the end of the discharge hose, shall be completely filled and pre-cooled to liquid temperatures before a "zero" condition is established and prior to the start of a commercial delivery. Means shall be provided to fill the discharge hose with liquid prior to the start of a delivery.

UR.2.3.   Vapor Equalization Line. – A vapor equalization line shall not be used during a metered delivery unless the quantity of vapor displaced from the buyer's tank to the seller's tank is deducted from the metered quantity. The appropriate correction values shall apply as specified in Table N.4.4.

UR.2.4.   Temperature or Density Compensation.

UR.2.4.1.   Use of Automatic Temperature or Density Compensators. – Devices equipped with an automatic temperature or density compensator shall have the compensator connected, operable, and in use at all times. Such automatic temperature or density compensator may not be removed.

UR.2.4.2.   Tickets or Invoices. – Any written invoice or printed ticket based on a reading of a device that is equipped with an automatic temperature or density compensator shall have shown thereon that the quantity delivered has been temperature or density compensated.

UR.2.5.   Ticket in Printing Device. – A ticket shall not be inserted into a device equipped with a ticket printer until immediately before a delivery is begun, and in no case shall a ticket be in the device when the vehicle is in motion while on a public street, highway, or thoroughfare.

UR.2.6.   Sale by Weight. – All quantity determinations shall be made by means of an approved and sealed weighing or measuring device. All sales shall be stated in kilograms or pounds.

THIS PAGE INTENTIONALLY LEFT BLANK

Table of Contents

Section 3.39.   Hydrogen Gas-Measuring Devices – Tentative Code

This tentative code has a trial or experimental status and is not intended to be enforced. The requirements are designed for study prior to the development and adoption of a final code. Requirements that apply to wholesale applications are under study and development by the U.S. National Work Group for the Development of Commercial Hydrogen Measurement Standards. Officials wanting to conduct an official examination of a device or system are advised to see paragraph G-A.3. Special and Unclassified Equipment.

(Tentative Code Added 2010)

A. Application

A.1.   General. – This code applies to devices that are used for the measurement of hydrogen gas in the vapor state used as a vehicle fuel.

A.2.   Exceptions. – This code does not apply to:

(a) Devices used solely for dispensing a product in connection with operations in which the amount dispensed does not affect customer charges.

(b) The wholesale delivery of hydrogen gas.

(c) Devices used for dispensing a hydrogen gas with a hydrogen fuel index lower than 99.97 % and concentrations of specified impurities that exceed level limits.

(d) Systems that measure pressure, volume, and temperature with a calculating device to determine the mass of gas accumulated in or discharged from a tank of known volume.

A.3.   Additional Code Requirements. – In addition to the requirements of this code, Hydrogen Gas-Measuring Devices shall meet the requirements of Section 1.10. General Code.

A.4.   Type Evaluation. – The National Type Evaluation Program (NTEP) will accept for type evaluation only those devices that comply with all requirements of this code.

S. Specifications

S.1.   Indicating and Recording Elements.

S.1.1.   Indicating Elements. – A measuring assembly shall include an indicating element that continuously displays measurement results relative to quantity and total price. Indications shall be clear, definite, accurate, and easily read under normal conditions of operation of the device.

S.1.2.   Vehicle Fuel Dispensers. – A hydrogen gas dispenser used to fuel vehicles shall be of the computing type and shall indicate the mass, the unit price, and the total price of each delivery.

S.1.3.   Units.

S.1.3.1.   Units of Measurement. – Deliveries shall be indicated and recorded in kilograms and decimal subdivisions thereof.

S.1.3.2.   Numerical Value of Quantity-Value Divisions. – The value of an interval (i.e., increment or scale division) shall be equal to:

(a) 1, 2, or 5; or

(b) a decimal multiple or submultiple of 1, 2, or 5.

   Examples: quantity-value divisions may be 10, 20, 50, 100; or 0.01, 0.02, 0.05; or 0.1, 0.2, or 0.5 etc.

S.1.3.3.   Maximum Value of Quantity-Value Divisions. – The maximum value of the quantity-value division shall not be greater than 0.5% of the minimum measured quantity.

S.1.3.4.   Values Defined. – Indicated values shall be adequately defined by a sufficient number of figures, words, symbols, or combinations thereof. A display of "zero" shall be a zero digit for all displayed digits to the right of the decimal mark and at least one to the left.

S.1.4.   Value of Smallest Unit. – The value of the smallest unit of indicated delivery, and recorded delivery if the device is equipped to record, shall not exceed the equivalent of:

(a) 0.001 kg on devices with a marked maximum flow rated of 30 kg/min or less; or

(b) 0.01 kg on devices with a marked maximum flow rate of more than 30 kg/min.

S.2.   Operating Requirements.

S.2.1.   Return to Zero.

(a) The primary indicating and the primary recording elements, if the device is equipped to record, shall be provided with a means for readily returning the indication to zero either automatically or manually.

(b) It shall not be possible to return primary indicating elements, or primary recording elements, beyond the correct zero position.

S.2.2.   Indicator Reset Mechanism. – The reset mechanism for the indicating element shall not be operable during a delivery. Once the zeroing operation has begun, it shall not be possible to indicate a value other than the latest measurement, or "zeros" when the zeroing operation has been completed.

S.2.3.   Provision for Power Loss.

S.2.3.1.   Transaction Information. – In the event of a power loss, the information needed to complete any transaction in progress at the time of the power loss (such as the quantity and unit price, or sales price) shall be determinable for at least 15 minutes at the dispenser or at the console if the console is accessible to the customer.

S.2.3.2.   User Information. – The device memory shall retain information on the quantity of fuel dispensed and the sales price totals during power loss.

S.2.4.   Display of Unit Price and Product Identity.

S.2.4.1.   Unit Price. – A computing or money-operated device shall be able to display on each face the unit price at which the device is set to compute or to dispense.

S.2.4.2.   Product Identity. – A device shall be able to conspicuously display on each side the identity of the product being dispensed.

S.2.4.3.   Selection of Unit Price. – When a product is offered for sale at more than on unit price through a computing device, the selection of the unit price shall be made prior to delivery using controls on the device or other customer-activated controls. A system shall not permit a change to the unit price during delivery of a product.

S.2.4.4.   Agreement Between Indications. – All quantity, unit price, and total price indications within a measuring system shall agree for each transaction.

S.2.5.   Money-Value Computations. – A computing device shall compute the total sales price at any single-purchase unit price for which the product being measured is offered for sale at any delivery possible within either the measurement range of the device or the range of the computing elements, whichever is less.

S.2.5.1.   Auxiliary Elements. – If a system is equipped with auxiliary indications, all indicated money value and quantity divisions of the auxiliary element shall be identical with those of the primary element.

S.2.5.2.   Display of Quantity and Total Price. – When a delivery is completed, the total price and quantity for that transaction shall be displayed on the face of the dispenser for at least 5 minutes or until the next transaction is initiated by using controls on the device or other user-activated controls.

S.2.6.   Recorded Representations, Point of Sale Systems. – A printed receipt shall be available through a built-in or separate recording element for transactions conducted with point-of-sale systems or devices activated by debit cards, credit cards, and/or cash. The printed receipt shall contain the following information for products delivered by the dispenser:

(a) the total mass of the delivery;

(b) the unit price;

(c) the total computed price; and

(d) the product identity by name, symbol, abbreviation, or code number.

S.2.7.   Indication of Delivery. – The device shall automatically show on its face the initial zero condition and the quantity delivered (up to the nominal capacity).

S.3.   Design of Measuring Elements and Measuring Systems.

S.3.1.   Maximum and Minimum Flow-Rates. – The ratio of the maximum to minimum flow-rates specified by the manufacturer for devices measuring gases shall be 10:1 or greater.

S.3.2.   Adjustment Means. – An assembly shall be provided with means to change the ratio between the indicated quantity and the quantity of gas measured by the assembly. A bypass on the measuring assembly shall not be used for these means.

S.3.2.1. Discontinuous Adjusting Means. – When the adjusting means changes ratio between the indicated quantity and the quantity of measured gas in a discontinuous manner, the consecutive values of the ratio shall not differ by more than 0.1 %.

S.3.3.   Provision for Sealing. – Adequate provision shall be made for an approved means of security (e.g., data change audit trail) or physically applying security seals in such a manner that no adjustment may be made of:

(a) each individual measurement element;

(b) any adjustable element for controlling delivery rate when such rate tends to affect the accuracy of deliveries;

(c) the zero adjustment mechanism; and

(d) any metrological parameter that detrimentally affects the metrological integrity of the device or system. When applicable, the adjusting mechanism shall be readily accessible for purposes of affixing a security seal. Audit trails shall use the format set forth in Table S.3.3. Categories of Device and Methods of Sealing.

Table S.3.3. Categories of Device and Methods of Sealing	
Categories of Device	**Method of Sealing**
Category 1: No remote configuration capability.	Seal by physical seal or two event counters: one for calibration parameters and one for configuration parameters.
Category 2: Remote configuration capability, but access is controlled by physical hardware. The device shall clearly indicate that it is in the remote configuration mode and record such message if capable of printing in this mode or shall not operate while in this mode.	The hardware enabling access for remote communication must be on-site. The hardware must be sealed using a physical seal or an event counter for calibration parameters and an event counter for configuration parameters. The event counters may be located either at the individual measuring device or at the system controller; however, an adequate number of counters must be provided to monitor the calibration and configuration parameters of the individual devices at a location. If the counters are located in the system controller rather than at the individual device, means must be provided to generate a hard copy of the information through an on-site device.
Category 3: Remote configuration capability access may be unlimited or controlled through a software switch (e.g., password). The device shall clearly indicate that it is in the remote configuration mode and record such message if capable of printing in this mode or shall not operate while in this mode.	An event logger is required in the device; it must include an event counter (000 to 999), the parameter ID, the date and time of the change, and the new value of the parameter. A printed copy of the information must be available on demand through the device or through another on-site device. The information may also be available electronically. The event logger shall have a capacity to retain records equal to 10 times the number of sealable parameters in the device, but not more than 1000 records are required. (**Note:** Does not require 1000 changes to be stored for each parameter.)

(Amended 2016)

S.3.4. Automatic Density Correction.

(a) An automatic means to determine and correct for changes in product density shall be incorporated in any hydrogen gas-measuring system where measurements are affected by changes in the density of the product being measured.

(b) Volume-measuring devices with automatic temperature compensation used to measure hydrogen gas as a vehicle fuel shall be equipped with an automatic means to determine and correct for changes in product density due to changes in the temperature, pressure, and composition of the product.

S.3.5. Pressurizing the Discharge Hose. – The discharge hose for hydrogen gas shall automatically pressurize to a pressure equal to or greater than the receiving vessel prior to the device beginning to register the delivery. The indications shall not advance as a result of the initial pressurization or the purging/bleeding of the discharge hose.

S.3.6.   Zero-Set-Back Interlock, Retail Vehicle Fuel Devices.

(a) A device shall be constructed so that:

(1) when the device is shut-off at the end of a delivery an automatic interlock prevents a subsequent delivery until the indicating element and recording elements, if the device is equipped and ac-tivated to record, have been returned to their zero positions; and

(2) it shall not be possible to return the discharge nozzle to its start position unless the zero set back interlock is engaged or becomes engaged.

(b) For systems with more than one:

(1) dispenser supplied by a single measuring element, an effective automatic control valve in each dispenser prevents product from being delivered until the indicating elements on that dispenser are in a correct zero position; or

(2) hose supplied by a single measuring element, effective automatic means must be provided to prevent product from being delivered until the indicating element(s) corresponding to each hose are in a correct zero position.

S.4.   Discharge Lines and Valves.

S.4.1.   Diversion of Measured Product. – No means shall be provided by which any measured product can be diverted from the measuring device.

S.4.2.   Directional Flow Valves. – If a reversal of flow could result in errors that exceed the tolerance for the minimum measured quantity, a valve or valves or other effective means, automatic in operation (and equipped with a pressure limiting device, if necessary) to prevent the reversal of flow shall be properly installed in the system. (Also see N.1. Minimum Measured Quantity.)

S.4.3.   Other Valves. – Check valves and closing mechanisms that are not used to define the measured quantity shall have relief valves (if necessary) to dissipate any abnormally high pressure that may arise in the measuring assembly.

S.5.   Markings. – A measuring system shall be conspicuously, legibly, and indelibly marked with the following information:

(a) pattern approval mark (i.e., type approval number);

(b) name and address of the manufacturer or his trademark and, if required by the weights and measures authority, the manufacturer's identification mark in addition to the trademark;

(c) model designation or product name selected by the manufacturer;

(d) nonrepetitive serial number;

(e) the accuracy class of the device as specified by the manufacturer consistent with Table T.2. Accuracy Classes and Tolerances for Hydrogen-Gas Measuring Devices;

(f) maximum and minimum flow rates in kilograms per unit of time;

(g) maximum working pressure;

(h) applicable range of ambient temperature if other than $-10 \,°C$ to $+50 \,°C$;

(i) minimum measured quantity; and

(j) product limitations (such as fuel quality), if applicable.

S.5.1.   Location of Marking Information; Hydrogen-Fuel Dispensers. – The marking information required in General Code, paragraph G S.1. Identification shall appear as follows:

(a) within 60 cm (24 in) to 150 cm (60 in) from the base of the dispenser;

(b) either internally and/or externally provided the information is permanent and easily read; and accessible for inspection; and

(c) on a portion of the device that cannot be readily removed or interchanged (i.e., not on a service access panel).

Note: The use of a dispenser key or tool to access internal marking information is permitted for retail hydrogen-measuring devices.

S.6.   Printer. – When an assembly is equipped with means for printing the measured quantity, the printed information must agree with the indications on the dispenser for the transaction and the printed values shall be clearly defined.

S.6.1.   Printed Receipt. – Any delivered, printed quantity shall include an identification number, the time and date, and the name of the seller. This information may be printed by the device or pre-printed on the ticket.

S.7.   Totalizers for Vehicle Fuel Dispensers. – Vehicle fuel dispensers shall be equipped with a nonresettable totalizer for the quantity delivered through each separate measuring device.

S.8.   Minimum Measured Quantity. – The minimum measured quantity shall satisfy the conditions of use of the measuring system as follows:

(a) Measuring systems having a maximum flow rate less than or equal to 4 kg/min shall have a minimum measured quantity not exceeding 0.5 kg.

(b) Measuring systems having a maximum flow rate greater than 4 kg/min but not greater than 12 kg/min shall have a minimum measured quantity not exceeding 1.0 kg.

N. Notes

N.1.   Minimum Measured Quantity. – The minimum measured quantity shall be specified by the manufacturer.

N.2.   Test Medium. – The device shall be tested with the product commercially measured except that, in a type evaluation examination, hydrogen gas as specified in NIST Handbook 130 shall be used.

Note: Corresponding requirements are under development and this paragraph will be revisited.

N.3.   Test Drafts. – The minimum test shall be one test draft at the declared minimum measured quantity and one test draft at approximately ten times the minimum measured quantity or 1 kg, whichever is greater. More tests may be performed over the range of normal quantities dispensed. (Also see T.3. Repeatability.)

The test draft shall be made at flows representative of that during normal delivery. The pressure drop between the dispenser and the proving system shall not be greater than that for normal deliveries. The control of the flow (e.g., pipework or valve(s) size, etc.) shall be such that the flow of the measuring system is maintained within the range specified by the manufacturer.

N.4. Tests.

N.4.1. Master Meter (Transfer) Standard Test. – When comparing a measuring system with a calibrated transfer standard, the minimum test shall be one test draft at the declared minimum measured quantity and one test draft at approximately ten times the minimum measured quantity or 1 kg, whichever is greater. More tests may be performed over the range of normal quantities dispensed.

N.4.1.1. Verification of Master Metering Systems. – A master metering system used to verify a hydrogen gas-measuring device shall be verified before and after the verification process. A master metering system used to calibrate a hydrogen gas-measuring device shall be verified before starting the calibration and after the calibration process.

N.4.2. Gravimetric Tests. – The weight of the test drafts shall be equal to at least the amount delivered by the device at the declared minimum measured quantity and one test draft at approximately ten times the minimum measured quantity or 1 kg, whichever is greater. More tests may be performed over the range of normal quantities dispensed.

N.4.3. PVT Pressure Volume Temperature Test. – The minimum test with a calibrated volumetric standard shall be one test draft at the declared minimum measured quantity and one test draft at approximately ten times the minimum measured quantity or 1 kg, whichever is greater. More tests may be performed over the range of normal quantities dispensed.

N.5. Minimum Measured Quantity. – The device shall be tested for a delivery equal to the declared minimum measured quantity when the device is likely to be used to make deliveries on the order of the declared minimum measured quantity.

N.6. Testing Procedures.

N.6.1. General. – The device or system shall be tested under normal operating conditions of the dispenser.

The test draft shall be made at flows representative of that during normal delivery. The pressure drop between the dispenser and the proving system shall not be greater than that for normal deliveries. The control of the flow (e.g., pipework or valve(s) size, etc.) shall be such that the flow of the measuring system is maintained within the range specified by the manufacturer.

N.6.1.1. Repeatability Tests. – Tests for repeatability should include a minimum of three consecutive test drafts of approximately the same size and be conducted under controlled conditions where variations in factors are reduced to minimize the effect on the results obtained.

N.7. Density. – Temperature and pressure of hydrogen gas shall be measured during the test for the determination of density or volume correction factors when applicable. For the thermophysical properties of hydrogen the following publications shall apply: for density calculations at temperatures above 255 K and pressures up to 120 MPa, a simple relationship may be used that is given in the publication of Lemmon et al., J. Res. NIST, 2008. Calculations for a wider range of conditions and additional thermophysical properties of hydrogen are available free of charge online at the "NIST Chemistry WebBook" **http://webbook.nist.gov/chemistry**, or available for purchase from NIST as the computer program NIST Standard Reference Database 23 "NIST Reference Fluid Thermodynamic and Transport Properties Database (REFPROP): Version 8.0" **http://www.nist.gov/srd/nist23.cfm**. These calculations are based on the reference Leachman, J.W., Jacobsen, R.T, Lemmon, E.W., and Penoncello, S.G. "Fundamental Equations of State for Parahydrogen, Normal Hydrogen, and Orthohydrogen" to be published in the Journal of Physical and Chemical Reference Data (**http://www.nist.gov/manuscript-publication-search.cfm?pub_id=832374**). More information may be obtained from NIST online at **http://www.boulder.nist.gov/div838/Hydrogen/Index.htm**.

T. Tolerances

T.1. Tolerances, General.

(a) The tolerances apply equally to errors of underregistration and errors of overregistration.

(b) The tolerances apply to all products at all temperatures measured at any flow rate within the rated measuring range of the device.

T.2. Tolerances.
– The tolerances for hydrogen gas measuring devices are listed in Table T.2. Accuracy Classes and Tolerances for Hydrogen Gas-Measuring Devices. (Proposed tolerance values are based on previous work with compressed gas products and will be confirmed based on performance data evaluated by the U.S. National Work Group.)

Table T.2. Accuracy Classes and Tolerances for Hydrogen Gas-Measuring Devices			
Accuracy Class	Application or Commodity Being Measured	Acceptance Tolerance	Maintenance Tolerance
7.0	Hydrogen gas as a vehicle fuel	5.0 %	7.0 %

(Amended 2016)

T.3. Repeatability.
– When multiple tests are conducted at approximately the same flow rate and draft size, the range of the test results for the flow rate shall not exceed 40 % of the absolute value of the maintenance tolerance and the results of each test shall be within the applicable tolerance. (Also see N.6.1.1. Repeatability Tests.)

T.4. Tolerance Application on Test Using Transfer Standard Test Method.
– To the basic tolerance values that would otherwise be applied, there shall be added an amount equal to two times the standard deviation of the applicable transfer standard when compared to a basic reference standard.

T.5. Tolerance Application in Type Evaluation Examinations for Devices.
– For type evaluation examinations, the tolerance values shall apply under the following conditions:

(a) at any temperature and pressure within the operating range of the device, and

(b) for all quantities greater than the minimum measured quantity.

UR. User Requirements

UR.1. Selection Requirements.

UR.1.1. Computing-Type Device; Retail Dispenser.
– A hydrogen gas dispenser used to refuel vehicles shall be of the computing type and shall indicate the mass, the unit price, and the total price of each delivery.

UR.1.2. Discharge Hose-Length.
– The length of the discharge hose on a retail fuel dispenser:

(a) shall not exceed 4.6 m (15 ft) unless it can be demonstrated that a longer hose is essential to permit deliveries to be made to receiving vehicles or vessels;

(b) shall be measured from its housing or outlet of the discharge line to the inlet of the discharge nozzle; and

(c) shall be measured with the hose fully extended if it is coiled or otherwise retained or connected inside a housing.

An unnecessarily remote location of a device shall not be accepted as justification for an abnormally long hose.

UR.1.3.   Minimum Measured Quantity.

(a) The minimum measured quantity shall be specified by the manufacturer.

(b) The minimum measured quantity appropriate for a transaction may be specified by the weights and measures authority. A device may have a declared minimum measured quantity smaller than that specified by the weights and measures authority; however, the device must perform within the performance requirements for the declared or specified minimum measured quantity up to deliveries at the maximum measurement range.

(c) The minimum measured quantity shall satisfy the conditions of use of the measuring system as follows:

  (1) measuring systems having a maximum flow rate less than or equal to 4 kg/min shall have a minimum measured quantity not exceeding 0.5 kg; and

  (2) measuring systems having a maximum flow rate greater than 4 kg/min but not greater than 12 kg/min shall have a minimum measured quantity not exceeding 1.0 kg.

UR.2.   Installation Requirements.

UR.2.1.   Manufacturer's Instructions. – A device shall be installed in accordance with the manufacturer's instructions, and the installation shall be sufficiently secure and rigid to maintain this condition.

UR.2.2.   Discharge Rate. – A device shall be installed so that after initial equalization the actual maximum discharge rate will not exceed the rated maximum discharge rate. Automatic means of flow regulation shall be incorporated in the installation if necessary.

UR.2.3.   Low-Flow Cut-Off Valve. – If a measuring system is equipped with a programmable or adjustable "low-flow cut-off" feature:

(a) the low-flow cut-off value shall not be set at flow rates lower than the minimum operating flow rate specified by the manufacturer on the measuring device; and

(b) the system shall be equipped with flow control valves which prevent the flow of product and stop the indicator from registering product flow whenever the product flow rate is less than the low-flow cut-off value.

UR.3.   Use of Device.

UR.3.1.   Unit Price and Product Identity for Retail Dispensers. – The unit price at which the dispenser is set to compute shall be conspicuously displayed or posted on the face of a retail dispenser used in direct sale.

UR.3.2.   Vehicle-mounted Measuring Systems Ticket Printer.

UR.3.2.1. Customer Ticket. – Vehicle-mounted measuring systems shall be equipped with a ticket printer which shall be used for all sales where product is delivered through the device. A copy of the ticket issued by the device shall be left with the customer at the time of delivery or as otherwise specified by the customer.

UR.3.2.2. Ticket in Printing Device. – A ticket shall not be inserted into a device equipped with a ticket printer until immediately before a delivery is begun, and in no case shall a ticket be in the device when the vehicle is in motion while on a public street, highway, or thoroughfare.

UR.3.3. Printed Ticket. – The total price, the total quantity of the delivery, and the price per unit shall be printed on any ticket issued by a device of the computing type and containing any one of these values.

UR.3.4. Steps After Dispensing. – After delivery to a customer from a retail dispenser:

(a) the device shall be shut-off at the end of a delivery, through an automatic interlock that prevents a subsequent delivery until the indicating elements and recording elements, if the device is equipped and activated to record, have been returned to their zero positions; and

(b) the discharge nozzle shall not be returned to its start position unless the zero set-back interlock is engaged or becomes engaged by the act of disconnecting the nozzle or the act of returning the discharge nozzle.

UR.3.5. Return of Indicating and Recording Elements to Zero. – The primary indicating elements (visual), and the primary recording elements shall be returned to zero immediately before each delivery.

UR.3.6. Return of Product to Storage, Retail Hydrogen Gas Dispensers. – Provisions at the site shall be made for returning product to storage or disposing of the product in a safe and timely manner during or following testing operations. Such provisions may include return lines, or cylinders adequate in size and number to permit this procedure.

UR.3.7. Conversion Factors. – Established correction values. (Also see references in N.7. Density.) shall be used whenever measured hydrogen gas is billed. All sales shall be based on kilograms.

Appendix D. Definitions

The specific code to which the definition applies is shown in [brackets] at the end of the definition. Definitions for the General Code [1.10] apply to all codes in Handbook 44.

A

audit trail. – An electronic count and/or information record of the changes to the values of the calibration or configuration parameters of a device. [1.10, 2.20, 2.21, 2.24, 3.30, 3.37, 3.39, 5.56(a)]

automatic temperature or density compensation. – The use of integrated or ancillary equipment to obtain from the output of a volumetric meter an equivalent mass, or an equivalent liquid volume at the assigned reference temperature below and a pressure of 14.696 lb/in^2 absolute.

Cryogenic liquids	21 °C (70 °F) [3.34]
Hydrocarbon gas vapor	15 °C (60 °F) [3.33]
Hydrogen gas	21 °C (70 °F) [3.39]
Liquid carbon dioxide	21 °C (70 °F) [3.38]
Liquefied petroleum gas (LPG) and Anhydrous ammonia	15 °C (60 °F) [3.32]
Petroleum liquid fuels and lubricants	15 °C (60 °F) [3.30]

[3.39]

C

calibration parameter. – Any adjustable parameter that can affect measurement or performance accuracy and, due to its nature, needs to be updated on an ongoing basis to maintain device accuracy (e.g., span adjustments, linearization factors, and coarse zero adjustments). [2.20, 2.21, 2.24, 3.30, 3.37, 3.39, 5.56(a)]

D

discharge hose. – A flexible hose connected to the discharge outlet of a measuring device or its discharge line. [3.30, 3.31, 3.32, 3.34, 3.37, 3.38, 3.39]

discharge line. – A rigid pipe connected to the outlet of a measuring device. [3.30, 3.31, 3.32, 3.34, 3.37, 3.39]

E

event counter. – A non-resettable counter that increments once each time the mode that permits changes to sealable parameters is entered and one or more changes are made to sealable calibration or configuration parameters of a device. [2.20, 2.21, 3.30, 3.37, 3.39, 5.54, 5.56(a), 5.56(b), 5.57]

event logger. – A form of audit trail containing a series of records where each record contains the number from the event counter corresponding to the change to a sealable parameter, the identification of the parameter that was changed, the time and date when the parameter was changed, and the new value of the parameter. [2.20, 2.21, 3.30, 3.37, 3.39, 5.54, 5.56(a), 5.56(b), 5.57]

I

indicating element. – An element incorporated in a weighing or measuring device by means of which its performance relative to quantity or money value is "read" from the device itself as, for example, an index-and-graduated-scale combination, a weighbeam-and-poise combination, a digital indicator, and the like. (Also see "primary indicating or recording element.") [1.10]

M

minimum measured quantity (MMQ). – The smallest quantity delivered for which the measurement is to within the applicable tolerances for that system. [3.37, 3.39]

N

non-resettable totalizer. – An element interfaced with the measuring or weighing element that indicates the cumulative registration of the measured quantity with no means to return to zero. [3.30, 3.37, 3.39]

P

point-of-sale system. – An assembly of elements including a weighing or measuring element, an indicating element, and a recording element (and may also be equipped with a "scanner") used to complete a direct sales transaction. [2.20, 3.30, 3.32, 3.37, 3.39]

R

remote configuration capability. – The ability to adjust a weighing or measuring device or change its sealable parameters from or through some other device that is not itself necessary to the operation of the weighing or measuring device or is not a permanent part of that device. [2.20, 2.21, 2.24, 3.30, 3.37, 3.39, 5.56(a)]

retail device. – A measuring device primarily used to measure product for the purpose of sale to the end user. [3.30, 3.32, 3.37, 3.39]

W

wet hose. – A discharge hose intended to be full of product at all times. (Also see "wet-hose type.") [3.30, 3.31, 3.38, 3.39]

wet-hose type. – A type of device designed to be operated with the discharge hose full of product at all times. (Also see "wet hose.") [3.30, 3.32, 3.34, 3.37, 3.38, 3.39]

Table of Contents

Section 3.40.   Electric Vehicle Fueling Systems – Tentative Code

This tentative code has a trial or experimental status and is not intended to be enforced. The requirements are designed for study prior to the development and adoption of a final code. Officials wanting to conduct an official examination of an Electric Vehicle Supply Equipment (EVSE) or system are advised to see paragraph G-A.3. Special and Unclassified Equipment.

(Tentative Code Added 2015)

A. Application

A.1.   General. – This code applies to devices, accessories, and systems used for the measurement of electricity dispensed in vehicle fuel applications wherein a quantity determination or statement of measure is used wholly or partially as a basis for sale or upon which a charge for service is based.

A.2.   Exceptions. – This code does not apply to:

(a)   The use of any measure or measuring device owned, maintained, and used by a public utility or municipality only in connection with measuring electricity subject to the authority having jurisdiction such as the Public Utilities Commission.

(b)   Electric Vehicle Supply Equipment (EVSEs) used solely for dispensing electrical energy in connection with operations in which the amount dispensed does not affect customer charges or compensation.

(c)   The wholesale delivery of electricity.

A.3.   Additional Code Requirements. – In addition to the requirements of this code, Electric Fueling Systems shall meet the requirements of Section 1.10. General Code.

A.3.1.   Electric Vehicle Supply Equipment (EVSE) with Integral Time-Measuring Devices. – An EVSE that is used for both the sale of electricity as vehicle fuel and used to measure time during which services (e.g., vehicle parking) are received. These devices shall also meet the requirements of Section 5.55. Timing Devices.

A.4.   Type Evaluation. – The National Type Evaluation Program (NTEP) will accept for type evaluation only those EVSEs that comply with all requirements of this code and have received safety certification by a nationally recognized testing laboratory (NRTL).

S. Specifications

S.1.   Primary Indicating and Recording Elements.

S.1.1.   Electric Vehicle Supply Equipment (EVSE). – An EVSE used to charge electric vehicles shall be of the computing type and shall indicate the electrical energy, the unit price, and the total price of each transaction.

(a)   EVSEs capable of applying multiple unit prices over the course of a single transaction shall also be capable of indicating the start and stop time, the total quantity of energy delivered, the unit price, and the total price for the quantity of energy delivered during each discrete phase corresponding to one of the multiple unit prices.

(b)   EVSEs capable of applying additional fees for time-based and other services shall also be capable of indicating the total time measured; the unit price(s) for the additional time based service(s); the total computed price(s) for the time measured; and the total transaction price, including the total price for the energy and all additional fees.

S.1.2.   EVSE Indicating Elements. – An EVSE used to charge electric vehicles shall include an indicating element that accumulates continuously and displays, for a minimum of 15 seconds at the activation by the user and at the start and end of the transaction, the correct measurement results relative to quantity and total price. Indications shall be clear, definite, accurate, and easily read under normal conditions of operation of the device. All indications and representations of electricity sold shall be clearly identified and separate from other time-based fees indicated by an EVSE that is used for both the sale of electricity as vehicle fuel and the sale of other separate time-based services (e.g., vehicle parking).

S.1.2.1.   Multiple EVSEs Associated with a Single Indicating Element. – A system with a single indicating element for two or more EVSEs shall be provided with means to display information from the individual EVSE(s) selected or displayed, and shall be provided with an automatic means to indicate clearly and definitely which EVSE is associated with the displayed information.

S.1.3.   EVSE Units.

S.1.3.1.   EVSE Units of Measurement. – EVSE units used to charge electric vehicles shall be indicated and recorded in megajoules (MJ) or kilowatt-hours (kWh) and decimal subdivisions thereof.

S.1.3.2.   EVSE Value of Smallest Unit. – The value of the smallest unit of indicated delivery by an EVSE, and recorded delivery if the EVSE is equipped to record, shall be 0.005 MJ or 0.001 kWh.

S.1.3.3.   Values Defined. – Indicated values shall be adequately defined by a sufficient number of figures, words, symbols, or combinations thereof. An indication of "zero" shall be a zero digit for all displayed digits to the right of the decimal mark and at least one to the left.

S.2.   EVSE Operating Requirements.

S.2.1.   EVSE Return to Zero.

(a)   The primary indicating and the primary recording elements of an EVSE used to charge electric vehicles, if the EVSE is equipped to record, shall be provided with a means for readily returning the indication to zero either automatically or manually.

(b)   It shall not be possible to return primary indicating elements, or primary recording elements, beyond the correct zero position.

S.2.2.   EVSE Indicator Zero Reset Mechanism. – The reset mechanism for the indicating element of an EVSE used to charge electric vehicles shall not be operable during a transaction. Once the zeroing operation has begun, it shall not be possible to indicate a value other than: the latest measurement; "all zeros;" blank the indication; or provide other indications that cannot be interpreted as a measurement during the zeroing operation.

S.2.3.   EVSE Provision for Power Loss.

S.2.3.1.   Transaction Information. – In the event of a power loss, the information needed to complete any transaction (i.e., delivery is complete and payment is settled) in progress at the time of the power loss (such as the quantity and unit price, or sales price) shall be determinable through one of the means listed below or the transaction shall be terminated without any charge for the electrical energy transfer to the vehicle:

(a)   at the EVSE;

(b)   at the console, if the console is accessible to the customer;

(c) via on site internet access; or

(d) through toll-free phone access.

For EVSEs in parking areas where vehicles are commonly left for extended periods, the information needed to complete any transaction in progress at the time of the power loss shall be determinable through one of the above means for at least eight hours.

S.2.3.2.   Transaction Termination. – In the event of a power loss, either:

(a) the transaction shall terminate at the time of the power loss; or

(b) the EVSE may continue charging without additional authorization if the EVSE is able to determine it is connected to the same vehicle before and after the supply power outage.

In either case, there must be a clear indication on the receipt provided to the customer of the interruption, including the date and time of the interruption along with other information required under S.2.6. EVSE Recorded Representations.

S.2.3.3.   User Information. – The EVSE memory, or equipment on the network supporting the EVSE, shall retain information on the quantity of fuel dispensed and the sales price totals during power loss.

S.2.4.   EVSE Indication of Unit Price and Equipment Capacity and Type of Voltage.

S.2.4.1.   Unit Price. – An EVSE shall be able to indicate on each face the unit price at which the EVSE is set to compute or to dispense at any point in time during a transaction.

S.2.4.2.   Equipment Capacity and Type of Voltage. – An EVSE shall be able to conspicuously indicate on each face the maximum rate of energy transfer (i.e., maximum power) and the type of current associated with each unit price offered (e.g., 7 kW AC, 25 kW DC, etc.).

S.2.4.3.   Selection of Unit Price. – When electrical energy is offered for sale at more than one unit price through an EVSE, the selection of the unit price shall be made prior to delivery through a deliberate action of the purchaser to select the unit price for the fuel delivery. Except when the conditions for variable price structure have been approved by the customer prior to the sale, a system shall not permit a change to the unit price during delivery of electrical energy.

Note: When electrical energy is offered at more than one unit price, selection of the unit price may be through the deliberate action of the purchaser: 1) using controls on the EVSE; 2) through the purchaser's use of personal or vehicle-mounted electronic equipment communicating with the system; or 3) verbal instructions by the customer.

S.2.4.4.   Agreement Between Indications. – All quantity, unit price, and total price indications within a measuring system shall agree for each transaction.

S.2.5.   EVSE Money-Value Computations. – An EVSE shall compute the total sales price at any single-purchase unit price for which the electrical energy being measured is offered for sale at any delivery possible within either the measurement range of the EVSE or the range of the computing elements, whichever is less.

S.2.5.1.   Money-Value Divisions Digital. – An EVSE with digital indications shall comply with the requirements of paragraph G-S.5.5. Money-Values, Mathematical Agreement, and the total price computation shall be based on quantities not exceeding 0.5 MJ or 0.1 kWh.

S.2.5.2.   Auxiliary Elements. – If a system is equipped with auxiliary indications, all indicated money value and quantity divisions of the auxiliary element shall be identical to those of the primary element.

S.2.6.   EVSE Recorded Representations. – A receipt, either printed or electronic, providing the following information shall be available at the completion of all transactions:

(a) the total quantity of the energy delivered with unit of measure;

(b) the total computed price of the energy sale;

(c) the unit price of the energy, and for systems capable of applying multiple unit prices for energy during a single transaction, the following additional information is required:

    (1) the start and stop time of each phase during which one of the multiple unit prices was applied;

    (2) the unit price applied during each phase;

    (3) the total quantity of energy delivered during each phase;

    (4) the total purchase price for the quantity of energy delivered during each phase;

(d) the maximum rate of energy transfer (i.e., maximum power) and type of current (e.g., 7 kW AC, 25 kW DC, etc.);

(e) any additional separate charges included in the transaction (e.g., charges for parking time) including:

    (1) the time and date when the service begins and the time and date when the service ends; or the total time interval purchased, and the time and date that the service either begins or ends;

    (2) the unit price applied for the time-based service;

    (3) the total purchase price for the quantity of time measured during the complete transaction;

(f) the final total price of the complete transaction including all items;

(g) the unique EVSE identification number;

(h) the business name; and

(i) the business location.

S.2.7.   Indication of Delivery. – The EVSE shall automatically show on its face the initial zero condition and the quantity delivered (up to the capacity of the indicating elements).

S.3.   Design of Measuring Elements and Measuring Systems.

S.3.1.   Metrological Components. – An EVSE measuring system shall be designed and constructed so that metrological components are adequately protected from environmental conditions likely to be detrimental to accuracy. The system shall be designed to prevent undetected access to adjustment mechanisms and terminal blocks by providing for application of a physical security seal or an audit trail.

S.3.2.   Terminals. – The terminals of the EVSE system shall be arranged so that the possibility of short circuits while removing or replacing the cover, making connections, or adjusting the system, is minimized.

S.3.3.   Provision for Sealing. – Adequate provision shall be made for an approved means of security (e.g., data change audit trail) or physically applying security seals in such a manner that no adjustment may be made of:

(a) each individual measurement element;

(b) any adjustable element for controlling voltage or current when such control tends to affect the accuracy of deliveries;

(c) any adjustment mechanism that corrects or compensates for energy loss between the system and vehicle connection; and

(d) any metrological parameter that detrimentally affects the metrological integrity of the EVSE or system.

When applicable, the adjusting mechanism shall be readily accessible for purposes of affixing a security seal. Audit trails shall use the format set forth in Table S.3.3. Categories of Device and Methods of Sealing.

Table S.3.3. Categories of Device and Methods of Sealing	
Categories of Device	**Method of Sealing**
Category 1: No remote configuration capability.	Seal by physical seal or two event counters: one for calibration parameters and one for configuration parameters.
Category 2: Remote configuration capability, but access is controlled by physical hardware. The device shall clearly indicate that it is in the remote configuration mode and record such message if capable of printing in this mode or shall not operate while in this mode.	The hardware enabling access for remote communication must be on-site. The hardware must be sealed using a physical seal or an event counter for calibration parameters and an event counter for configuration parameters. The event counters may be located either at the individual measuring EVSE or at the system controller; however, an adequate number of counters must be provided to monitor the calibration and configuration parameters of the individual EVSEs at a location. If the counters are located in the system controller rather than at the individual EVSE, means must be provided to generate a hard copy of the information through an on-site device.
Category 3: Remote configuration capability access may be unlimited or controlled through a software switch (e.g., password). The device shall clearly indicate that it is in the remote configuration mode and record such message if capable of printing in this mode or shall not operate while in this mode.	An event logger is required in the device; it must include an event counter (000 to 999), the parameter ID, the date and time of the change, and the new value of the parameter. A printed copy of the information must be available through the EVSE or through another on-site device. The event logger shall have a capacity to retain records equal to 10 times the number of sealable parameters in the EVSE, but not more than 1000 records are required. (**Note:** Does not require 1000 changes to be stored for each parameter.)

S.3.4.   Data Storage and Retrieval.

(a) EVSE data accumulated and indicated shall be unalterable and accessible.

(b) Values indicated or stored in memory shall not be affected by electrical, mechanical, or temperature variations, radio-frequency interference, power failure, or any other environmental influences to the extent that accuracy is impaired.

(c) Memory and/or display shall be recallable for a minimum of three years. A replaceable battery shall not be used for this purpose.

S.3.5.   Temperature Range for System Components. – EVSEs shall be accurate and correct over the temperature range of – 40 °C to + 85 °C (– 40 °F to 185 °F). If the system or any measuring system components are not capable of meeting these requirements, the temperature range over which the system is capable shall be stated on the NTEP CC, marked on the EVSE, and installations shall be limited to the narrower temperature limits.

S.4.   Connections.

S.4.1.   Diversion of Measured Electricity. – No means shall be provided by which any measured electricity can be diverted from the measuring device.

S.4.1.1.   Unauthorized Disconnection. – Means shall be provided to automatically terminate the transaction in the event that there is an unauthorized break in the connection with the vehicle.

S.4.2.   Directional Control. – If a reversal of energy flow could result in errors that exceed the tolerance for the minimum measured quantity, effective means, automatic in operation to prevent or account for the reversal of flow shall be properly installed in the system. (See N.3. Minimum Test Draft [Size])

S.5.   Markings. – The following identification and marking requirements are in addition to the requirements of Section 1.10. General Code, paragraph G-S.1. Identification.

S.5.1.   Location of Marking Information; EVSE. – The marking information required in General Code, paragraph G-S.1. Identification shall appear as follows:

(a) within 60 cm (24 in) to 150 cm (60 in) from ground level; and

(b) on a portion of the EVSE that cannot be readily removed or interchanged (e.g., not on a service access panel).

S.5.2.   EVSE Identification and Marking Requirements. – In addition to all the marking requirements of Section 1.10. General Code, paragraph G-S.1. Identification, each EVSE shall have the following information conspicuously, legibly, and indelibly marked:

(a) voltage rating;

(b) maximum current deliverable;

(c) type of current (AC or DC or, if capable of both, both shall be listed);

(d) minimum measured quantity (MMQ); and

(e) temperature limits, if narrower than and within – 20 °C to + 50 °C (– 4 °F to 122 °F).

S.5.3.   Abbreviations and Symbols. – The following abbreviations or symbols may appear on an EVSE system.

(a) VAC = volts alternating current;

(b) VDC = volts direct current;

(c) MDA = maximum deliverable amperes;

(d) J = joule.

S.6.   Printer. – When a system is equipped with means for printing the measured quantity, the printed information must agree with the indications on the EVSE for the transaction and the printed values shall be clearly defined.

S.6.1.   Printed Receipt. – Any delivered, printed quantity shall include an EVSE identification number that uniquely identifies the EVSE from all other EVSEs within the seller's facility, the time and date, and the name of the seller. This information may be printed by the EVSE system or pre-printed on the ticket.

S.7.   Totalizers for EVSE Systems. – EVSE systems shall be designed with a nonresettable totalizer for the quantity delivered through each separate measuring device. Totalizer information shall be adequately protected and unalterable. Totalizer information shall be provided by the system and readily available on site or via on site internet access.

S.8.   Minimum Measured Quantity (MMQ). – The minimum measured quantity shall satisfy the conditions of use of the measuring system as follows:

(a) Measuring systems shall have a minimum measured quantity not exceeding 2.5 MJ or 0.5 kWh.

N. Notes

N.1.   No Load Test. – A no load test may be conducted on an EVSE measuring system by applying rated voltage to the system under test and no load applied.

N.2.   Starting Load Test. – A system starting load test maybe conducted by applying rated voltage and 0.5-ampere load.

N.3.   Minimum Test Draft (Size). – Full and light load tests shall require test of the EVSE System for a delivery of the minimum measured quantity as declared by the manufacturer.

N.4.   EVSE System Test Loads. – EVSE measuring system testing shall be accomplished by connecting the test load and test standard at the point where the fixed cord is connected to the vehicle. Losses in the cord between the EVSE under test and the test standard should be automatically corrected for in the EVSE quantity indication for direct comparison to the test standard and also while the EVSE is in normal operation. For EVSEs that require a customer-supplied cord, system testing shall be accomplished by connecting the test load and test standard at the point where the customer's cord is connected to the EVSE.

N.5.   Test of an EVSE System.

N.5.1.   Performance Verification in the Field. – Testing in the field is intended to validate the transactional accuracy of the EVSE system. The following testing is deemed sufficient for a field validation.

N.5.2.   Accuracy Testing. – The testing methodology compares the total energy delivered in a transaction and the total cost charged as displayed/reported by the EVSE with that measured by the measurement standard.

(a) For AC systems:

(1) Accuracy test of the EVSE system at a load of not less than 85 % of the maximum deliverable amperes (expressed as MDA) as determined from the pilot signal for a total energy delivered of at least twice the minimum measured quantity (MMQ). If the MDA would result in maximum deliverable power of greater than 7.2 kW, then the test may be performed at 7.2 kW.

(2) Accuracy test of the EVSE system at a load of not greater than 10 % of the maximum deliverable amperes (expressed as MDA) as determined from the pilot signal for a total energy delivered of at least the minimum measured quantity (MMQ).

(b) For DC systems (see note):

(1) Accuracy test of the EVSE system at a load of not less than 85 % of the maximum deliverable amperes current (expressed as MDA) as determined from the digital communication message from the DC EVSE to the test standard for a total energy delivered of at least twice the minimum measured quantity (MMQ).

(2) Accuracy test of the EVSE system at a load of not more than 10 % of the maximum deliverable amperes (expressed as MDA) as determined from the digital communication message from the DC EVSE to the test standard for a total energy delivered of at least the minimum measured quantity (MMQ).

Note: For DC systems it is anticipated that an electric vehicle may be used as the test load. Under that circumstance, testing at the load presented by the vehicle shall be sufficient.

N.6.   Repeatability Tests. – Tests for repeatability shall include a minimum of three consecutive tests at the same load, similar time period, etc., and be conducted under conditions where variations in factors are reduced to minimize the effect on the results obtained.

T. Tolerances

T.1.   Tolerances, General.

(a) The tolerances apply equally to errors of underregistration and errors of overregistration.

(b) The tolerances apply to all deliveries measured at any load within the rated measuring range of the EVSE.

(c) Where instrument transformers or other components are used, the provisions of this section shall apply to all system components.

T.2.   Load Test Tolerances.

T.2.1.   EVSE Load Test Tolerances. – The tolerances for EVSE load tests are:

(a) Acceotance Tolerance: 1.0 %; and

(b) Maintenance Tolerance: 2.0 %.

T.3.   Repeatability. – When multiple load tests are conducted at the same load condition, the range of the load test results shall not exceed 25 % of the absolute value of the maintenance tolerance and the results of each test shall be within the applicable tolerance.

T.4.   Tolerance Application in Type Evaluation Examinations for EVSEs. – For type evaluation examinations, the acceptance tolerance values shall apply under the following conditions:

(a) at any temperature, voltage, load, and power factor within the operating range of the EVSE, and

(b) regardless of the influence factors in effect at the time of the conduct of the examination, and

(c) for all quantities greater than the minimum measured quantity.

T.5.   No Load Test. – An EVSE measuring system shall not register when no load is applied.

T.6.   Starting Load. – An EVSE measuring system shall register a starting load test at a 0.5 ampere (A) load.

UR. User Requirements

UR.1. Selection Requirements.

UR.1.1. Computing-Type Device; Retail EVSE. – An EVSE used to charge electric vehicles shall be of the computing type and shall indicate the electrical energy, the unit price, and the total price of each delivery.

UR.1.2. Connection Cord-Length. – An adequate means for cord management shall be in use when the cord exceeds 25 ft in length.

UR.2. Installation Requirements.

UR.2.1. Maximum Deliverable Current. – The marked maximum deliverable current shall not exceed the total capacity in amperes of the EVSE or the thermal overload protectors of the installation site.

UR.2.2. Manufacturer's Instructions. – An EVSE shall be installed in accordance with the manufacturer's instructions, and the installation shall be sufficiently secure and rigid to maintain this condition.

UR.2.3. Load Range. – An EVSE shall be installed so that the current and voltage will not exceed the rated maximum values over which the EVSE is designed to operate continuously within the specified accuracy. Means to limit current and/or voltage shall be incorporated in the installation if necessary.

UR.2.4. Regulation Conflicts and Permit Compliance. – If any provision of Section UR.2. Installation Requirements is less stringent than that required of a similar installation by the serving utility, the installation shall be in accordance with those requirements of the serving utility.

The installer of any EVSE shall obtain all necessary permits.

UR.2.5. Responsibility, Unattended EVSE. – An unattended EVSE shall have clearly and conspicuously displayed thereon, or immediately adjacent thereto, adequate information detailing the name, address, and phone number of the local responsible party for the device.

UR.3. Use of EVSE.

UR.3.1. Unit Price for Retail EVSE Devices. – The unit price at which the EVSE is set to compute shall be conspicuously displayed or posted on the face of the retail EVSE used in direct sale.

UR.3.2. Return of Indicating and Recording Elements to Zero. – The primary indicating elements (visual) and the primary recording elements shall be returned to zero immediately before each transaction.

UR.3.3. EVSE Recorded Representations. – A receipt, either printed or electronic, providing the following information shall be available at the completion of all transactions:

(a) the total quantity of the energy delivered with unit of measure;

(b) the total computed price of the energy sale;

(c) the unit price of the energy; and for systems capable of applying multiple unit prices for energy during a single transaction, the following additional information is required:

   (1) the start and stop time of each phase during which one of the multiple unit prices was applied;

   (2) the unit price applied during each phase;

   (3) the total quantity of energy delivered during each phase;

(4) the total purchase price for the quantity of energy delivered during each phase;

(d) the maximum rate of energy transfer (i.e., maximum power) and type of current (e.g., 7 kW AC, 25 kW DC, etc.);

(e) any additional separate charges included in the transaction (e.g., charges for parking time) including:

(1) the time and date when the service begins and the time and date when the service ends; or the total time interval purchased, and the time and date that the service either begins or ends;

(2) the unit price applied for the time-based service;

(3) the total purchase price for the quantity of time measured during the complete transaction;

(f) the final total price of the complete transaction including all items;

(g) the unique EVSE identification number;

(h) the business name; and

(i) the business location.

UR.3.4.   EVSE in Operation. – The EVSE shall be permanently, plainly, and visibly identified so that it is clear which EVSE and connector is in operation.

UR.3.5.   Steps After Charging. – After delivery to a customer from a retail EVSE:

(a) the EVSE shall be shut-off at the end of a charge, through an automatic interlock that prevents subsequent charging until the indicating elements and recording elements, if the EVSE is equipped and activated to record, have been returned to their zero positions; and

(b) the vehicle connector shall not be returned to its starting position unless the zero set-back interlock is engaged or becomes engaged by the act of disconnecting from the vehicle or the act of returning the connector to the starting position.

Appendix D. Definitions

The following includes new definitions to address Electric Vehicle Fueling Systems. Also included are those definitions currently found in Appendix D that are intended to apply to these systems. The specific code(s) to which each definition applies is shown in [brackets] at the end of the definition. Definitions for the General Code [1.10] apply to all codes in Handbook 44.

A

alternating current (AC). – An electric current that reverses direction in a circuit at regular intervals. [3.40]

ampere. – The practical unit of electric current. It is the quantity of current caused to flow by a potential difference of one volt through a resistance of one ohm. One ampere (A) is equal to the flow of one coulomb of charge per second. One coulomb (C) is the unit of electric charge equal in magnitude to the charge of 6.24×10^{18} electrons. [3.40]

audit trail. – An electronic count and/or information record of the changes to the values of the calibration or configuration parameters of a device. [1.10, 2.20, 2.21, 2.24, 3.30, 3.37, 3.39, 3.40, 5.56(a)]
(Added 1993)

C

calibration parameter. – Any adjustable parameter that can affect measurement or performance accuracy and, due to its nature, needs to be updated on an ongoing basis to maintain device accuracy (e.g., span adjustments, linearization factors, and coarse zero adjustments). [2.20, 2.21, 2.24, 3.30, 3.37, 3.39, 3.40, 5.56(a)]
(Added 1993)

configuration parameter. – Any adjustable or selectable parameter for a device feature that can affect the accuracy of a transaction or can significantly increase the potential for fraudulent use of the device and, due to its nature, needs to be updated only during device installation or upon replacement of a component (e.g., division value [increment], sensor range, and units of measurement). [2.20, 2.21, 2.24, 3.30, 3.37, 3.40, 5.56(a)]
(Added 1993)

creep. – A continuous apparent measurement of energy indicated by a system with operating voltage applied and no power consumed (load terminals open circuited). [3.40]

current. – The rate of the flow of electrical charge past any one point in a circuit. The unit of measurement is amperes (A) or coulombs (C) per second. [3.40]

D

direct current (DC). – An electric current that flows in one direction. [3.40]

E

electric vehicle, plug-in. – A vehicle that employs electrical energy as a primary or secondary mode of propulsion. Plug-in electric vehicles may be all-electric vehicles (EV's) or plug-in hybrid electric vehicles (PHEV's). All-electric vehicles are powered by an electric motor and battery at all times. All-electric vehicles may also be called battery-electric vehicles (BEV's). Plug-in hybrid electric vehicles employ both an electric motor and an internal combustion engine that consumes either conventional or alternative fuel or a fuel cell. In a parallel type hybrid-electric vehicle, either the electric motor or the engine may propel the vehicle. In a series type hybrid-electric vehicle, the engine or fuel cell generates electricity that is then used by the electric motor to propel the vehicle. EV's, BEV's, and PHEV's are capable of receiving and storing electricity via connection to an external electrical supply. Not all hybrid-electric vehicles are of the plug-in type. Hybrid-electric vehicles that do not have the capability to receive electrical energy from an external supply (HEV's) generate electrical energy onboard with the internal combustion engine, regenerative braking, or both. [3.40]

electric vehicle supply equipment (EVSE). – A device or system designed and used specifically to transfer electrical energy to an electric vehicle, either as charge transferred via physical or wireless connection, by loading a fully charged battery, or by other means. [3.40]

electricity as vehicle fuel. – Electrical energy transferred to and/or stored onboard an electric vehicle primarily for the purpose of propulsion. [3.40]

energy. – The integral of active power with respect to time. [3.40]

energy flow. – The flow of energy between line and load terminals (conductors) of an electricity system. Flow from the line to the load terminals is considered energy delivered. Energy flowing in the opposite direction (i.e., from the load to line terminals) is considered as energy received. [3.40]

equipment, commercial. – Weights, measures, and weighing and measuring devices, instruments, elements, and systems or portion thereof, used or employed in establishing the measurement or in computing any basic charge or payment for services rendered on the basis of weight or measure. As used in this definition, measurement includes the determination of size, quantity, value, extent, area, composition (limited to meat and poultry), constituent value (for grain), or measurement of quantities, things, produce, or articles for distribution or consumption, purchased, offered, or submitted for sale, hire, or award. [1.10, 2.20, 2.21, 2.22, 2.24, 3.30, 3.31, 3.32, 3.33, 3.34, 3.35, 3.38, 3.40, 4.40, 5.51, 5.56.(a), 5.56.(b), 5.57, 5.58, 5.59]
(Added 2008)

event counter. – A nonresettable counter that increments once each time the mode that permits changes to sealable parameters is entered and one or more changes are made to sealable calibration or configuration parameters of a device. [2.20, 2.21, 3.30, 3.37, 3.39, 3.40, 5.54, 5.56(a), 5.56(b), 5.57]
(Added 1993)

event logger. – A form of audit trail containing a series of records where each record contains the number from the event counter corresponding to the change to a sealable parameter, the identification of the parameter that was changed, the time and date when the parameter was changed, and the new value of the parameter. [2.20, 2.21, 3.30, 3.37, 3.39, 3.40, 5.54, 5.56(a), 5.56(b), 5.57]
(Added 1993)

EVSE field reference standard. – A portable apparatus that is traceable to NIST and is used as a standard to test EVSEs in commercial applications. This instrument is also known as a portable standard or working standard. [3.40]

F

face. – That portion of a computing-type pump or dispenser which displays the actual computation of price per unit, delivered quantity, and total sale price. In the case of some electronic displays, this may not be an integral part of the pump or dispenser. [3.30, 3.40]
(Added 1987)

H

hertz (Hz). – Frequency or cycles per second. One cycle of an alternating current or voltage is one complete set of positive and negative values of the current or voltage. [3.40]

J

megajoule (MJ). – An SI unit of energy equal to 1 000 000 joules (J). [3.40]

K

kilowatt (kW). – A unit of power equal to 1000 watts (W). [3.40]

kilowatt-hour (kWh). – A unit of energy equal to 1000 watthours (W h). [3.40]

L

load, full. – A test condition with rated voltage, current at 100 % of test amps level, and power factor of 1.0. [3.40]

load, light. – A test condition with rated voltage, current at 10 % of test amps level, and power factor of 1.0. [3.40]

M

master meter, electric. – An electric watthour meter owned, maintained, and used for commercial billing purposes by the serving utility. All the electric energy served to a submetered service system is recorded by the master meter. [3.40]

meter, electricity. – An electric watthour meter. [3.40]

metrological components. – Elements or features of a measurement device or system that perform the measurement process or that may affect the final quantity determination or resulting price determinations. This includes accessories that can affect the validity of transactions based upon the measurement process. The measurement process includes determination of quantities; the transmission, processing, storage, or other corrections or adjustments of measurement data or values; and the indication or recording of measurement values or other derived values such as price or worth or charges. [3.40]

N

nationally recognized testing laboratory (NRTL). – A laboratory that conducts testing and certification that is recognized by the Occupational Safety and Health Administration (OSHA). [3.40]

nonresettable totalizer. – An element interfaced with the measuring or weighing element that indicates the cumulative registration of the measured quantity with no means to return to zero. [3.30, 3.37, 3.39, 3.40]

O

ohm (Ω). – The practical unit of electric resistance that allows one ampere of current to flow when the impressed potential is one volt. [3.40]

P

percent registration. – Percent registration is calculated as follows:

$$Percent\ \text{Registration} = \frac{Wh\ measured\ by\ EVSE}{\textbf{Wh measured by STANDARD}}\ x\ 100$$

[3.40]

power factor. – The ratio of the active power to the apparent power in an AC circuit. The power factor is a number between 0 and 1 that is equal to 1 when the voltage and current are in phase (load is entirely resistive). [3.40]

primary indicating or recording elements. – The term "primary" is applied to those principal indicating (visual) elements and recording elements that are designed to, or may, be used by the operator in the normal commercial use of a device. The term "primary" is applied to any element or elements that may be the determining factor in arriving at the sale representation when the device is used commercially. (Examples of primary elements are the visual indicators for meters or scales not equipped with ticket printers or other recording elements and both the visual indicators and the ticket printers or other recording elements for meters or scales so equipped.) The term "primary" is not applied to such auxiliary elements as, for example, the totalizing register, or predetermined-stop mechanism on a meter or the means for producing a running record of successive weighing operations, these elements being supplementary to those that are the determining factors in sales representations of individual deliveries or weights. (See "indicating element" and "recording element."). [1.10, 3.40]

R

recorded representation. – The printed, electronically recorded, or other representation that retains a copy of the quantity and any other required information generated by a weighing or measuring device. [1.10, 3.40]

recording element. – An element incorporated, connected to, or associated with a weighing or measuring device by means of which its performance relative to quantity or money value is permanently recorded in a printed or electronic form. [1.10, 3.40]

remote configuration capability. – The ability to adjust a weighing or measuring device or change its sealable parameters from or through some other device that is not itself necessary to the operation of the weighing or measuring device or is not a permanent part of that device. [2.20, 2.21, 2.24, 3.30, 3.37, 3.39, 3.40, 5.56(a)]

(Added 1993)

retail device. – A measuring device primarily used to measure electrical energy for the purpose of sale to the end user. [3.40]

S

serving utility. – The utility distribution company that owns the master meter and sells electric energy to the owner of a submeter system. [3.40]

starting load. – The minimum load above which the device will indicate energy flow continuously. [3.40]

submeter. – A system furnished, owned, installed, and maintained by the customer who is served through a utility owned master meter. [3.40]

T

test accuracy – in-service. – The device accuracy determined by a test made during the period that the system is in service. It may be made on the customer's premises without removing the system from its mounting or by removing the EVSE for testing either on the premises or in a laboratory or shop. [3.40]

test amperes (TA). – The full load current (amperage) specified by the EVSE manufacturer for testing and calibration adjustment. (Example: TA 30). [3.40]

thermal overload protector. – A circuit breaker or fuse that automatically limits the maximum current in a circuit. [3.40]

U

unit price. – The price at which the electrical energy is being sold and expressed in whole units of measurement. [1.10, 3.30, 3.40]

(Added 1992)

V

vehicle connector. – A device that by insertion into a vehicle inlet, establishes an electrical connection to the electric vehicle for the purpose of providing power and information exchange, with means for attachment of an electric vehicle cable. This device is a part of the vehicle coupler. [3.40]

vehicle coupler. – A means enabling the connection, at will, of an electric vehicle cable to the equipment. It consists of a vehicle connector and a vehicle inlet. [3.40]

vehicle inlet. – The part incorporated in, or fixed to the vehicle, which receives power from a vehicle connector. [3.40]

volt. – The practical unit of electromotive force. One volt will cause one ampere to flow when impressed across a resistance of one ohm. [3.40]

W

watt. – The practical unit of electric power. In an alternating-current (AC) circuit, the power in watts is volts times amperes multiplied by the circuit power factor. [3.40]

watthour (Wh). – The practical unit of electric energy that is expended in one hour when the average power consumed during the hour is one watt. [3.40]

THIS PAGE INTENTIONALLY LEFT BLANK

Section 4

Table of Contents

Page

THIS PAGE INTENTIONALLY LEFT BLANK

Table of Contents

THIS PAGE INTENTIONALLY LEFT BLANK

Section 4.40. Vehicle Tanks Used as Measures

A. Application

A.1.   General. – This code applies to vehicle tanks when these are used as commercial measures.

A.2.   Exceptions. – This code does not apply to the following devices:

(a) Devices used solely for dispensing a product in connection with operations in which the amount dispensed does not affect customer charges.

(b) Meters mounted on vehicle tanks (for which see Section 3.31. Code for Vehicle-Tank Meters).

A.3.   Additional Code Requirements. – In addition to the requirements of this code, Vehicle Tanks Used as Measures shall meet the requirements of Section 1.10. General Code.

S. Specifications

S.1.   Design of Compartments.

S.1.1.   Compartment Distortion. – The shell and bulkheads of a vehicle tank shall be so constructed that under any condition of liquid lading they will not become distorted sufficiently to cause a change in the capacity of any compartment (as determined by volumetric test) equal to more than 0.25 L per 750 L (0.5 pt per 200 gal), or fraction thereof, of the nominal compartment capacity, or to more than 0.5 L (1 pt), whichever is greater. (This specification prescribes a limit on permissible distortion only, and is not to be construed as setting up a secondary tolerance on compartment capacities to be added to the values given in tolerance paragraph T.2. Tolerance Values.)

S.1.2.   Venting. – During filling operations, effective venting of a compartment shall be provided to permit air to escape from all spaces designed to be filled with liquid and to permit the influx of air to the compartment during the discharge of liquid therefrom. Venting shall prevent any formation of air pockets.

S.1.3.   Completeness of Delivery. – A tank shall be so constructed that, when it is standing on a level surface, complete delivery can be made from any compartment through its delivery faucet or valve whether other compartments are full or empty, and whether or not the delivery is through a manifold.

S.1.4.   Fill or Inspection Opening. – The fill or inspection opening of a compartment shall be of such size and location that it can readily be determined by visual inspection that the compartment has been properly filled or completely emptied and shall be so positioned with respect to the ends of the compartment that the indicator may be positioned as required. In no case shall the opening, if circular, have a diameter of less than 20 cm (7⅝ in), or, if other than circular, have an effective area of less than 290 cm^2 (45 in^2).

S.1.5.   Dome Flange and Baffle Plates. – Any dome flange extending into a compartment shall be provided with sufficient perforations or openings flush with the compartment shell to prevent any trapping of air. All baffle plates in a compartment shall be so cut away at top and bottom, and elsewhere as necessary, as to facilitate loading and unloading.

S.1.6.   Compartment and Piping Capacities and Emergency Valve. – If a compartment is equipped with an emergency (or safety) valve, this shall be positioned at the lowest point of outlet from the compartment, and the compartment capacity or capacities shall be construed as excluding the capacity of the piping leading therefrom. However, the capacity of the piping leading from such a compartment shall be separately determined and reported, and may be separately marked as specified in S.4. Marking of Compartments.

S.1.6.1.   On Vehicle Tanks Equipped for Bottom Loading. – On equipment designed for bottom loading, the compartment capacity shall include the piping of a compartment to the valve located on the upstream side of the manifold and immediately adjacent thereto or, if not manifolded, to the outlet valve, provided that on or immediately adjacent to the marking as specified in S.4. Marking of Compartments the following words or a statement of similar meaning shall be affixed: "Warning: Emergency valves must be opened before checking measurement."

S.1.7.   Expansion Space. – When a compartment is filled to the level of the highest indicator in the compartment, there shall remain an expansion space of at least 0.75 % of the nominal compartment capacity as defined by that indicator.

S.2.   Design of Compartment Indicators.

S.2.1.   General. – An indicator shall be so designed that it will distinctly and unmistakably define a capacity point of its compartment when liquid is in contact with the lowest portion of the indicator.

S.2.2.   Number of Indicators. – In no case shall a compartment be provided with more than five indicators.
(Amended 1972)

S.2.3.   Identification of Multiple Indicators. – If a compartment is provided with multiple indicators, each such indicator shall be conspicuously marked with an identifying letter or number.

S.2.4.   Location. – An indicator shall be located:

(a) midway between the sides of its compartment;

(b) as nearly as practicable midway between the ends of its compartment, and in no case offset by more than 10 % of the compartment space or 15 cm (6 in), whichever is less;
(Amended 1972)

(c) so that it does not extend into, nor more than 15 cm (6 in) from that section of the compartment defined by a vertical projection of the fill opening;
(Amended 1974)

(d) at a depth, measuring from the top of the dome opening, not lower than 46 cm (18 in) for fill openings of less than 38 cm (15 in) in diameter, or, if other than circular, an effective area of less than 1130 cm^2 (175 in^2), and not lower than 61 cm (24 in) for larger fill openings; and

(e) to provide a clearance of not less than 5 cm (2 in) between indicators.

S.2.5.   Permanence. – Any indicator that is not intended to remain adjustable and all brackets or supports shall be securely welded in position.

S.2.6.   Adjustable Indicators. – Adequate provision shall be made for conveniently affixing a security seal or seals:

(a) to any indicator intended to remain adjustable, so that no adjustment of the indicator can be made without mutilating or destroying the seal; and

(b) to any removable part to which an indicator may be attached, so that the part cannot be removed without mutilating or destroying the seal.

S.2.7.   Sensitiveness. – The position of any indicator in its compartment shall be such that at the level of the indicator a change of 1.0 mm (0.04 in) in the height of the liquid surface will represent a volume change of not more than the value of the tolerance for the nominal compartment capacity as defined by that indicator.

S.3.   Design of Compartment Discharge Manifold. – When two or more compartments discharge through a common manifold or other single outlet, effective means shall be provided to ensure that:

(a)   liquid can flow through the delivery line leading from only one compartment at one time and that flow of liquid from one compartment to any other is automatically prevented; or

(b)   all compartments will discharge simultaneously.

If the discharge valves from two or more compartments are automatically controlled so that they can only be operated together, thus effectively connecting these compartments to one another, such compartments shall, for purposes of this paragraph, be construed to be one compartment.

S.4.   Marking of Compartments.

S.4.1.   Compartment Identification. – Each compartment of a multiple-compartment tank shall be conspicuously identified by a letter or number marked on the dome or immediately below the fill opening. Such letters or numbers shall be in regular sequence from front to rear, and the delivery faucets or valves shall be marked to correspond with their respective compartments.

S.4.2.   Compartment Capacity, Single Indicator. – A compartment provided with a single indicator shall be clearly, permanently, and conspicuously marked with a statement of its capacity as defined by its indicator.

S.4.3.   Compartment Capacity, Multiple Indicators. – A compartment provided with two or more indicators shall be clearly, permanently, and conspicuously marked with a statement identifying:

(a)   each indicator by a letter or number; and

(b)   immediately adjacent to each letter or number, the capacity of the compartment as defined by the particular indicator.

N. Notes

N.1.   Test Liquid. – Water or light fuel oil shall be used as the test liquid for a vehicle-tank compartment.

N.2.   Evaporation and Volume Change. – Care shall be exercised to reduce to a minimum, evaporation losses and volume changes resulting from changes in temperature of the test liquid.

N.3.   To Deliver. – A vehicle-tank compartment shall be gauged "to deliver." If the compartment is gauged by measuring the test liquid into the tank, the inside tank walls shall first be thoroughly wetted.

N.4.   Gauging of Compartments. – When a compartment is gauged to determine the proper position for an indicator or to determine what a capacity marking should be, whether on a new vehicle tank or following repairs or modifications that might affect compartment capacities, tolerances are not applicable, and the indicator shall be set and the compartment capacity shall be marked as accurately as practicable.

N.5.   Adjustment and Remarking. – When a compartment is found upon test to have an error in excess of the applicable tolerance, the capacity of the compartment shall be adjusted to agree with its marked capacity, or its marked capacity shall be changed to agree with its capacity as determined by the test.

T. Tolerances

T.1. Application.

T.1.1.   To Excess and to Deficiency. – The tolerances hereinafter prescribed shall be applied to errors in excess and in deficiency.

T.2.   Tolerance Values. – Maintenance and acceptance tolerances shall be as shown in Table 1. Maintenance and Acceptance Tolerances on Vehicle-Tank Compartments.

Table 1. Maintenance and Acceptance Tolerances on Vehicle-Tank Compartments		
Nominal Capacity of Compartment	**Maintenance and Acceptance Tolerances**	
Gallons	**Expressed in Quarts**	**Expressed in Gallons**
200 or less	2	0.5
201 to 400, inclusive	3	0.75
401 to 600, inclusive	4	1.0
601 to 800, inclusive	5	1.25
801 to 1000, inclusive	6	1.50
over 1000	Add 1 quart per 200 gallons or fraction thereof	Add 0.25 gallon per 200 gallons or fraction thereof

UR.   User Requirements

UR.1.   Conditions of Use.

UR.1.1.   Filling. – A vehicle shall stand upon a level surface during the filling of a compartment.

UR.1.2.   Delivering. – During a delivery, a vehicle shall be so positioned as to assure complete emptying of a compartment. Each compartment shall be used for an individual delivery only; that is, an individual delivery shall consist of the entire contents of a compartment or compartments.

(Amended 1976)

Table of Contents

THIS PAGE INTENTIONALLY LEFT BLANK

Section 4.41.   Liquid Measures

A. Application

A.1.   General. – This code applies to liquid measures; that is, to rigid measures of capacity designed for general and repeated use in the measurement of liquids.

A.2.   Exceptions. – The code does not apply to test measures or other volumetric standards.

A.3.   Additional Code Requirements. – In addition to the requirements of this code, Liquid Measures shall meet the requirements of Section 1.10. General Code.

S. Specifications

S.1.   Units.

(a) The capacity of a liquid measure shall be 0.1 L, 0.2 L, 0.5 L, 1 L, 2 L, 5 L, or a multiple of 5 L, and the measure shall not be subdivided.

(b) The capacity of a liquid measure shall be 1 gill, ½ liq pt, 1 liq pt, 1 liq qt, ½ gal, 1¼ gal, 1½ gal, or a multiple of 1 gal, and the measure shall not be subdivided. However, 3 pt and 5 pt brick molds and 2½ gal (10 qt) cans shall be permitted when used exclusively for ice cream.

S.2.   Material. – Measures shall be made of metal, glass, earthenware, enameled ware, composition, or similar and suitable material. If made of metal, the thickness of the metal shall not be less than the appropriate value given in Table 1. Minimum Thickness of Metal for Liquid Measures.

Table 1. Minimum Thickness of Metal for Liquid Measures		
Nominal Capacity	**Minimum Thickness**	
	For Iron or Steel, Plated, or Unplated (inch)	**For Copper or Aluminum (inch)**
1 pint or less	0.010	0.020
1 quart, ½ gallon, 1 gallon	0.014	0.028
Over 1 gallon	0.016	0.032

S.3.   Capacity Point. – The capacity of a measure shall be determined to a definite edge, or to the lowest portion of a plate, bar, or wire, at or near the top of the measure, and shall not include the capacity of any lip or rim that may be provided.

S.4.   Reinforcing Rings. – Reinforcing rings, if used, shall be attached to the outside of the measure and shall show no divisions or lines on the inside surface of the measure.

S.5.   Discharge. – A measure equipped with a discharge faucet or valve shall be susceptible to complete discharge through the faucet or valve when the measure is standing on a level surface.

S.6.   Marking Requirements. – A measure shall be marked on its side with a statement of its capacity. If the capacity is stated in terms of the pint or quart, the word "Liquid" or the abbreviation "Liq" shall be included.

T. Tolerances

T.1. – Maintenance tolerances in excess and in deficiency shall be as shown in Table 2. Maintenance Tolerances, in Excess and in Deficiency, for Liquid Measures. Acceptance tolerances shall be one-half the maintenance tolerances.

Table 2. Maintenance Tolerances, in Excess and in Deficiency, for Liquid Measures				
Nominal Capacity	Tolerance			
	In Excess		In Deficiency	
	fluid drams	cubic inches	fluid drams	cubic inches
½ pt or less	2.0	0.4	1.0	0.2
1 pt	3.0	0.7	1.5	0.3
1 qt	4.0	0.9	2.0	0.5
½ gal	6.0	1.4	3.0	0.7
	fluid ounces	cubic inches	fluid drams	cubic inches
1 and 1¼ gal	1.0	1.8	4.0	0.9
1½ gal	1.5	2.7	6.0	1.4
	fluid ounces	cubic inches	fluid ounces	cubic inches
2 gal	2.0	3.5	1.0	1.8
3 and 4 gal	4.0	7.0	2.0	3.6
5 gal	6.0	11.0	3.0	5.4
10 gal	10.0	18.0	5.0	9.0

Table of Contents

Section 4.42. Farm Milk Tanks

A. Application

A.1. General. – This code applies to farm milk tanks on the premises of producers when these are used, or are to be used, for the commercial measurement of milk.

A.2. Exceptions. – This code does not apply to tanks mounted on highway vehicles.

A.3. Additional Code Requirements. – In addition to the requirements of this code, Farm Milk Tanks shall meet the requirements of Section 1.10. General Code.

S. Specifications

S.1. Components. – A farm milk tank, whether stationary or portable, shall be considered suitable for commercial use only when it comprises:

(a) a vessel, whether or not it is equipped with means for cooling its contents;

(b) a means for reading the level of liquid in the tank, such as a removable gauge rod or surface gauge; and

(c) a chart for converting level-of-liquid readings to volume.

Each compartment of a subdivided tank shall, for the purposes of this code, be construed to be a farm milk tank.

S.2. Design of Tank.

S.2.1. Level. – A farm milk tank shall be designed to be in normal operating position when it is in level. The tank shall be so constructed that it will maintain its condition of level under all normal conditions of lading.

S.2.2. Level-Indicating Means. – A tank shall be permanently equipped with sensitive means by which the level of the tank can be determined.

S.2.2.1. On a Stationary Tank. – A stationary tank shall be provided with such level-indicating means as a two-way or circular level, a plumb bob, two-way leveling lugs, or the like; or the top edge or edges of the tank shall be so constructed throughout as to provide an accurate reference for level determinations; provided, that when leveling lugs or the top edge or edges of the tank are used as the reference for level determinations, there shall be supplied with the tank a sensitive spirit level of appropriate dimensions, and the positions where such level is intended to be used shall be permanently marked on the reference surface of the tank; and provided further, that when leveling lugs are used they shall be so designed, constructed, and installed at the factory that any alteration of the original position or condition, such as by hammering or filing, would be difficult and would become obvious. A stationary tank with a nominal capacity of 2 000 L or 500 gal, or greater shall be provided with at least two similar level-indicating means, and these shall be located in opposite and distant positions from each other to facilitate an accurate level determination in both directions of the tank's horizontal plane.
(Amended 1980)

S.2.2.2. On a Portable Tank. – A portable tank shall be provided with either a two-way or a circular level.

S.2.3. Portable Tank. – A portable tank shall be of the center-reading type; that is, it shall be so designed that the gauge rod or surface gauge, when properly positioned for use, will be approximately in the vertical axis of the tank, centrally positioned with respect to the tank walls.

*S.2.4.   Capacity. – A farm milk tank shall be clearly and permanently marked on a surface visible after installation with its capacity as determined by the manufacturer. The capacity shall not exceed an amount that can be agitated without overflowing and that can be measured accurately with the liquid at rest.
[Nonretroactive as of January 1, 1979]*

S.3.   Design of Indicating Means.

*S.3.1.   General. – A tank shall include indicating means and shall be calibrated over the entire range of the volume of the tank from 5 % of capacity or 2 m³ (500 gal) whichever is less, to its maximum capacity.
[Nonretroactive as of January 1, 1986]*

(Added 1985)

S.3.2.   Gauge-Rod Bracket or Supports. – If a tank is designed for use with a gauge rod, a substantial and rigid gauge-rod bracket or other suitable supporting elements for positioning the gauge rod shall be provided. A gauge rod and its brackets or other supporting elements shall be so constructed that, whenever the rod is placed in engagement with the bracket or supports and released, the rod will automatically seat itself at a fixed height and in a vertical position. When a gauge rod is properly seated on its brackets or supports, there shall be a clearance of at least 7.5 cm (3 in) between the graduated face of the rod and any tank wall or other surface that it faces.

S.3.3.   Gauge Rod. – When properly seated in position, a rod shall not touch the bottom of the tank unless this is required by the design of the supporting elements. The rod shall be graduated throughout an interval corresponding to the volume range within which readings of liquid level are to be made.

S.3.4.   Surface-Gauge Bracket or Supports. – If a tank is designed for use with a surface gauge, a substantial and rigid surface-gauge bracket or other suitable supporting elements for positioning the surface gauge shall be provided. A surface gauge and its brackets or other supporting elements shall be so constructed that, whenever the gauge assembly is placed in engagement with the bracket or supports, the indicator, if not permanently mounted on the tank, will automatically seat itself in correct operating position, and the graduated element will be vertically positioned and will be securely held at any height to which it may be manually set.

S.3.5.   Surface Gauge. – When properly engaged with its bracket and set to its lowest position, a surface gauge shall not touch the bottom of the tank. The gauge shall be graduated throughout an interval corresponding to the volume range within which readings of liquid level are to be made.

S.3.6.   External Gauge Assemblies.

S.3.6.1.   Design and Installation. – The gauge assembly shall be designed to meet sanitary requirements and shall be readily accessible for cleaning purposes. The gauge assembly shall be mounted in a vertical position and equipped with a sliding mechanism to assist in determining the liquid level.

S.3.6.2.   Gauge Tube. – The gauge tube shall be borosilicate glass or approved rigid plastic or rigidly supported flexible tubing with a uniform internal diameter not less than 2 cm (¾ in). It shall be designed and constructed so that all product in the gauge can be discarded in such a manner that no product in the gauge tube will enter the discharge line or tank.

(Amended 1983)

S.3.6.3.   Scale Plate. – The scale plate shall be mounted adjacent to and parallel with the gauge tube and be no more than 7 mm (¼ in) from the tube.

S.3.6.4.   Scale Graduations. – The graduation lines shall be clear and easily readable and shall comply with the requirements of paragraphs included under S.3.7. Graduations.

S.3.6.5.   Venting. – An external gauge tube shall be adequately vented at the top, open to the atmosphere. Any attachment to the gauge tube shall not adversely affect the operation of this vent.

(Added 1984)

(Added 1977)

S.3.7.   Graduations.

S.3.7.1.   Spacing and Width of Graduations. – On a gauge rod or surface gauge, the spacing of the graduations, center to center, shall be not more than 1.6 mm (0.0625 in or $^1/16$ in) and shall not be less than 0.8 mm (0.03125 in or $^1/32$ in). The graduations shall not be less than 0.12 mm (0.005 in) in width, and the clear interval between adjacent edges of successive graduations shall be not less than 0.4 mm (0.015625 in or $^1/64$ in).

S.3.7.2.   Values of Graduations. – On a gauge rod or surface gauge, the graduations may be designated in inches or in centimeters and fractions thereof, or may be identified in a numerical series without reference to inches or centimeters or fractions thereof. In either case, a volume chart shall be provided for each such rod or gauge and each tank with which it is associated, showing values in terms of the graduation on the rod or gauge. If a rod or gauge is associated with but one tank, in lieu of linear or numerical series graduations and volume chart, values in terms of volume of liquid in the tank may be shown directly on the rod or gauge.

S.3.7.3.   Value of Graduated Interval. – The value of a graduated interval on a gauge rod or surface gauge (exclusive on the interval from the bottom of the tank to the lowest graduation) shall not exceed:

(a)   2 L for a tank of a nominal capacity of 1000 L or less; ½ gal for a tank of a nominal capacity of 250 gal or less;

(b)   4 L for a tank of a nominal capacity of 1001 L to 2000 L, inclusive; 1 gal for a tank of a nominal capacity of 251 gal to 500 gal, inclusive;

(c)   6 L for a tank of a nominal capacity of 2001 L to 6000 L, inclusive; 1½ gal for a tank of a nominal capacity of 501 gal to 1500 gal, inclusive;

(d)   8 L for a tank of a nominal capacity of 6001 L to 10 000 L, inclusive; 2 gal for a tank of a nominal capacity of 1501 gal to 2500 gal, inclusive; or

(e)   8 L plus 4 L for each additional 10 000 L or fraction thereof, for tanks of nominal capacity above 10 000 L or 2 gal plus 1 gal for each additional 2500 gal or fraction thereof, for tanks with nominal capacity above 2500 gal.

(Amended 1980)

S.3.8.   Design of Indicating Means on Tanks with a Capacity Greater than 8000 Liters or 2000 Gallons. *– Any farm milk tank with a capacity greater than 8000 L, or 2000 gal, shall be equipped with an external gauge assembly.*
[Nonretroactive and applicable only to tanks manufactured after January 1, 1981]

(Added 1980)

S.4.   Design of Volume Chart.

S.4.1.   General. – A volume chart shall show volume values only, *over the entire range of the volume of the tank from 5 % of capacity or 2 m³ (500 gal) whichever is less, to its maximum capacity.* All letters and figures on the chart shall be distinct and easily readable. The chart shall be substantially constructed, and the face of the chart shall be so protected that its lettering and figures will not tend easily to become obliterated or illegible.
*[*Nonretroactive as of January 1, 1986]*

(Amended 1985)

S.4.2.   For a Tank of 1 000 Liters, or 250 Gallons, or Less. – The volume chart for a tank of nominal capacity of 1000 L, or 250 gal, or less shall show values at least to the nearest 1 L, or ¼ gal.

S.4.3.   For a Tank of 1 001 Liters to 2000 Liters, or 251 to 500 Gallons. – The volume chart for a tank of nominal capacity of 1001 L to 2000 L, or 251 gal to 500 gal, inclusive, shall show values at least to the nearest 2 L, or ½ gal.

S.4.4.   For a Tank of Greater than 2000 Liters, or 500 Gallons. – The volume chart for a tank of nominal capacity of greater than 2000 L, or 500 gal, shall show values at least to the nearest gallon, or 4 L.
(Amended 1980)

S.5.   Gauging.

S.5.1.   Level. – A farm milk tank shall be level, as shown by the level-indicating means, during the original gauging operation.

S.5.2.   To Deliver. – A farm milk tank shall be originally gauged "to deliver." If the tank is gauged by measuring the test liquid into the tank, the inside tank walls shall first be thoroughly wetted and the tank shall then be drained for 30 seconds after the main drainage flow has ceased.

S.5.3.   Preparation of Volume Chart. – When a tank is gauged for the purposes of preparing a volume chart, tolerances are not applicable, and the chart shall be prepared as accurately as practicable.

S.6.   Identification. – A tank and any gauge rod, surface gauge, spirit level, and volume chart intended to be used therewith shall be mutually identified, as by a common serial number, in a prominent and permanent manner.

N. Notes

N.1.   Test Liquid. – Water shall be used as the test liquid for a farm milk tank.

N.2.   Evaporation and Volume Change. – Care shall be exercised to reduce to a minimum, evaporation losses and volume changes resulting from changes in temperature of the test liquid.

N.3.   To Deliver. – A farm milk tank shall be tested "to deliver." If the tank is gauged by measuring the test liquid delivered into the tank, the inside tank walls shall first be thoroughly wetted and the tank then shall be drained for 30 seconds after the main drainage flow has ceased.

N.4.   Level. – A farm milk tank shall be level, as shown by the level-indicating means, during gauging and testing.

N.5.   Test Methods. – Acceptance tests of milk tanks may be of either the prover method or the master meter method provided that the master metering system is capable of operating within 25 % of the applicable tolerance found in T.3. Basic Tolerance Values. Subsequent tests may be of either the prover method or the master meter method provided that the master metering system is capable of operating within 25 % of the applicable tolerance found in T.4. Basic Tolerance Values, Master Meter Method.

N.5.1.   Verification of Master Metering Systems. – A master metering system used to gauge a milk tank shall be verified before and after the gauging process. A master metering system used to calibrate a milk tank shall be verified before starting the calibration and re-verified at least every quarter of the tank capacity, or every 2000 L (500 gal), whichever is greater. The above process of re-verifying the master metering system may be waived if the system is verified using a NIST traceable prover with a minimum of two tests immediately before and one test immediately after the gauging process and that each test result is within 25 % of T.3. Basic Tolerance Values.
(Added 2001) (Amended 2012)

N.5.2.   Temperature Changes in Water Supply. – When using a master metering system to gauge or calibrate a milk tank, the official shall monitor the temperature of the water before and after changing sources of supply.

If the water temperature of the new source changes by more than 2.8 °C (5 °F) from the previous supply, the official shall re-verify the accuracy of the master metering system as soon as possible after the system reaches temperature equilibrium with the new supply source.

(Added 2001)

N.6.   Reading the Meniscus. – When a reading or setting is to be obtained from a meniscus formed by milk or other opaque liquid, the index or reading line is the position of the highest point of the center of the meniscus. When calibrating a device with water and the device is to be used with an opaque liquid, the reading should be obtained accordingly; that is, the position of the highest point of the center of the meniscus.

(Added 1984)

T. Tolerances

T.1.   Application. – The tolerances hereinafter prescribed shall be applied equally to errors in excess and errors in deficiency.

T.2.   Minimum Tolerance Values. – On a particular tank, the maintenance and acceptance tolerance applied shall be not smaller than the volume corresponding to the graduated interval at the point of test draft on the indicating means or 2 L (½ gal), whichever is greater.

(Amended 1980)

T.3.   Basic Tolerance Values. – The basic maintenance and acceptance tolerance shall be 0.2 % of the volume of test liquid in the tank at each test draft.

(Amended 1975)

T.4.   Basic Tolerance Values, Master Meter Method. – The basic maintenance and acceptance tolerance for tanks tested by the master meter method shall be 0.4 % of the volume of test liquid in the tank at each test draft.

(Added 1975)

UR.   User Requirements

UR.1.   Installation. – A stationary tank shall be rigidly installed in level without the use of removable blocks or shims under the legs. If such tank is not mounted permanently in position, the correct position on the floor for each leg shall be clearly and permanently defined.

UR.2.   Level Condition.

 UR.2.1.   Stationary Tank. – A stationary farm milk tank shall be maintained in level.

  UR.2.1.1.   Leveling Lugs. – If leveling lugs are provided on a stationary tank, such lugs shall not be hammered or filed to establish or change a level condition of the tank.

 UR.2.2.   Portable Tank. – On a portable tank, measurement readings shall be made only when the tank is approximately level; that is, when it is not out of level by more than 5 % or approximately three degrees in any direction.

UR.3.   Weight Chart. – An auxiliary weight chart may be provided, on which shall be prominently displayed the weight per unit volume value used to derive the weight values from the official volume chart.

UR.4.   Use. – A farm milk tank shall not be used to measure quantities greater than an amount that can be agitated without overflowing.

THIS PAGE INTENTIONALLY LEFT BLANK

Table of Contents

THIS PAGE INTENTIONALLY LEFT BLANK

Section 4.43.   Measure-Containers

A. Application

A.1.   General. – This code applies to measure-containers, including lids or closures if such are necessary to provide total enclosure of the measured commodity, as follows:

    (a) Retail measure-containers intended to be used only once to determine at the time of retail sale, and from bulk supply, the quantity of commodity on the basis of liquid measure. The retail measure-container serves as the container for the delivery of the commodity.

    (b) Prepackaged measure-containers intended to be used only once to determine in advance of sale the quantity of a commodity (such as ice cream, ice milk, or sherbet) on the basis of liquid measure. The prepackaged measure-container serves as the container for the delivery of the commodity, in either a wholesale or a retail marketing unit.

A.2.   Exceptions. – This code does not apply to rigid containers used for milk, cream, or other fluid dairy products, which are covered by packaging requirements.

A.3.   Additional Code Requirements. – In addition to the requirements of this code, Measure-Containers shall meet the requirements of Section 1.10. General Code.

S. Specifications

S.1.   Units. – The capacity of a measure-container shall be a multiple of or a binary submultiple of a quart or a liter, and the measure shall not be subdivided. However, for prepackaged measure-containers, any capacity less than ¼ L or ½ liq pt shall be permitted.

(Amended 1979)

S.2.   Capacity Point. – The capacity of a measure-container shall be sharply defined by:

    (a) the top edge;

    (b) a line near the top edge; or

    (c) the horizontal cross-sectional plane established by the bottom surface of the removable lid or cap when seated in the container.

S.3.   Shape. – A measure-container shall be designed as some suitable geometrical shape, and its capacity shall be determined without distortion from its normal assembled shape.

S.4.   Marking.

    S.4.1.   Capacity Point. – If the capacity point of a measure-container is defined by a line, the container shall be marked conspicuously on its side with a suitable statement clearly identifying this line as the capacity point.

    S.4.2.   Capacity Statement. – A measure-container shall be clearly and conspicuously marked with a statement of its capacity in terms of one of the units prescribed in Section S.1. Units.

N. Notes

N.1.   Test Liquid. – Water shall be used as the test liquid for a measure-container.

N.2.   Preparation of Container for Test.

N.2.1.   General. – Before an actual test is begun, a measure-container shall, if necessary, be so restrained that it will maintain its normal assembled shape and that its sides will not bulge when it is filled with water.

N.2.2.   Restraining Form for Test.

N.2.2.1.   For Rectangular Containers of One Liter, One Quarter Less. – Bulging of the sides of a rectangular measure-container of 1 L (1 qt) capacity or less may be controlled by holding against each side of the container, with a cord, rubber bands, or tape, a metal plate or a piece of heavy cardboard slightly smaller than the side of the container.

(Amended 1979)

N.2.2.2.   For Rectangular Prepackaged Measure-container of Two Quarts or Two Liters or Greater. – A rectangular prepackaged measure-container of 2 L (2 qt) capacity or greater shall be supported during a test by a rigid restraining form. This form shall restrain not less than the entire area of the central two-thirds of each side of the container, measured from bottom to top. The inside width dimension of any side panel of the restraining form shall be 1.6 mm ($\frac{1}{16}$ in) greater than the corresponding outside dimension of the container. (The outside width dimension of any side panel of the container shall be established by adding to the inner side center-of-score to center-of-score dimension two thicknesses of the board used, and the sum thus obtained shall be rounded off to the nearest 0.4 mm ($\frac{1}{64}$ in).)

(Amended 1979)

T. Tolerances

T.1.   Tolerances on an Individual Measure. – The acceptance tolerances in excess and in deficiency on an individual measure shall be as shown in Table 1. Acceptance Tolerances, in Excess and in Deficiency, for Measure-Containers.

T.2.   Tolerance on Average Capacity. – The average capacity on a random sample of 10 measures selected from a lot of 25 or more shall be equal to or greater than the nominal capacity.

(Amended 1979)

UR.   User Requirements

UR.1.   Limitation of Use. – The use of a measure-container with a rectangular cross section of a capacity of 2 L (2 qt) or greater shall be limited to the packaging, in advance of sale, of ice cream, sherbet, or other similar frozen desserts.

(Amended 1979)

Table 1. Acceptance Tolerances, in Excess and in Deficiency, for Measure-Containers				
Nominal Capacity	**Tolerance**			
	In Excess		**In Deficiency**	
	milliliters		**milliliters**	
¼ liter or less	10		5.0	
½ liter	15		7.5	
1 liter	20		10.0	
Over 1 liter	Add per liter 10 milliliters		Add per liter 5.0 milliliters	
	fluid drams	**cubic inches**	**fluid drams**	**cubic inches**
½ pint or less	3	0.6	1.5	0.3
1 pint	4	1.0	2.0	0.5
1 quart	6	1.4	3.0	0.7
2 quarts	9	2.0	4.5	1.0
3 quarts	10	2.4	5.0	1.2
4 quarts	12	2.8	6.0	1.2
Over 4 quarts	Add per quart 3 fluid drams	Add per quart 0.7 cubic inch	Add per quart 1.5 fluid drams	Add per quart 0.35 cubic inch

THIS PAGE INTENTIONALLY LEFT BLANK

Table of Contents

THIS PAGE INTENTIONALLY LEFT BLANK

Section 4.44. Graduates

A. Application

A.1.   General. – This code applies to subdivided glass measures of capacity, either cylindrical or conical in shape.

A.2.   Additional Code Requirements. – In addition to the requirements of this code, Graduates shall meet the requirements of Section 1.10. General Code.

S. Specifications

S.1.   Units. – Nominal capacities, graduation ranges, values of graduated intervals, and numbered graduations applicable to single-scale graduates and to the appropriate portions of double scale graduates shall be as shown in Table 1. Design Details for Graduates.

Table 1. Design Details for Graduates				
Nominal Capacity	**To be Graduated Between**		**Value of Graduated Intervals**	**Number at Each Graduation Divisible by**
milliliters	milliliters		milliliters	milliliters
5	1 and	5	½	1
10	2 and	10	1	2
25	5 and	25	5	5
50	10 and	50	5	10
100	20 and	100	10	20
500	100 and	500	25	50
1000	200 and	1000	50	100
minims	minims		minims	minims
60	15 and	60	5	10[a]
120	30 and	120	10	20[b]
fluid drams	fluid drams		fluid drams	fluid drams
4	1 and	4	½	1
8	2 and	8	1	2
fluid ounces	fluid ounces		fluid ounces	fluid ounces
2	½ and	2	¼	½
4	1 and	4	½	1
8	2 and	8	½	1
16	4 and	16	1	2
32	8 and	32	2	4
[a] And, in addition, at the first (15-minim) graduation.				
[b] And, in addition, at the first (30-minim) graduation.				

S.2.   Initial Interval. – A graduate shall have an initial interval that is not subdivided, equal to not less than one-fifth and not more than one-fourth of the capacity of the graduate.

S.3.   Shape. – A graduate of a capacity of more than 15 mL (4 fl dr) may be of either the cylindrical or circular conical type. A graduate of a capacity of 15 mL (4 fl dr) or less shall be of the single-scale cylindrical type.

S.4.   Material. – A graduate shall be made of good-quality, thoroughly annealed, clear, transparent glass, free from bubbles and streaks that might affect the accuracy of measurement. The glass shall be uniform in thickness and shall not be excessively thick.

S.5.   Dimensional Proportions.

S.5.1.   On a Circular Conical Graduate. – The inside measurement from the bottom of a circular conical graduate to the capacity graduation shall be not less than two times the inside diameter at the capacity graduation. The inside measurement from the bottom of the graduate to the point representing one-fourth of the capacity shall be not less than the inside diameter at that point.

S.5.2.   On a Cylindrical Graduate. – The inside measurement from the bottom of a cylindrical graduate to the capacity graduation shall be not less than five times the inside diameter at the capacity graduation.

S.6.   Base. – The base of the graduate shall be perpendicular to the vertical axis of the graduate. The diameter of the base shall be of such size that the empty graduate will remain standing on an inclined surface of 25 %, or approximately 15 degrees, from the horizontal.

S.7.   Design of Graduations.

S.7.1.   General. – Graduations shall be perpendicular to the vertical axis of the graduate and parallel to each other. Graduations shall be continuous, of uniform thickness not greater than 0.4 mm (0.015 in), clearly visible, permanent, and indelible under normal conditions of use.

(Amended 1977)

S.7.2.   On a Single-Scale Graduate. – On a single-scale graduate, the main graduations shall completely encircle the graduate and subordinate graduations shall extend at least one-half the distance around the graduate.

S.7.3.   On a Double-Scale or a Duplex Graduate. – On a double-scale or duplex graduate, there shall be a clear space between the ends of the main graduations on the two scales, and this space shall be approximately 90 degrees from the lip of the graduate and shall conform to the requirements of Table 2. Clear Space Between Ends of Main Graduations on Double Scale Graduates.

Table 2. Clear Space Between Ends of Main Graduations on Double Scale Graduates	
Inside Diameter of Graduate at the Graduations (inches)	**Clear Space Between Ends of Main Graduations (inch)**
Less than 1.5	$\frac{1}{8}$ to $\frac{1}{4}$
1.5 to 3, inclusive	$\frac{1}{4}$ to $\frac{1}{2}$
Over 3	$\frac{3}{8}$ to $\frac{5}{8}$

S.8.   Basis of Graduation. – A graduate shall be graduated "to deliver" when the temperature of the graduate is 20 °C (68 °F), and shall be marked accordingly in a permanent and conspicuous manner.

S.9.   Marking Requirements. – Each main graduation shall be marked to show its value. Intermediate graduations shall not be marked. Value figures shall be uniformly positioned either directly upon or immediately above the graduations to which they refer. Figures placed upon graduations shall be set in from the ends of the graduations a sufficient distance to allow the ends of the graduations to be used in making a setting.

N. Notes

N.1.   Test Liquid. – Water shall be used as the test liquid for graduates.

N.2.   Temperature Control. – During the test of a graduate, appropriate precautions shall be exercised to reduce any detrimental temperature effects to the practicable minimum.

T. Tolerances

T.1. – Maintenance and acceptance tolerances in excess and in deficiency shall be as shown in Table 3. Maintenance and Acceptance Tolerances, in Excess and Deficiency, for Graduates for graduates that are graduated "to contain" or "to deliver." (The tolerance to be applied at any graduation is determined by the inside diameter of the graduate at the graduation in question.)

Table 3. Maintenance and Acceptance Tolerances, in Excess and in Deficiency, for Graduates					
Inside Diameter of Graduate		Tolerance	Inside Diameter of Graduate		Tolerance
From	To but Not Including		From	To but Not Including	
millimeters		milliliters	inches		minims
0	16	0.1	0	$9/16$	2
16	21	0.2	$9/16$	$13/16$	3
21	26	0.4	$13/16$	$1^1/16$	6
26	31	0.6	$1^1/16$	$1^5/16$	10
31	36	0.8	$1^5/16$	$1^9/16$	15
36	41	1.1	$1^9/16$	$1^{13}/16$	20
41	46	1.4	$1^{13}/16$	$2^1/16$	30
46	51	1.8	$2^1/16$	$2^5/16$	40
51	56	2.2	$2^5/16$	$2^9/16$	50
56	61	2.8	$2^9/16$	$2^{13}/16$	65
61	66	3.4	$2^{13}/16$	$3^1/16$	80
66	71	4.1	$3^1/16$	$3^5/16$	95
71	76	4.8	$3^5/16$	$3^9/16$	110
76	81	5.6	$3^9/16$	$3^{13}/16$	130
81	86	6.4	$3^3/16$	$4^1/16$	150
86	91	7.2			
91	96	8.1			
96	101	9.0			

(Amended 1974)

THIS PAGE INTENTIONALLY LEFT BLANK

Table of Contents

THIS PAGE INTENTIONALLY LEFT BLANK

Section 4.45.   Dry Measures

A.   Application

A.1.   General. – This code applies to rigid measures of capacity designed for general and repeated use in the measurement of solids, including the capacities of ½ bu or more.

A.2.   Exceptions.

   (a)   This code does not apply to "standard containers" used for the measurement of fruits and vegetables and as shipping containers thereof.

   (b)   This code does not apply to berry baskets and boxes.   (Also see Section 4.46. Code for Berry Baskets and Boxes.)
   (Added 1976)

A.3.   Additional Code Requirements. – In addition to the requirements of this code, Dry Measures shall meet the requirements of Section 1.10. General Code.

S.   Specifications

S.1.   Units. – The capacity of a measure shall be 1 bu, a multiple of the bushel, or a binary submultiple of the bushel, and the measure shall not be subdivided or double-ended.

S.2.   Material. – A dry measure shall be made of any suitable material that will retain its shape during normal usage.

S.3.   Shape. – A measure, other than a basket, of a capacity of ½ bu or less, shall be cylindrical or conical in shape. The top diameter shall in no case be less than the appropriate minimum diameter shown in Table 1. Minimum Top Diameters for Dry Measure other than Baskets.   The bottom of a measure, other than a basket, shall be perpendicular to the vertical axis of the measure and shall be flat, except that a metal bottom may be slightly corrugated.   The bottom of a measure shall not be adjustable or movable.

Table 1. Minimum Top Diameters for Dry Measures other than Baskets	
Nominal Capacity	**Minimum Top Diameter Inches**
1 pint	4
1 quart	$5^3/_8$
2 quarts	$6^5/_8$
½ peck	8½
1 peck	$10^7/_8$
½ bushel	13¾

S.3.1.   Conical Dry Measure. – If conical, the top diameter shall exceed the bottom diameter by not more than 10 % of the bottom diameter.

S.4.   Capacity Point. – The capacity of a measure shall be determined by the top edge of the measure.

S.5.   Top Reinforcement. – The top edge of a measure shall be reinforced. On a wooden measure other than a basket, of a capacity of 1 qt or more, this reinforcement shall be in the form of a firmly attached metal band.

S.6.   Marking Requirements. – A measure shall be conspicuously marked on its side with a statement of its capacity. If the capacity is stated in terms of the pint or quart, the word "Dry" shall be included. The capacity statement shall be in letters of the following dimensions:

(a)   At least ½ in high and ¼ in wide on a measure of any capacity between ½ pt and 1 pk.

(b)   At least 1 in high and ½ in wide on a measure of a capacity of ½ bu or more.

(c)   On a measure of a capacity of ¼ pt or less, the statement shall be as prominent as practicable, considering the size and design of such measure.

N. Notes

N.1.   Testing Medium.

N.1.1.   Watertight Dry Measures. – Water shall be used as the testing medium for watertight dry measures.

N.1.2.   Non-Watertight Dry Measures. – A dry measure shall be tested either volumetrically using rapeseed as a testing medium or geometrically through inside measurement and calculation.
(Amended 1988)

T. Tolerances

T.1. – Maintenance tolerances in excess and in deficiency shall be as shown in Table 2. Maintenance Tolerances, in Excess and in Deficiency, for Dry Measure. Acceptance tolerances shall be one-half the maintenance tolerances.

Table 2. Maintenance Tolerances, in Excess and in Deficiency, for Dry Measures		
Nominal Capacity	**Tolerance**	
	In Excess Cubic Inches	**In Deficiency Cubic Inches**
¹/₃₂ pint or less	0.1	0.05
¹/₁₆ pint	0.15	0.1
⅛ pint	0.25	0.15
¼ pint	0.5	0.3
½ pint	1.0	0.5
1 pint	2.0	1.0
1 quart	3.0	1.5

Table 2. Maintenance Tolerances, in Excess and in Deficiency, for Dry Measures		
Nominal Capacity	**Tolerance**	
	In Excess Cubic Inches	**In Deficiency Cubic Inches**
2 quarts	5.0	2.5
½ peck	10.0	5.0
1 peck	16.0	8.0
½ bushel	30.0	15.0
1 bushel	50.0	25.0

THIS PAGE INTENTIONALLY LEFT BLANK

Table of Contents

THIS PAGE INTENTIONALLY LEFT BLANK

Section 4.46.   Berry Baskets and Boxes

A. Application

A.1.   General. – This code applies to baskets and boxes for berries and small fruits in capacities of 1 dry quart and less.

A.2.   Additional Code Requirements. – In addition to the requirements of this code, berry baskets and boxes shall meet the requirements of Section 1.10. General Code.

S. Specifications

S.1.   Units. – The capacity of a berry basket or box shall be ½ dry pint, 1 dry pint, or 1 dry quart.

S.2.   Materials. – A berry basket or box shall be made of any suitable materials that will retain its shape during normal filling, storage, and handling.

S.3.   Capacity Point. – The capacity of a berry basket or box shall be determined by its top edges.

N. Notes

N.1.   Method of Test. – A berry basket or box may be tested either volumetrically, using rape seed as the testing medium, or geometrically through accurate inside dimension measurement and calculation.

T. Tolerances

T.1.   Tolerances on Individual Measures. – Maintenance and acceptance tolerances in excess and deficiency on an individual measure shall be as shown in Table 1. Maintenance and Acceptance Tolerances in Excess and in Deficiency.

T.2.   Tolerances on Average Capacity. – The average capacity on a random sample of 10 measures selected from a lot of 25 or more shall be equal to or greater than the nominal capacity.
(Amended 1979)

Table 1. Maintenance and Acceptance Tolerances in Excess and in Deficiency		
Nominal Capacity	Tolerance	
	In Excess Cubic Inches	In Deficiency Cubic Inches
½ pint	1	0.5
1 pint	2	1.0
1 quart	3	1.5

THIS PAGE INTENTIONALLY LEFT BLANK

Section 5

Table of Contents

THIS PAGE INTENTIONALLY LEFT BLANK

Table of Contents

THIS PAGE INTENTIONALLY LEFT BLANK

Section 5.50. Fabric-Measuring Devices

A. Application

A.1. General. – This code applies only to mechanisms and machines designed to indicate automatically (with or without value-computing capabilities) the length of fabric passed through the measuring elements.

A.2. Devices Used to Measure Other Similar Material in Sheet, Roll, or Bolt Form. – Insofar as they are clearly appropriate, the requirements and provisions of this code apply also to devices designed for the commercial measurement of other material similar to fabrics, in sheet, roll, or bolt form.

A.3. Additional Code Requirements. – In addition to the requirements of this code, Fabric-Measuring Devices shall meet the requirements of Section 1.10. General Code.

S. Specifications

S.1. Units. – A fabric-measuring device shall indicate lengths as follows:

(a) For devices indicating in metric units, lengths shall be indicated in terms of 5 cm; 10 cm; 25 cm; or 50 cm and meters.

In addition, lengths may be indicated in terms of any or all of the following subdivisions: millimeters and centimeters.

(b) For devices indicating in U.S. customary units, lengths shall be indicated in terms of $^1/_8$ yd; $^1/_4$ yd; or $^1/_2$ yd; and yards.

In addition, lengths may be indicated in terms of any or all of the following subdivisions: $^1/_8$ yd; $^1/_{16}$ yd; inches and feet.

Digital indicators may indicate values in decimal fractions.
(Amended 1977)

S.2. Design of Indicating Elements.

S.2.1. Graduations.

S.2.1.1. Length. – Graduations shall be so varied in length that they may be conveniently read.

S.2.1.2. Width. – In any series of graduations, the width of a graduation shall in no case be greater than the width of the minimum clear interval between graduations, and the width of main graduations shall be not more than 50 % greater than the width of subordinate graduations. Graduations shall in no case be less than 0.2 mm (0.008 in) in width.

S.2.1.3. Clear Interval between Graduations. – The clear interval between graduations shall be at least 6 mm for cm graduations ($^1/_4$ in for $^1/_8$ yd graduations), and 3 mm for 20 cm graduations ($^1/_8$ in for 1 in graduations).

S.2.2. Indicator.

S.2.2.1. Symmetry. – The index of an indicator shall be symmetrical with respect to the graduations, at least throughout that portion of its length associated with the graduations.

S.2.2.2.   Length. – The index of an indicator shall reach to the finest graduations with which it is used, unless the indicator and the graduations are in the same plane, in which case the distance between the end of the indicator and the ends of the graduations, measured along the line of the graduations, shall be not more than 1.0 mm (0.04 in).

S.2.2.3.   Width. – The index of an indicator shall not be wider than the narrowest graduations with which it is used, and shall in no case exceed 0.4 mm (0.015 in).

S.2.2.4.   Clearance. – The clearance between the index of an indicator and the graduations shall in no case be more than 1.5 mm (0.06 in).

S.2.2.5.   Parallax. – Parallax effects shall be reduced to the practicable minimum.

S.2.3.   Money-Value Computations.

S.2.3.1.   Full-Computing Type. – In this type, the money value at each of a series of unit prices shall be computed automatically for every length within the range of measurement of the fabric-measuring device. Value graduations shall be provided and shall be accurately positioned. The value of each graduated interval shall be 1 cent at all prices per yard of 30 cents and less, and shall not exceed 2 cents at higher prices per yard. Five-cent intervals may be represented in the two-cent range by special graduations, but these shall not be positioned in the clear intervals between graduations of the regular series.

S.2.3.2.   Limited-Computing Type. – In this type, the money value at each of a series of unit prices shall be computed automatically only for lengths corresponding to a definite series of length graduations. There shall be no value graduations. At no position that the chart can assume shall two value figures at the same price per yard be completely and clearly exposed to view at one time. Money values shown shall be mathematically accurate, except that a fraction of less than ½ cent shall be dropped and the next higher cent shall be shown in the case of a fraction of ½ cent or more. One of the following requirements shall be met:

(a) There shall be a money-value computation for each length graduation within the range of measurement of the device.

(b) No money-value computation shall be exposed to view except at such times as the device shows a length indication for which a corresponding series of value indications is computed.

(c) Each column or row of money-value computations shall be marked to show the length to which the computations correspond, the device shall be marked to show the character and limitations of the computations, and there shall be computations corresponding to at least 10 cm (⅛ yd) throughout the range of measurement of the device.

S.2.4.   Return to Zero. – Primary indicating elements shall be readily returnable to a definite zero indication. Means shall be provided to prevent the return of the indicating elements beyond their correct zero positions.

S.3.   Marking Requirements. – If a device will not accurately measure all fabrics, it shall be marked to indicate clearly its limitations.

S.4.   Design Accuracy. – Indications of length and money value shall be accurate whether the values of the indications are being increased or decreased.

N. Notes

N.1.   Testing Medium. – A fabric-measuring device shall be tested with a suitable testing tape approximately 7.62 cm (3 in) wide and with a graduated length of at least 11 m (12 yd), made from such material and having such surface finish as to provide dimensional stability and reduce slippage to the practicable minimum.

T. Tolerances

T.1.   Tolerance Values. – Maintenance and acceptance tolerances shall be as shown in Table 1. Maintenance and Acceptance Tolerances for Fabric-Measuring Devices.

Indication of Device (yards)	Maintenance Tolerance		Acceptance Tolerance	
	On Under-registration (inches)	On Over-registration (inches)	On Under-registration (inches)	On Over-registration (inches)
2 or less	3/8	1/4	1/4	1/8
3	3/8	5/16	1/4	5/32
4	1/2	5/16	1/4	5/32
5	5/8	3/8	5/16	3/16
6	3/4	3/8	3/8	3/16
7 and 8	1	1/2	1/2	1/4
9	1 1/4	5/8	5/8	5/16
10 and 11	1 1/2	3/4	3/4	3/8
12 and 13	1 3/4	7/8	7/8	7/16
14 and 15	2	1	1	1/2
Over 15	Add 1/8 inch per indicated yard	Add 1/16 inch per indicated yard	Add 1/16 inch per indicated yard	Add 1/32 inch per indicated yard

Table 1. Maintenance and Acceptance Tolerances for Fabric-Measuring Devices

UR.   User Requirements

UR.1.   Installation Requirements.

UR.1.1.   Installation. – A fabric-measuring device shall be securely supported and firmly fixed in position.

UR.2.   Use Requirements.

UR.2.1.   Limitation of Use. – A fabric-measuring device shall be used to measure only those fabrics that it was designed to measure, and in no case shall it be used to measure a fabric that a marking on the device indicates should not be measured.

UR.2.2.   Return of Indicating Elements to Zero. – The primary indicating elements shall be returned to zero before each measurement.

THIS PAGE INTENTIONALLY LEFT BLANK

Table of Contents

THIS PAGE INTENTIONALLY LEFT BLANK

Section 5.51.   Wire- and Cordage-Measuring Devices

A. Application

A.1.   General. – This code applies to mechanisms and machines designed to indicate automatically the length of cordage, rope, wire, cable, or similar flexible material passed through the measuring elements.

A.2.   Additional Code Requirements. – In addition to the requirements of this code, Wire- and Cordage-Measuring Devices shall meet the requirements of Section 1.10. General Code.

S. Specifications

S.1.   Units. – A wire- or cordage-measuring device shall indicate lengths in terms of feet, yards, or meters, or combinations of units of the same measurement system, and shall have minimum increments with values that do not exceed the equivalent of 0.1 meter or 0.1 yard.

(Amended 1989)

S.2.   Design of Indicating Elements.

S.2.1.   Graduations.

S.2.1.1.   Length. – Graduations shall be so varied in length that they may be conveniently read.

S.2.1.2.   Width. – In any series of graduations, the width of a graduation shall in no case be greater than the width of the minimum clear interval between graduations, and the width of main graduations shall be not more than 50 % greater than the width of subordinate graduations. Graduations shall in no case be less than 0.2 mm (0.008 in), nor more than 1.0 mm (0.04 in), in width.

S.2.1.3.   Clear Interval between Graduations. – The clear interval between graduations shall be at least as wide as the widest graduation, and in no case less than 0.8 mm (0.03 in).

S.2.2.   Indicator.

S.2.2.1.   Symmetry. – The index of an indicator shall be symmetrical with respect to the graduations, at least throughout that portion of its length associated with the graduations.

S.2.2.2.   Length. – The index of an indicator shall reach to the finest graduations with which it is used, unless the indicator and the graduations are in the same plane, in which case the distance between the end of the indicator and the ends of the graduations, measured along the line of the graduations, shall be not more than 1.0 mm (0.04 in).

S.2.2.3.   Width. – The index of an indicator shall not be wider than the narrowest graduations with which it is used, and shall in no case exceed 0.4 mm (0.015 in).

S.2.2.4.   Clearance. – The clearance between the index of an indicator and the graduations shall in no case be more than 1.5 mm (0.06 in).

S.2.2.5. Parallax. – Parallax effects shall be reduced to the practicable minimum.

S.2.3.   Zero Indication. – Primary indicating elements shall be readily returnable to a definite zero indication.

S.3.   Design of Measuring Elements.

S.3.1.   Sensitiveness. – If the most sensitive element of the indicating system utilizes an indicator and graduations, the relative movement of these parts corresponding to a measurement of 30 cm (1 ft) shall be not less than 6 mm (¼ in).

S.3.2.   Slippage. – The measuring elements of a wire- or cordage-measuring device shall be so designed and constructed as to reduce to the practicable minimum any slippage of material being measured and any lost motion in the measuring mechanism.

S.3.3.   Accessibility. – A wire- or cordage-measuring device shall be so constructed that the measuring elements are readily visible and accessible, without disassembly of any supporting frame or section of the main body, for purposes of cleaning or removing any foreign matter carried into the mechanism by the material being measured.

S.4.   Marking Requirements.

S.4.1.   Limitation of Use. – If a device will measure accurately only certain configurations, diameters, types, or varieties of materials, or with certain accessory equipment, all limitations shall be clearly and permanently stated on the device.

S.4.2.   Operating Instructions. – Any necessary operating instructions shall be clearly stated on the device.

S.4.3.   Indications. – Indicating elements shall be identified by suitable words or legends so that the values of the indications will be unmistakable.

S.5.   Design Accuracy. – Indications of length shall be accurate whether the values of the indications are being increased or decreased.

N. Notes

N.1. Testing Medium. – Wherever feasible, a wire- or cordage-measuring device shall be tested with a steel tape not less than 10 mm (⅜ in) in width and at least 15 m (50 ft) in length. When a device cannot be tested in this manner because of the design of the device, it shall be tested with a dimensionally stable material appropriately marked and compared at frequent periodic intervals with a steel tape in order to assure that any marked interval is not in error by more than ⅓ of the tolerance of the device at that particular interval.
(Amended 1981)

N.2. Minimum Test. – Tests shall be conducted at a minimum initial increment of 5 m (20 ft) and appropriate increments up to at least 15 m (50 ft).

T. Tolerances

T.1. Tolerance Values. – Maintenance and acceptance tolerances shall be as shown in Table 1. Maintenance and Acceptance Tolerances for Wire- and Cordage-Measuring Devices.

Table 1. Maintenance and Acceptance Tolerances for Wire- and Cordage-Measuring Devices		
Indication of Device (feet)	**Acceptance and Maintenance Tolerances**	
	On Underregistration (inches)	**On Overregistration (inches)**
20	6	3
Over 20 to 30	8	4
Over 30 to 40	10	5
Over 40 to 50	12	6
Over 50	Add 2 inches per indicated 10 feet	Add 1 inch per indicated 10 feet

UR. User Requirements

UR.1.   Installation Requirements.

UR.1.1.   Installation. – A wire- or cordage-measuring device shall be securely supported and firmly fixed in position.

UR.2.   Use Requirements.

UR.2.1.   Limitation of Use. – A wire- or cordage-measuring device shall be used to measure only those materials that it was designed to measure, and in no case shall it be used to measure a material that a marking on the device indicates should not be measured.

UR.2.2.   Return to Zero. – The primary indicating elements of a wire- or cordage-measuring device shall be returned to zero before each measurement.

UR.2.3.   Operation of Device. – A wire- or cordage-measuring device shall not be operated in such a manner as to cause slippage or inaccurate measurement.

UR.2.4.   Cleanliness. – The measuring elements of a wire- or cordage-measuring device shall be kept clean to prevent buildup of dirt and foreign material that would adversely affect the measuring capability of the device.

THIS PAGE INTENTIONALLY LEFT BLANK

Table of Contents

THIS PAGE INTENTIONALLY LEFT BLANK

Section 5.52.   Linear Measures

A. Application

A.1.   General. – This code applies to any linear measure or measure of length, whether flexible or inflexible, permanently installed or portable.

A.2.   Additional Code Requirements. – In addition to the requirements of this code, Linear Measures shall meet the requirements of Section 1.10. General Code.

S. Specifications

S.1.M.   Units. – A linear measure may be in total length, and the total length may be subdivided in any or all of the following:

(a) centimeters and tenths of the centimeter;

(b) meters; and

(c) multiples of meters.

An one-meter measure may be graduated, in addition, to show 0.1 m and multiples of 0.1 m subdivisions.

S.1.   Units. – A linear measure may be in total length, and the total length may be subdivided in any or all of the following:

(a) inches and binary submultiples of the inch;

(b) feet;

(c) yards and multiples of yards.

A 1-yard measure may be graduated, in addition, to show ⅓ yd and ⅔ yd subdivisions. A flexible tape may be graduated in tenths or hundredths of a foot, or both tenths and hundredths of a foot. (Any other subdivisions are allowable only on measures of special purposes and when required for such purposes.)

S.2.   Material.

S.2.1.   Flexible Tape. – A flexible tape shall be made of metal.

S.2.2.   End Measure. – If an end measure is made of material softer than brass, the ends of the measure shall be protected by brass (or other metal at least equally hard) securely attached.

S.3.   Finish. – Measures shall be smoothly finished.

S.4.   Design.

S.4.1.   Rigid Measure. – A rigid measure shall be straight.

S.4.2.   Folding Measure. – A folding measure shall open to a definite stop, and when so opened shall be straight.

S.5.   Graduations.

    S.5.1.   General. – Graduations shall be perpendicular to the edge of the measure.

    S.5.2.   Width. – The width of the graduations on any measure shall not exceed one-half the width of the smallest graduated interval on the measure, and in no case shall be wider than 0.75 mm (0.03 in).
(Amended 1982)

T. Tolerances

T.1.   For Measures Except Metal Tapes. – Maintenance tolerances in excess and in deficiency for measures except metal tapes shall be as shown in Table 1. Maintenance Tolerances, in Excess and in Deficiency, for Linear Measures Except Metal Tapes. Acceptance tolerances shall be one-half the maintenance tolerances.

Table 1. Maintenance Tolerances, in Excess and in Deficiency, for Linear Measures Except Metal Tapes	
Nominal Interval from Zero	Tolerance
Feet	Inch
½ or less	$1/64$
1	$1/32$
2	$1/16$
3	$3/32$
4	$1/8$
5	$5/32$
6	$3/16$

T.2.   For Metal Tapes. – Maintenance and acceptance tolerances in excess and in deficiency for metal tapes shall be as shown in Table 2. Maintenance and Acceptance Tolerances, in Excess and in Deficiency, for Metal Tapes. Tapes of 10 m (25 ft) or over shall be tested at a tension resulting from a load of 5 kg (10 lb). Tapes less than 10 m (25 ft) shall be tested at a tension resulting from a load of 2.5 kg (5 lb). However, flexible metal tapes of 10 m (25 ft) or less that are not normally used under tension shall be tested with no tension applied. All tapes shall be supported throughout on a horizontal flat surface whenever tested.
(Amended 1972)

Table 2. Maintenance and Acceptance Tolerances, in Excess and in Deficiency, for Metal Tapes	
Nominal Interval from Zero	Tolerance
Feet	Inch
6 or less	$1/32$
7 to 30, inclusive	$1/16$
31 to 55, inclusive	$1/8$
56 to 80, inclusive	$3/16$
81 to 100, inclusive	$1/4$

Table of Contents

THIS PAGE INTENTIONALLY LEFT BLANK

Section 5.53. Odometers

A. Application

A.1. General. – This code applies to odometers that are used or are to be used to determine the charges for rent or hire of passenger vehicles and trucks and buses. (When official examinations are undertaken on odometers that form the basis for the payment of fees or taxes to, or the preparation of reports for, governmental agencies, and in similar cases, the requirements of this code shall be applied insofar as they are applicable and appropriate to the conditions of such special uses.)
(Amended 1977)

A.2. Exceptions. – This code does not apply to taximeters (for which see Section 5.54. Code for Taximeters).
(Amended 1977)

A.3. Additional Code Requirements. – In addition to the requirements of this code, Odometers shall meet the requirements of Section 1.10. General Code.

S. Specifications

S.1. Design of Indicating Elements.

    S.1.1. General. – The primary indicating element of an odometer may be:

        (a) the distance-traveled portion of the "speedometer" assembly of a motor vehicle;

        (b) a special cable-driven distance-indicating device; or

        (c) a hub odometer attached to the hub of a wheel on a motor vehicle.
(Amended 1977)

    S.1.2. Units. – An odometer shall indicate in terms of miles or kilometers.
(Amended 1977)

    S.1.3. Minimum Indicated Value. – The value of the interval of indicated distance shall be:

        (a) for odometers indicating in kilometers, 0.1 km; or

        (b) for odometers indicating in miles, 0.1 mi.
(Amended 1977)

    S.1.4. Advancement of Indicating Elements. – The most sensitive indicating elements of an odometer may advance continuously or intermittently; all other elements shall advance intermittently. Except when the indications are being returned to zero, the indications of an installed odometer shall be susceptible to advancement only by the rotation of the vehicle wheel or wheels.
(Amended 1977)

    S.1.5. Readability. – Distance figures and their background shall be of sharply contrasting colors. Figures indicating tenth units shall be differentiated from other figures with different colors, or with a decimal point, or by other equally effective means. Except during the period of advance of any decade to the next higher indication, only one figure in each decade shall be exposed to view. Any protective covering intended to be transparent shall be in such condition that it can be made transparent by ordinary cleaning of its exposed surface.
(Amended 1977)

S.1.6.   Digital Indications and Representation. – Digital indicating odometers (discontinuous registration) shall "round off" indications to the nearest minimum division or truncate indications to the lower minimum division.

(Added 1990)

N. Notes

N.1.   Testing Procedures.

N.1.1.   Test Methods. – To determine compliance with distance tolerances, a distance test of an odometer shall be conducted using one or more of the following test methods:

(a) **Road Test.** – A road test consists of driving the vehicle over a precisely measured road course.

(b) **Fifth-Wheel Test.** – A fifth-wheel test consists of driving the vehicle over any reasonable road course and determining the distance actually traveled through the use of a mechanism known as a "fifth wheel" that is attached to the vehicle and that independently measures and indicates the distance.

(c) **Simulated-Road Test.** – A simulated-road test consists of determining the distance traveled by use of a roller device, or by computation from rolling circumference and wheel-turn data.

(Amended 1977)

N.1.2.   Test Runs. – Not less than two test runs shall be conducted. Acceleration and deceleration shall be carefully controlled to avoid spinning or skidding the wheels.

(Amended 1977)

N.1.2.1.   For Devices Indicating in Miles. – The test runs shall be 2 mi in length, shall start from, and finish at, a dead stop with a minimum of 80 % of the run between 30 mi/h and 45 mi/h.

(Added 1977)

N.1.2.2.   For Devices Indicating in Kilometers. – The test runs shall be 3 km in length, shall start from, and finish at, a dead stop with a minimum of 80 % of the run between 50 km/h and 75 km/h.

(Added 1977)

N.1.3.   Test Conditions.

N.1.3.1.   Tire Stabilization. – Road tests or fifth-wheel tests shall be preceded by a run of at least 8 km or 5 mi, for the purpose of stabilizing tire pressures. Simulated road tests on a roller device shall be made at stable tire pressures.

(Amended 1977)

N.1.3.2.   Tire Pressure. – At the completion of the test run or runs, the tires of the vehicle under test shall be checked to determine that the tire pressure is that operating tire pressure posted in the vehicle. If not, the tire pressure should be adjusted to the posted tire pressure and further tests may be conducted to determine the operating characteristics of the odometer.

(Amended 1977)

N.1.3.3.   Vehicle Loading.

(a) **Passenger Load.** – During the distance test of an odometer, the vehicle may carry two persons.

(b) **Truck Cargo Load.** – Truck odometers shall be tested by one of the following methods:

    (1) the truck is loaded with one-half of the maximum cargo load; or

    (2) unloaded if unloaded test tolerances are applied.

(Amended 1977 and 1987)

T. Tolerances

T.1.   To Underregistration and to Overregistration. – The tolerances hereinafter prescribed shall be applied to errors of underregistration and errors of overregistration.

T.2.   Tolerance Values. – Except for unloaded trucks, maintenance and acceptance tolerances on odometers shall be 4 % of the interval under test.

(Amended 1977and 1987)

T.2.1.   Tolerances for Unloaded Trucks. – Maintenance and acceptance tolerances on truck odometers shall be 5 % for underregistration and 3 % for overregistration of the interval under test.

(Added 1987)

UR. User Requirements

UR.1.   Inflation of Vehicle Tires. – The operational tire pressure of passenger vehicle and truck tires shall be posted in the vehicle and tires shall be maintained at the posted pressure.

(Amended 1977)

THIS PAGE INTENTIONALLY LEFT BLANK

Table of Contents

THIS PAGE INTENTIONALLY LEFT BLANK

Section 5.54.   Taximeters

A. Application

A.1.   General. – This code applies to taximeters; that is, to devices that automatically calculate at a predetermined rate or rates and indicate the charge for hire of a vehicle.

A.2.   Exceptions. – This code does not apply to:

(a) Odometers on vehicles that are rented on a distance basis (for which see Section 5.53. Code for Odometers).

(b) Devices that only display a flat rate or negotiated rate.

(Amended 1977 and 2016)

A.3.   Additional Code Requirements. – In addition to the requirements of this code, Taximeters shall meet the requirements of Section 1.10. General Code.

S. Specifications

S.1.   Design of Indicating and Recording Elements.

S.1.1.   General. – A taximeter shall be equipped with a primary indicating element.
(Amended 1988 and 2015)

> ***S.1.1.1. Recording Elements.*** *– A receipt providing information as required in S.1.9. Recorded Representations shall be available from a taximeter or taximeter system through an integral or separate recording element for all transactions conducted.*
> *[Nonretroactive January 1, 2016]*
> (Added 2015)

S.1.2.   Advancement of Indicating Elements. – Except when a taximeter is being cleared, the primary indicating and recording elements shall be susceptible of advancement only by the movement of the vehicle or by the time mechanism.

At the conclusion of a transaction (e.g., following the totalizing of all accrued charges and having a customer receipt made available), no other advancement of fare, extras, or other charges shall occur until the taximeter has been cleared.
[Nonretroactive as of January 1, 2017

Where permitted, a flat rate or negotiated rate shall be displayed in the "fare" indicating mechanism, provided that once a flat rate or negotiated rate is entered the fare may no longer be advanced by movement of the vehicle or the time mechanism.

(Amended 1988 and 2016)

S.1.3.   Visibility of Indications. – The indications of fare, including extras, and the mode of operation, such as "time" or "hired," shall be constantly displayed whenever the meter is in operation. All indications of passenger interest shall be easily read from a distance of 1.2 m (4 ft) under any condition of normal operation.

(Amended 1977, 1986, and 1988)

S.1.3.1.   Minimum Height of Figures, Words, and Symbols. – The minimum height of the figures used to indicate the fare shall be 10 mm and for extras, 8 mm. The minimum height of the figures, words, or symbols used for other indications, including those used to identify or define, shall be 3.5 mm.
(Added 1986)

S.1.3.2.   Lighting of Indications. – *Integral lighting shall be provided to illuminate the fare, extras, the rate or rate code, and the taximeter status (i.e., vacant, hired, and time off).*
[Nonretroactive as of January 1, 1989]
(Added 1988) (Amended 1990)

S.1.3.3. Passenger's Indications. – *A supplementary indicating element installed in a taxi to provide information regarding the taxi service to the passenger, shall clearly display the current total of all charges incurred for the transaction. The accruing total of all charges must remain clearly visible on the passenger's display (unless disabled by the passenger) at all times during the transaction.*
[Nonretroactive as of January 1, 2016]
(Added 2015)

S.1.3.3.1.   Additional Information. – *Additional information shall be displayed or made available through a passenger's indicating element (as described in S.1.3.3. Passenger's Indications) and shall be current and reflect any charges that have accrued. This additional information shall include:*

(a) an itemized account of all charges incurred including fare, extras, and other additional charges; and

(b) the rate(s) in use at which any fare is calculated.

Any additional information made available must not obscure the accruing total of charges for the taxi service. This additional information may be made accessible through clearly identified operational controls (e.g., keypad, button, menu, touch-screen).
[Non retroactive as of January 1, 2016]
(Added 2015)

S.1.3.3.2.   Fare and Extras Charges. – *The indication of fare and extras charges on a passenger's indicating element shall agree with similar indications displayed on all other indicating elements in the system.*
[Nonretroactive as of January 1, 2016]
(Added 2015)

S.1.4.   Actuation of Fare-Indicating Mechanism. – When a taximeter designed to calculate fares upon the basis of a combination of distance traveled and time elapsed is operative with respect to fare indication, the fare-indicating mechanism shall be actuated by the distance mechanism whenever the vehicle is in motion at such a speed that the rate of distance revenue equals or exceeds the time rate, and may be actuated by the time mechanism whenever the vehicle speed is less than this and when the vehicle is not in motion. Means shall be provided for the vehicle operator to render the time mechanism either operative or inoperative with respect to the fare-indicating mechanism.
(Amended 1977)

S.1.5.   Operating Condition.

S.1.5.1.   General. – When a taximeter is cleared, the indication "Not Registering," "Vacant," or an equivalent expression shall be shown. Whenever a taximeter is set to register charges, it shall indicate

"Registering," "Hired," or an equivalent expression and the rate at which it is set shall be automatically indicated (Rate 1 or Rate A, for example).

(Amended 1988)

S.1.5.2.   Time not Recording. – When a taximeter is set for fare registration with the time mechanism inoperative, it shall indicate "Time Not Recording" or an equivalent expression.

(Amended 1988)

S.1.6.   Fare Identification. – Fare indications shall be identified by the word "Fare" or by an equivalent expression. Values shall be defined by suitable words or monetary signs.

S.1.7.   Extras. – Extras shall be indicated as a separate item and shall not be included in the fare indication. They shall be identified by the word "Extras" or by an equivalent expression. Values shall be defined by suitable words or monetary signs. Means may be provided to totalize the fare and extras if the totalized amount returns to separate indications of fare and extras within 5 seconds or less.

(Amended 1988)

S.1.7.1.   Nonuse of Extras. – If and when taximeter extras are prohibited by legal authority or are discontinued by a vehicle operator, the extras mechanisms shall be rendered inoperable or the extras indications shall be effectively obscured by permanent means.

S.1.8.   Protection of Indications. – All indications of fare and extras shall be protected from unauthorized alteration or manipulation.

(Amended 2015)

S.1.9.   Recorded Representation. – A printed receipt issued from a taximeter, whether through an integral or separate recording element, shall include as a minimum, the following information when processed through the taximeter system:

(a)   date;

(b)   unique vehicle identification number, such as the medallion number, taxi number, vehicle identification number (VIN), permit number, or other identifying information as specified by the statutory authority; *

(c)   start and end time of the trip; *

(d)   distance traveled, maximum increment of 0.1 km (0.1 mi); *

(e)   fare in $;

(f)   each rate at which the fare was computed and the associated fare at that rate; *

(g)   additional charges (in $) where permitted such as extras, any surcharges, telecommunication charges, and taxes shall be identified and itemized; *

(h)   total charge for service in $ (inclusive of fare, extras, and all additional charges); *

(i)   trip number, if available; ** and

(j)   telephone number (or other contract information) for customer assistance. **

Note: When processed through the taximeter or taximeter system, any adjustments (in $) to the total charge for service including discounts, credits, and tips shall also be included on the receipt.**
[Nonretroactive as of January 1, 1989] *[Nonretroactive as of January 1, 2000]
**[Nonretroactive as of January 1, 2016]
(Added 1988) (Amended 1999 and 2015)

S.1.9.1.   Multiple Recorded Representations.

S.1.9.1.1.   Duplicate Receipts. *– A recording element may produce a duplicate receipt for the previous transaction provided the information printed is identical to the original with the exception of time issued. The duplicate receipt shall include the words "duplicate" or "copy." The feature to print a duplicate receipt shall be deactivated at the time the meter is hired for the next fare.*
[Nonretroactive as of January 1, 2000]
(Added 1999)

S.1.10.   Non-fare Information. *– The fare and extras displays may be used to display auxiliary information provided the meter is in the vacant condition and such information is only displayed for 10 seconds, or less. If the information consists of a list of information, the list may be displayed one item after another, provided that each item is displayed for 10 seconds, or less.*
[Nonretroactive as of January 1, 2002]
(Added 2000)

S.2.   Basis of Fare Calculations. – A taximeter shall calculate fares only upon the basis of:

(a) distance traveled;

(b) time elapsed; or

(c) a combination of distance traveled and time elapsed.

A taximeter may utilize more than one rate to calculate the fare during a trip. Any change in the applied rate must occur at the completion of the current interval.
(Amended 1977 and 2016)

S.2.1.   Initial Time and Distance Intervals. – The time and distance intervals of a taximeter shall be directly proportional as expressed in the following formula:

$$\frac{Seconds\ of\ Initial\ Time\ Interval}{Seconds\ per\ Non-Initial\ Time\ Interval} = \frac{Distance\ of\ Initial\ Mileage\ Interval}{Distance\ per\ Non-Initial\ Mileage\ Interval}$$

(Added 1990)

S.3.   Design of Operating Control.

S.3.1.   Positions of Control. – The several positions of the operating controls shall be clearly defined and shall be so constructed that accidental or inadvertent changing of the operating condition of the taximeter is improbable. Movement of the operating controls to an operating position immediately following movement to the cleared position shall be delayed enough to permit the taximeter to come to a complete rest in the cleared position.
(Amended 1988)

S.3.2.   Control for Extras Mechanism. – The knob, handle, or other means provided to actuate the extras mechanism shall be inoperable whenever the taximeter is cleared.

S.4.   Interference. – The design of a taximeter shall be such that there will be no interference between the time and the distance portions of the mechanism device at any speed of operation.
(Amended 1977 and 1988)

S.5.   Provision for Security Seals. – Adequate provision shall be made to provide security for a taximeter. Security may be provided either by:

  (a) Affixing security seals to the taximeter and to all other components required for service operation of a complete installation on a vehicle, so that no adjustments, alterations, or replacements affecting accuracy or indications of the device or the assembly can be made without mutilating the seal or seals; or

  (b) Using a combination of security seals described in paragraph (a) and, in the case of a component that may be removed from a vehicle (e.g., slide mounting the taximeter), providing a physical or electronic link between components affecting accuracy or indications of the device to ensure that its performance is not affected and operation is permitted only with those components having the same unique properties.

The sealing means shall be such that it is not necessary to disassemble or remove any part of the device or of the vehicle to apply or inspect the seals.
(Amended 1988 and 2000)

S.6.   Power Interruption, Electronic Taximeters.

  (a) After a power interruption of 3 seconds or less, the fare and extras indications shall return to the previously displayed indications and may be susceptible to advancement without the taximeter being cleared.

  (b) After a power interruption exceeding 3 seconds, the fare and extras indications shall return to the previously displayed indications and shall not be susceptible to advancement until the taximeter is cleared.

After restoration of power following an interruption exceeding 3 seconds, the previously displayed fare shall be displayed for a maximum of 1 minute at which time the fare shall automatically clear and the taximeter shall return to the vacant condition. *
[*Nonretroactive as of January 1, 2002]
(Added 1988) (Amended 1989, 1990, and 2000)

S.7.   Anti-Fraud Provisions, Electronic Taximeters. – An electronic taximeter may have provisions to detect and eliminate distance input that is inconsistent with output of the vehicle's distance sensor. When a taximeter equipped with this feature detects input inconsistent with the distance sensor:

  (a) The meter shall either filter out the inconsistent distance input signals or cease to increment fare based on distance until the distance input signal returns to normal. If the meter ceases to increment fare based on distance, the taximeter may continue to increment fare based on elapsed time;

  (b) The taximeter shall provide a visible or audible signal that inconsistent input signals are being detected; and

  (c) The taximeter shall record the occurrence in an event logger. The event logger shall include an event counter (000 to 999), the date, and the time of at least the last 1000 occurrences.
(Added 2001)

N. Notes

N.1.   Distance Tests.

  N.1.1.   Test Methods. – To determine compliance with distance tolerances, a distance test of a taximeter shall be conducted utilizing one or more of the following test methods:

(a) **Road Test.** – A road test consists of driving the vehicle over a precisely measured road course.

(b) **Fifth-Wheel Test.** – A fifth-wheel test consists of driving the vehicle over any reasonable road course and determining the distance actually traveled through the use of a mechanism known as a "fifth wheel" that is attached to the vehicle and that independently measures and indicates the distance.

(c) **Simulated-Road Test.** – A simulated road test consists of determining the distance traveled by use of a roller device, or by computation from rolling circumference and wheel-turn data.

(Amended 1977)

N.1.2.   Test Procedures. – The distance test of a taximeter, whether a road test, a simulated-road test, or a fifth-wheel test, shall include at least duplicate runs of sufficient length to cover at least the third money drop or 1 mi, whichever is greater, and shall be at a speed approximating the average speed traveled by the vehicle in normal service. In the case of metric-calibrated taximeters, the test should cover at least the third money drop or 2 km, whichever is greater.

(Amended 1977)

N.1.3.   Test Conditions.

N.1.3.1.   Vehicle Lading. – During the distance test of a taximeter, the vehicle shall carry two persons, or in the case of a simulated-road test, 70 kg or 150 lb of test weights may be substituted in lieu of the second person.

N.1.3.2.   Tire Pressure. – At the completion of test run or runs, the tires of the vehicle under test shall be checked to determine that the tire pressure is that operating tire pressure posted in the vehicle. If not, the tire pressure should be adjusted to the posted tire pressure and further tests may be conducted to determine the operating characteristics of the taximeter.

(Amended 1977)

N.2.   Time Test. – If a taximeter is equipped with a timing device through which charges are made for time intervals, the timer shall be tested at the initial interval, four separate subsequent intervals, and an average time test of at least four consecutive subsequent time intervals.

(Amended 1988)

N.3.   Interference Test. – If a taximeter is equipped with a timing device through which charges are made for time intervals, a test shall be conducted to determine whether there is interference between the time and distance elements. During the interference test, the vehicle's operating speed shall be 3 km/h or 4 km/h, or 2 mi/h or 3 mi/h faster than the speed at which the basic distance rate equals the basic time rate. The basic rate per hour divided by the basic rate per mile is the speed (km/h or mi/h) at which the basic time rate and basic distance rate are equal.

(Amended 1988)

T. Tolerances

T.1.   Tolerance Values.

T.1.1.   On Distance Tests. – Maintenance and acceptance tolerances for taximeters shall be as follows:

(a) On Overregistration: 1 % of the interval under test.

(b) On Underregistration: 4 % of the interval under test, with an added tolerance of 30 m or 100 ft whenever the initial interval is included in the interval under test.

T.1.2.   On Time Tests.

T.1.2.1.   On Individual Time Intervals. – Maintenance and acceptance tolerances on individual time intervals shall be as follows:

(a) On Overregistration: 3 seconds per minute (5 %).

(b) On Underregistration: 9 seconds per minute (15 %) on the initial interval, and 6 seconds per minute (10 %) on subsequent intervals.

T.1.2.2.   On Average Time Interval Computed After the Initial Interval. – Except for the initial interval, maintenance and acceptance tolerances on the average time interval shall be as follows:

(a) On Overregistration: 0.2 second per minute (0.33 %).

(b) On Underregistration: 3 seconds per minute (5 %).

(Amended 1991)

T.1.3.   On Interference Tests.

T.1.3.1.   The registration of a taximeter in the "time on" position shall agree within 1 % of its performance in the "time off" position.

(Added 1988)

UR.   User Requirements

UR.1.   Inflation of Vehicle Tires. – The operational tire pressure of passenger vehicles and truck tires shall be posted in the vehicle and shall be maintained at the posted pressure.

(Amended 1977)

UR.2.   Position and Illumination of Taximeter. – A taximeter shall be so positioned and illuminated that its indications, operational markings, and controls of passenger interest can be conveniently read by a passenger seated in the back seat of the vehicle.

(Amended 1985 and 1986)

UR.3.   Statement of Rates. – The distance and time rates for which a taximeter is set, including the initial distance interval and the initial time interval, the local tax rate, and the schedule of extras when an extras indication is provided shall be conspicuously displayed inside the front and rear passenger compartments. The words "Rate," "Rates," or "Rates of Fare" shall precede the rate statement. The rate statement shall be fully informative, self-explanatory, and readily understandable by the ordinary passenger, and shall either be of a permanent character or be protected by glass or other suitable transparent material.

(Amended 1977, 1988, 1990, and 1999)

THIS PAGE INTENTIONALLY LEFT BLANK

Table of Contents

THIS PAGE INTENTIONALLY LEFT BLANK

Section 5.55.   Timing Devices

A.   Application

A.1.   General. – This code applies to devices used to measure time during which services are being dispensed (such as vehicle parking, laundry drying, and car washing). This code also applies to Electric Vehicle Supply Equipment (EVSE) when used to assess charges for time-based services in addition to those charged for electrical energy.
(Amended 2015)

A.2.   Additional Code Requirements. – In addition to the requirements of this code, Timing Devices shall meet the requirements of Section 1.10. General Code.

S.   Specifications

S.1.   Design of Indicating and Recording Elements and of Recorded Representations.

S.1.1.   Primary Elements.

S.1.1.1.   General. – A timing device shall be equipped with a primary indicating element, and may also be equipped with a primary recording element. A timing device incorporated into an Electric Vehicle Supply Equipment system for use in assessing charges for timing separate from charges for electrical energy shall be equipped with the capability to provide a recorded representation of the transaction through a built-in or separate recording element. A readily observable in-service light or other equally effective means that automatically indicates when laundry driers, vacuum cleaners, and car washes are in operation shall be deemed an appropriate primary indicating element.
(Amended 1979 and 2015)

S.1.1.2.   Units. – A timing device shall indicate and record, if the device is equipped to record, the time in terms of minutes for time intervals of 60 minutes or less and in hours and minutes for time intervals greater than 60 minutes.

S.1.1.3.   Value of Smallest Unit. – The value of the smallest unit of indicated time and recorded time, if the device is equipped to record, shall not exceed the equivalent of:

(a) one-half hour on parking meters indicating time in excess of two hours;

(b) six minutes on parking meters indicating time in excess of one but not greater than two hours; or

(c) five minutes on all other devices, except those equipped with an in-service light.
(Amended 1975)

S.1.1.4.   Advancement of Indicating and Recording Elements. – Primary indicating and recording elements shall be susceptible to advancement only during the mechanical operation of the device, except that clocks may be equipped to manually reset the time.

S.1.1.5.   Operation of In-Service Indicator Light. – For devices equipped with an in-service indicator light, the indicator shall be operative only during the time the device is in operation.
(Amended 2015)

S.1.1.6.   Discontinuous Indicating Parking Meters. – An indication of the time purchased shall be provided at the time the meter is activated in units of no more than one minute for times less than one hour

and not more than two minutes for times of one hour or more. Convenient means shall be provided to indicate to the purchaser the unexpired time.
(Added 1975) (Amended 1976)

S.1.2. Graduations.

S.1.2.1. **Length.** – Graduations shall be so varied in length that they may be conveniently read.

S.1.2.2. **Width.** – In any series of graduations, the width of a graduation shall in no case be greater than the width of the minimum clear interval between graduations and the width of main graduations shall be not more than 50 % greater than the width of subordinate graduations. Graduations shall in no case be less than 0.2 mm (0.008 in) in width.

S.1.2.3. **Clear Interval Between Graduations.** – The clear interval shall be not less than 0.75 mm (0.03 in). If the graduations are not parallel, the measurement shall be made:

(a) along the line of relative movement between the graduations at the end of the indicator; or

(b) if the indicator is continuous, at the point of widest separation of the graduations.

S.1.3. Indicators.

S.1.3.1. **Symmetry.** – The index of an indicator shall be symmetrical with respect to the graduations, at least throughout that portion of its length associated with the graduations.

S.1.3.2. **Length.** – The index of an indicator shall reach to the finest graduations with which it is used, unless the indicator and the graduations are in the same plane, in which case the distance between the end of the indicator and the ends of the graduations, measured along the line of the graduations, shall be not more than 1.0 mm (0.04 in).

S.1.3.3. **Width.** – The width of the index of an indicator in relation to the series of graduations with which it is used shall be not greater than:

(a) the width of the widest graduation; and

(b) the width of the minimum clear interval between the graduations.

S.1.3.4. **Parallax.** – Parallax effect shall be reduced to a practicable minimum.

S.1.4. Recorded Representations.

S.1.4.1. Recorded Representations, Electric Vehicle Supply Equipment (EVSE) Timing Devices. – A timing device incorporated into an EVSE for use in assessing charges for timing separate from charges for electrical energy shall issue a recorded representation itemizing the charges for these services as defined in Section 3.40. Electric Vehicle Fueling Systems.
(Added 2015)

S.1.4.1.1. **Duplicate Receipts.** – Duplicate receipts are permissible, provided the word "duplicate" or "copy" is included on the receipt.
(Added 2015)

S.1.4.2. Recorded Representations, All Other Timing Devices. – A printed ticket issued or stamped by a timing device shall have printed clearly thereon:

(a) the time and day when the service ends and the time and day when the service begins, except that a self-service money-operated device that clearly displays the time of day need not record the time and day when the service begins; or

(b) the time interval purchased, and the time and day that the service either begins or ends.

(Added 2015)

(Amended 1983 and 2015)

S.2.   Marking Requirements, Operating Instructions. – Operating instructions shall be clearly stated on the device.

S.3.   Interference. – The design of the EVSE shall be such that there will be no interference between the time and electrical energy measurement elements of the system.

(Added 2015)

S.4.   Provisions for Sealing. – Adequate provisions shall be made to provide security for the timing element.

(Added 2015)

S.5.   Power Interruption. – In the event of a power loss, the information needed to complete any transaction (i.e., delivery is complete and payment is settled) in progress at the time of the power loss (such as the quantity and unit price, or sales price) shall be determinable through one of the means listed below or the transaction shall be terminated without any charge for the electrical energy transfer to the vehicle:

(a) at the EVSE;

(b) at the console, if the console is accessible to the customer;

(c) via on site Internet access; or

(d) through toll-free phone access.

For EVSEs in parking areas where vehicles are commonly left for extended periods, the information needed to complete any transaction in progress at the time of the power loss shall be determinable through one of the above means for at least eight hours.

(Added 2015)

S.5.1.   Transaction Termination. – In the event of a power loss, either:

(a) the transaction shall terminate at the time of the power loss; or

(b) the EVSE may continue charging without additional authorization if the EVSE is able to determine it is connected to the same vehicle before and after the supply power outage.

In either case, there must be a clear indication on the receipt provided to the customer of the interruption, including the date and time of the interruption along with other information required under S.1.4.2. Recorded Representations, All Other Timing Devices.

(Added 2015)

S.5.2.   User Information. – The EVSE memory, or equipment on the network supporting the EVSE, shall retain information on the quantity of time and the sales price totals during a power loss.

(Added 2015)

N. Notes

N.1.   Test Method. – A timing device shall be tested with a timepiece with an error of not greater than plus or minus 15 seconds per 24-hour period. In the test of timing devices with a nominal capacity of 1 hour or less, stopwatches with a minimum division of not greater than one-fifth second shall be used. In the test of timing devices with a nominal capacity of more than one hour, the value of the minimum division on the timepiece shall be not greater than one second. Time pieces and stopwatches shall be calibrated with standard time signals as described in National Institute of Standards and Technology Special Publication 432, NIST Time and Frequency Dissemination Services, or any superseding publication.

(Amended 1978)

N.2.   Broadcast Times and Frequencies. – Time and frequency standards are broadcast by the stations listed in Table N.2. Broadcast Times and Frequencies.

Table N.2.* Broadcast Times and Frequencies			
Station	**Location, Latitude, Longitude**	**Frequency (MHz)**	**Times of Transmission (UTC)**
WWV	Fort Collins, Colorado 40E41' N 105E02' W	2.5 5.0 10.0 15.0 20.0	Continuous
WWVH	Kauai, Hawaii 21E59' N 159E46' W	2.5 5.0 10.0 15.0	Continuous
CHU	Ottawa, Canada 45E18' N 75E45' W	3.330 7.335 14.670 14.670	Continuous

*From NIST Special Publication 559, "Time and Frequency Users' Manual," 1990.

(Added 1988)

N.3. Interference Tests, EVSE. – On an EVSE equipped with a timing device used to calculate time-based charges in addition to any charges assessed for electrical energy, a test shall be conducted to ensure that there is no interference between time and electrical energy measuring elements.

(Added 2015)

T. Tolerances

T.1.   Tolerance Values. – Maintenance and acceptance tolerances for timing devices shall be as follows:

T.1.1.   For Timing Devices Other Than Those Specified in T.1.2. For Time Clocks and Time Recorders and T.1.3. On Parking Meters. – The maintenance and acceptance tolerances shall be:

(a)   On Overregistration: five seconds for any time interval of one minute or more; and

   (Amended 1986)

(b) On Underregistration: six seconds per indicated minute.

(Amended 1975 and 1986)

T.1.2.   For Time Clocks and Time Recorders. – The maintenance and acceptance tolerances on over-registration and underregistration shall be three seconds per hour, but not to exceed one minute per day.

(Amended 1975)

T.1.3.   On Parking Meters and Other Timing Devices Used to Assess Charges for Parking. – The maintenance and acceptance tolerances are shown in Table T.1.3. Maintenance and Acceptance Tolerances for Parking Meters and Other Timing Devices Used to Assess Charges For Parking.

(Amended 2015)

Table T.1.3. Maintenance and Acceptance Tolerances for Parking Meters and Other Timing Devices Used to Assess Charges for Parking		
Maintenance and Acceptance Tolerances		
Nominal Time Capacity	**On Overregistration**	**On Underregistration**
30 minutes or less	No tolerance	10 seconds per minute, but not less than 2 minutes
Over 30 minutes to and including 1 hour	No tolerance	5 minutes plus 4 seconds per minute over 30 minutes
Over 1 hour	No tolerance	7 minutes plus 2 minutes per hour over 1 hour

(Amended 2015)

T.2.   Tests Involving Digital Indications or Representations. – To the tolerances that would otherwise be applied, there shall be added an amount equal to one-half the minimum value that can be indicated or recorded.

UR.   User Requirements

UR.1.   Statement of Rates. – The following information shall be clearly, prominently, and conspicuously displayed:

(a) the price in terms of money per unit or units of time for the service dispensed; and

(b) for a timing device other than an EVSE, the number of coins the device will accept and be activated by at one time.

(Amended 1976 and 2015)

UR.2.   Time Representations. – Any time representation shall be within plus or minus two minutes of the correct time in effect in the area, except on an individual clock used only for "time out"; in addition, the time indication of the "time-out" clock shall be the same as or less than that of the "time-in" clock.

(Amended 1975)

THIS PAGE INTENTIONALL LEFT BLANK

Table of Contents

Section 5.56.(a)   Grain Moisture Meters

Section 5.56. was reorganized into two sections beginning with the 1997 Edition of NIST Handbook 44. This Section, 5.56.(a), applies to all NTEP grain moisture meters. It also applies to any grain moisture meter manufactured or placed into service after January 1, 1998.
(Code reorganized and renumbered 1996)

A. Application

A.1.   General Code. – This code applies to grain moisture meters, that is, devices used to indicate directly the moisture content of cereal grain and oil seeds. The code consists of general requirements applicable to all moisture meters and specific requirements applicable only to certain types of moisture meters. Requirements cited for "test weight per bushel" indications or recorded representations are applicable only to devices incorporating an automatic test weight per bushel measuring feature.
(Amended 2003)

A.2.   Exceptions. – This code does not apply to devices used for in-motion measurement of grain moisture content or seed moisture content.

A.3.   Type Evaluation. – The National Type Evaluation Program (NTEP) will accept for type evaluation only those devices that comply with this code. State enforcement will be based upon the effective dates identified with each requirement when specific dates are shown.
(Added 1993)

A.4.   Additional Code Requirements. – In addition to the requirements of this code, 5.56.(a) Grain Moisture Meters shall meet the requirements of Section 1.10. General Code.

S. Specifications

S.1.   Design of Indicating, Recording, and Measuring Elements.

S.1.1.   Digital Indications and Recording Elements.

(a) Meters shall be equipped with a digital indicating element.

(b) The minimum height for the digits used to display moisture content shall be 10 mm.

(c) Meters shall be equipped with a communication interface that permits interfacing with a recording element and transmitting the date, grain type, grain moisture results, test weight per volume results, and calibration version identification.

(d) A digital indicating element shall not display, and a recording element shall not record, any moisture content values or test weight per volume values before the end of the measurement cycle.

(e) Moisture content results shall be displayed and recorded as percent moisture content, wet basis. Test weight per bushel results shall be displayed and recorded as pounds per bushel. Subdivisions of these units shall be in terms of decimal subdivisions (not fractions).

(f) A meter shall not display or record any moisture content or test weight per volume values when the moisture content of the grain sample is beyond the operating range of the device, unless the moisture and test weight representations include a clear error indication (and recorded error message with the recorded representation).

(g) On multi-constituent meters (e.g., meters which also measure grain protein), provision shall be made for displaying and recording the constituent label (such as moisture, protein, etc.) to make it clear which constituent is associated with each of the displayed and recorded values.

(Added 1995)

(Amended 1993, 1994, 1995, 1996, and 2003)

S.1.2. Selecting or Recording Grain or Seed Type and Class. – Provision shall be made for selecting and recording the type and class or multi-class group (as appropriate) of grain or seed to be measured. The means to select the type and class or multi-class group of grain or seed shall be readily visible and the type and class or multi-class group of grain or seed selected shall be clearly and definitely identified. Abbreviations for grain types and multi-class groups indicated on the meter must meet the minimum acceptable abbreviations listed in Table S.1.2. Grain Types and Multi-Class Groups Considered for Type Evaluation and Calibration and Their Minimum Acceptable Abbreviations.

(Amended 1993, 1995, and 2007)

Table S.1.2. Grain Types and Multi-Class Groups Considered for Type Evaluation and Calibration and Their Minimum Acceptable Abbreviations		
Grain Type	*Grain Class*	*Minimum Acceptable Abbreviation*
Barley	All-Class Barley*	BARLEY
	Six-Rowed Barley	SRB
	Two-Rowed Barley	TRB
Corn	---	CORN
Grain Sorghum	---	SORG or MILO
Oats	---	OATS
Rice	All-Class Rough Rice*	RGHRICE
	Long Grain Rough Rice	LGRR
	Medium Grain Rough Rice	MGRR
Small Oil Seeds (under consideration)	---	---
Soybeans	---	SOYB
Sunflower seed (Oil)	---	SUNF
Wheat	All-Class Wheat*	WHEAT
	Durum Wheat	DURW
	Hard Red Spring Wheat	HRSW
	Hard Red Winter Wheat	HRWW
	Hard White Wheat	HDWW
	Soft Red Winter Wheat	SRWW
	Soft White Wheat	SWW
	Wheat Excluding Durum*	WHTEXDUR

Note: Grain Types marked with an asterisk (*) are "Multi-Class Calibrations."

[Nonretroactive as of January 1, 1998]

(Table Added 1993) (Amended 1995, 1998, and 2007)

S.1.3.   Operating Range. – A meter shall automatically and clearly indicate when the operating range of the meter has been exceeded. The operating range shall specify the following:

(a) **Temperature Range of the Meter.** – The temperature range over which the meter may be used and still comply with the applicable requirements shall be specified. The minimum temperature range shall be 10 °C to 30 °C. No moisture value may be displayed when the temperature range is exceeded. An appropriate message shall be displayed when the temperature of the meter is outside its specified operating range.

(b) **Temperature Range of each Grain or Seed.** – The temperature range for each grain or seed for which the meter is to be used shall be specified. The minimum temperature range for each grain shall be 0 °C to 40 °C. No moisture value may be displayed when the temperature range is exceeded. An appropriate error message shall be displayed when the temperature of the grain sample exceeds the specified temperature range for the grain.

(c) **Moisture Range of the Grain or Seed.** – The moisture range for each grain or seed for which the meter is to be used shall be specified. Moisture and test weight per bushel values may be displayed when the moisture range is exceeded if accompanied by a clear indication that the moisture range has been exceeded.

(Amended 2003)

(d) **Maximum Allowable Meter/Grain Temperature Difference.** – The maximum allowable difference in temperature between the meter and the sample for which an accurate moisture determination can be made shall be specified. The minimum temperature difference shall be 10 °C. No moisture value may be displayed when the maximum allowable temperature difference is exceeded. An appropriate error message shall be displayed when the difference in temperature between the meter and the sample exceeds the specified difference.

(Added 1993) (Amended 1995)

S.1.4.   Value of Smallest Unit. – The display shall permit moisture value determination to both 0.01 % and 0.1 % resolution. The 0.1 % resolution is for commercial transactions; the 0.01 % resolution is for type evaluation and calibration purposes only, not for commercial purposes. Test weight per bushel values shall be determined to the nearest 0.1 pound per bushel.

(Amended 2003)

S.1.5.   Operating Temperature.

(a) Warm-up period: When a meter is turned on it shall not display or record any usable values until the operating temperature necessary for accurate determination has been attained, or the meter shall bear a conspicuous statement adjacent to the indication stating that the meter shall be turned on for a time period specified by the manufacturer prior to use.

(b) A meter shall meet the requirements of T.2. Tolerances when operated in the temperature range of 10 °C to 30 °C (50 °F to 86 °F) or within the range specified by the meter manufacturer.

(c) If the manufacturer specifies a temperature range, the range shall be at least 20 °C (36 °F).

(Added 1993) (Amended 1995 and 1996)

S.2.   Design of Grain Moisture Meters.

S.2.1.   Minimum Sample Size. – Meters shall be designed to measure the moisture content of representative-size grain samples. The minimum allowable sample size used in analysis shall be 100 g or 400 kernels or seeds, whichever is smaller.

(Added 1993) (Amended 1995)

S.2.2.   Electric Power Supply.

S.2.2.1.   Power Supply, Voltage and Frequency.

(a) A meter that operates using alternating current must perform within the tolerances defined in Section T.2. Tolerances over the line voltage range 100 V to 130 V, or 200 V to 250 V rms as designed, and over the frequency range of 59.5 Hz to 60.5 Hz.

(b) Battery-operated instruments shall not indicate or record values outside the applicable tolerance limits when battery power output is excessive or deficient.

S.2.2.2.   Power Interruption. – A power interruption shall not cause an indicating or recording element to display or record any values outside the applicable tolerance limits.
(Added 1988)

S.2.3.   Level Indicating Means. – A meter shall be equipped with a level indicator and leveling adjustments if its performance is changed by an amount greater than the applicable tolerance when the meter is moved from a level position to a position that is out of level in any upright direction by up to 5 % (approximately three degrees). The level-indicating means shall be readable without removing any meter parts requiring a tool.
(Added 1988) (Amended 1994)

S.2.4.   Calibration Integrity.

S.2.4.1.   Calibration Version. – A meter must be capable of displaying either calibration constants, a unique calibration name, or a unique calibration version number for use in verifying that the latest version of the calibration is being used to make moisture content and test weight per bushel determinations.
(Added 1993) (Amended 1995 and 2003)

S.2.4.2.   Calibration Corruption. – If calibration constants are digitally stored in an electronically alterable form, the meter shall be designed to make automatic checks to detect corruption of calibration constants. An error message must be displayed if calibration constants have been electronically altered.
(Added 1993) (Amended 1995)

S.2.4.3.   Calibration Transfer. – *The instrument hardware/software design and calibration procedures shall permit calibration development and the transfer of calibrations between instruments of like models without requiring user slope or bias adjustments.*

[Note: Only the manufacturer or the manufacturer's designated service agency may make standardization adjustments on moisture meters. This does not preclude the possibility of the operator installing manufacturer-specified calibration constants under the instructions of the manufacturer or its designated service agency.] Standardization adjustments (not to be confused with grain calibrations) are those physical adjustments or software parameters which make meters of like type respond identically to the grain(s) being measured.
[Nonretroactive as of January 1, 1999]
(Added 1994) (Amended 1998)

S.2.5.   Provision for Sealing. – Provision shall be made for applying a security seal in a manner that requires the security seal to be broken, or for using other approved means of providing security (e.g., audit trail available at the time of inspection as defined in Table S.2.5. Categories of Device and Methods of Sealing) before any change that affects the metrological integrity of the device can be made to any mechanism.

Table S.2.5. Categories of Device and Methods of Sealing	
Categories of Device	**Methods of Sealing**
Category 1: No remote configuration capability.	Seal by physical seal or two event counters: one for calibration parameters (000 to 999) and one for configuration parameters (000 to 999). If equipped with event counters, the device must be capable of displaying, or printing through the device or through another on-site device, the contents of the counters.
Category 2: Remote configuration capability, but access is controlled by physical hardware. A device shall clearly indicate that it is in the remote configuration mode and shall not be capable of operating in the measure mode while enabled for remote configuration.	The hardware enabling access for remote communication must be at the device and sealed using a physical seal or two event counters: one for calibration parameters (000 to 999) and one for configuration parameters (000 to 999). If equipped with event counters, the device must be capable of displaying, or printing through the device or through another on-site device, the contents of the counters.
Category 3: Remote configuration capability access may be unlimited or controlled through a software switch (e.g., password). When accessed for the purpose of modifying sealable parameters, the device shall clearly indicate that it is in the configuration mode and shall not be capable of operating in the measuring mode.	An event logger is required in the device; it must include an event counter (000 to 999), the parameter ID, the date and time of the change, and the new value of the parameter (for calibration changes consisting of multiple constants, the calibration version number may be used rather than the calibration constants). A printed copy of the information must be available through the device or through another on-site device. The event logger shall have a capacity to retain records equal to 25 times the number of sealable parameters in the device, but not more than 1000 records are required. (Note: Does not require 1000 changes to be stored for each parameter.)
Category 3a: No remote capability, but operator is able to make changes that affect the metrological integrity of the device (e.g., slope, bias, etc.) in normal operation. *When accessed for the purpose of modifying sealable parameters, the device shall clearly indicate that it is in the configuration mode and shall not be capable of operating in the measuring mode.	Same as Category 3
Category 3b: No remote capability, but access to metrological parameters is controlled through a software switch (e.g., password). *When accessed for the purpose of modifying sealable parameters, the device shall clearly indicate that it is in the configuration mode and shall not be capable of operating in the measuring mode.	Same as Category 3

[Nonretroactive as of January 1, 1999]
[*Nonretroactive as of January 1, 2014]

(Amended 1998 and 2013)

Note: Zero-setting and test point adjustments are considered to affect metrological characteristics and must be sealed.
(Added 1993) (Amended 1995 and 1997)

S.2.6.   Determination of Quantity and Temperature. – The moisture meter system shall not require the operator to judge the precise volume or weight and temperature needed to make an accurate moisture determination. External grinding, weighing, and temperature measurement operations are not permitted. In addition, if the meter is capable of measuring test weight per bushel, determination of sample volume and weight for this measurement shall be fully automatic *and means shall be provided to ensure that measurements of test weight per volume are not allowed to be displayed or printed when an insufficient sample volume is available to provide an accurate measurement. [Nonretroactive as of January 1, 2004]*
(Added 1994) (Amended 1995 and 2003)

S.3.   Accessory Equipment. – When the operating instructions for a moisture meter require accessory equipment separate from and external to the moisture meter, such equipment shall be appropriate and complete for the measurement.

S.4.   Operating Instructions and Use Limitations. – The manufacturer shall furnish operating instructions for the device and accessories that include complete information concerning the accuracy, sensitivity, and use of accessory equipment necessary in obtaining a moisture content. Operating instructions shall include the following information:

(a) name and address or trademark of the manufacturer;

(b) the type or design of the device with which it is intended to be used;

(c) date of issue;

(d) the kinds or classes of grain or seed for which the device is designed to measure moisture content and test weight per bushel; and
(Amended 2003)

(e) the limitations of use, including but not confined to the moisture measurement range, grain or seed temperature, maximum allowable temperature difference between grain sample and meter, kind or class of grain or seed, moisture meter temperature, voltage and frequency ranges, electromagnetic interferences, and necessary accessory equipment.
(Added 1984)

N. Notes

N.1.   Testing Procedures. – Field evaluation of grain moisture meters shall be performed by one of the following methods:

N.1.1.   Air Oven Reference Method Transfer Standards. – Official grain samples shall be used as the official transfer standards with moisture content and test weight per bushel values assigned by the reference methods. The reference methods for moisture shall be the oven drying methods as specified by the USDA GIPSA. The test weight per bushel value assigned to a test weight transfer standard shall be the average of 10 test weight per bushel determinations using the quart kettle test weight per bushel apparatus as specified by the USDA GIPSA. Tolerances shall be applied to the average of at least three measurements on each official grain sample. Official grain samples shall be clean and naturally moist, but not tempered (i.e., water not added).
(Amended 1992, 2001, and 2003)

N.1.2. Minimum Test. – A minimum test of a grain moisture meter shall consist of tests using samples (need not exceed three) of each grain or seed type for which the device is used, and for each grain or seed type shall include the following:

(a) tests of moisture indications, using samples having at least two different moisture content values within the operating range of the device; and

(b) if applicable, tests of test weight per volume indications, with at least the lowest moisture samples used in (a) above.

(Added 2003)

(Amended 1986, 1989, and 2003)

N.1.3. Meter to Like-Type Meter Method Transfer Standards. – Properly standardized reference meters using National Type Evaluation Program approved calibrations shall be used as transfer standards. A reference meter shall be of the same type as the meter under test. Tests shall be conducted side-by-side using, as a comparison medium, grain samples that are clean and naturally moist, but not tempered (i.e., water not added).

(Added 2001)

T. Tolerances

T.1. To Underregistration and to Overregistration. – The tolerances hereinafter prescribed shall be applied to errors of underregistration and errors of overregistration.

T.2. Tolerances.

T.2.1. Air Oven Reference Method. – Maintenance and acceptance tolerances shall be as shown in Table T.2.1. Acceptance and Maintenance Tolerances Air Oven Reference Method. Tolerances are expressed as a fraction of the percent moisture content of the official grain sample, together with a minimum tolerance.

(Amended 2001)

Table T.2.1. Acceptance and Maintenance Tolerances Air Oven Reference Method		
Type of Grain, Class, or Seed	**Tolerance**	**Minimum Tolerance**
Corn, oats, rice, sorghum, sunflower	0.05 of the percent moisture content	0.8 % in moisture content
All other cereal grains and oil seeds	0.04 of the percent moisture content	0.7 % in moisture content

(Amended 2001)

T.2.2. Meter to Like-Type Meter Method. – Maintenance and acceptance tolerances shall be as shown in Table T.2.2. Acceptance and Maintenance Tolerances Meter to Like-Type Meter Method. The tolerances shall apply to all types of grain and seed.

(Added 2001)

Table T.2.2. Acceptance and Maintenance Tolerances Meter to Like-Type Meter Method	
Sample Reference Moisture	**Tolerance**
Up to 22 %	0.5 % in moisture content

(Added 2001)

T.3.   For Test Weight Per Bushel Indications or Recorded Representations. – The maintenance and acceptance tolerances on test weight per bushel indications or recorded representations shall be as shown in Table T.3. Acceptance and Maintenance Tolerances Test Weight per Bushel. Tolerances are (+) positive or (−) negative with respect to the value assigned to the official grain sample.

(Amended 1992 and 2003)

Table T.3. Acceptance and Maintenance Tolerances Test Weight per Bushel	
Type of Grain or Seed	**Tolerance (Pounds Per Bushel)**
Corn, oats	0.8
All wheat classes	0.5
Soybeans, all barley classes, all rice classes, sunflower, sorghum	0.7

(Added 2003)

UR.   User Requirements

UR.1.   Selection Requirements.

UR.1.1.   Value of the Smallest Unit on Primary Indicating and Recording Elements. – The resolution of the moisture meter display shall be 0.1 % moisture and 0.1 pounds per bushel test weight during commercial use. (Amended 2003)

UR.1.2.   See G-UR.1.2.   Environment.

UR.2.   Installation Requirements. – The grain moisture meter shall be installed in an environment within the range of temperature and/or other environmental factors specified in the operating instructions.

UR.3.   Use Requirements.

UR.3.1.   Operating Instructions. – The operating instructions for the use of the grain moisture meter shall be readily available to the user, service technician, and weights and measures official at the place of installation. It shall include a list of accessory equipment and the kinds of grain or seed to be measured with the moisture meter. (Amended 1988)

UR.3.2.   Other Devices Not Used for Commercial Measurement. – If there are other moisture meters on the premises not used for trade or determining other charges for services, these devices shall be clearly and conspicuously marked "Not for Use in Trade or Commerce."

UR.3.3.   Maintaining Integrity of Grain Samples. – Whenever there is a time lapse (temperature change) between taking the sample and testing the sample, means to prevent condensation of moisture or loss of moisture from grain samples shall be used. For example, a cold grain sample may be kept in a closed container in order to permit the cold grain to come to the operating temperature range of the meter before the grain moisture measurements are made.

UR.3.4.   Printed Tickets.

(a) Printed tickets shall be free from any previous indication of moisture content or type of grain or seed selected.

(b) The customer shall be given a printed ticket at the time of the transaction or as otherwise specified by the customer. The printed ticket shall include the date, grain type, grain moisture results, test weight per bushel, and calibration version identification. The ticket information shall be generated by the grain moisture meter system.

(Amended 1993, 1995, 2003, and 2013)

UR.3.5.   Accessory Devices. – Accessory devices, if necessary in the determination of a moisture content value, shall be in close proximity to the moisture meter and allow immediate use.

UR.3.6.   Sampling. – A grain sample shall be obtained by following appropriate sampling methods and equipment. These include, but are not limited to grain probes of appropriate length used at random locations in the bulk, the use of a pelican sampler, or other techniques and equipment giving equivalent results. The grain sample shall be taken such that it is representative of the lot.

UR.3.7.   Location. – Also see G-UR.3.3. Position of Equipment.

UR.3.8.   Level Condition. – If equipped with a level indicator, a meter shall be maintained in a level condition.
(Added 1988)

UR.3.9.   Current Calibration Data. – Grain moisture determinations shall be made using only the most recently published calibration data.
(Added 1988)

UR.3.10.   Posting of Meter Operating Range. – The operating range of the grain moisture meter shall be clearly and conspicuously posted in the place of business such that the information is readily visible from a reasonable customer position. The posted information shall include the following:

(a) The temperature range over which the meter may be used and still comply with the applicable requirements. If the temperature range varies for different grains or seed, the range shall be specified for each.

(b) The moisture range for each grain or seed for which the meter is to be used.

(c) The temperature range for each grain or seed for which the meter is to be used.

(d) The maximum allowable difference in temperature that may exist between the meter and the sample for which an accurate moisture determination can be made.

(Added 1988)

THIS PAGE INTENTIONALLY LEFT BLANK

Table of Contents

Section 5.56.(b)   Grain Moisture Meters

Section 5.56. was reorganized into two sections beginning with the 1997 Edition of NIST Handbook 44. This Section, 5.56.(b), applies to all non-NTEP grain moisture meters manufactured or placed into service before January 1, 1998.

(Code reorganized and renumbered 1996)

A. Application

A.1.   General. – This code applies to grain moisture meters; that is, devices used to indicate directly or through conversion and/or correction tables the moisture content of cereal grain and oil seeds. The code consists of general requirements applicable to all moisture meters and specific requirements applicable only to certain types of moisture meters.

A.2.   Exceptions. – This code does not apply to devices used for in-motion measurement of grain moisture content or seed moisture content.

A.3.   Additional Code Requirements. – In addition to the requirements of this code, 5.56.(b) Grain Moisture Meters shall meet the requirements of Section 1.10. General Code.

S. Specifications

S.1.   Design of Indicating and Recording Elements and of Recorded Representations.

S.1.1.   Primary Elements, General. – A meter shall be equipped with a primary indicating element and may also be equipped with a primary recording element. If the meter indicates directly and/or is equipped to record, the meter shall indicate and/or record its measurements in terms of percent moisture content, wet basis. Subdivisions of this unit shall be in terms of decimal subdivisions (not fractions). If the meter indicates in the conventional scale and requires conversion or correction tables, the resulting values after use of such tables shall be in terms of percent moisture content, wet basis. Subdivisions of this unit shall be in terms of decimal subdivisions (not fractions).

S.1.2.   Digital Indications.

S.1.2.1.   Measurement Completion. – A digital indicating element shall not display any values (either moisture content or conventional scale) before the end of the measurement cycle.

S.1.3.   Graduations.

S.1.3.1.   Length. – Graduations shall be so varied in length that they may be conveniently read.

S.1.3.2.   Width. – In any series of graduations, the width of a graduation shall in no case be greater than the width of the minimum clear interval between graduations, and the width of the main graduations shall be not more than 50 % greater than the width of subordinate graduations. Graduations shall in no case be less than 0.2 mm (0.008 in) in width.

S.1.3.3.   Clear Interval between Graduations. – The clear interval shall be not less than 0.75 mm (0.03 in) between graduations. If the graduations are not parallel, the measurement shall be made:

(a)   along the line of relative movement between the graduations at the end of the indicator; or

(b)   if the indicator is continuous, at the point of widest separation of the graduations.

S.1.4. Indicators.

S.1.4.1.   Symmetry. – The index of an indicator shall be symmetrical with respect to the graduations, at least throughout that portion of its length associated with the graduations.

S.1.4.2.   Length. – The index of an indicator shall reach to the finest graduations with which it is used, unless the indicator and the graduations are in the same plane, in which case the distance between the end of the indicator and the ends of the graduations, measured along the line of the graduations, shall be not more than 1.0 mm (0.04 in).

S.1.4.3.   Width. – The width of the index of an indicator in relation to the series of graduations with which it is used shall be not greater than:

(a) the width of the widest graduation; nor

(b) the width of the minimum clear interval between graduations.

When the index of an indicator extends along the entire length of a graduation, that portion of the index of the indicator that may be brought into coincidence with the graduation shall be of the same width as the graduation throughout the length of the index that coincides with the graduation.

S.1.4.4.   Clearance. – The clearance between the index of an indicator and the graduations shall in no case be more than 1.5 mm (0.06 in).

S.1.4.5.   Parallax. – Parallax effects shall be reduced to the practicable minimum.

S.1.5. Recording Elements.

S.1.5.1.   General. – If a meter is equipped with a recording element, it shall record in terms of percent moisture content, wet basis only, and not in terms of conventional scale.

S.1.5.2.   Measurement Completion. – A recording element shall not record any values before the end of the measurement cycle.

S.1.5.3.   Range of Moisture Content. – A recording element shall not record any values when the moisture content of the grain sample is beyond the operating range of the device.

S.1.6. Design of Direct Reading Grain Moisture Meters.

S.1.6.1.   Grain or Seed Kind and Class Selection and Recording. – Provision shall be made for selecting and recording, if equipped to record, the kind and class (as appropriate) of grain or seed to be measured. The means to select the kind and class of grain or seed shall be readily visible and the kind and class of grain or seed selected shall be clearly and definitely identified in letters (such as Wheat or WHT, HRWW, etc.).

S.1.6.2.   Operating Range. – *A meter shall automatically and clearly indicate when the operating range of the meter has been exceeded or the manufacturer shall:*

(a) *clearly and conspicuously mark the operating ranges on the meter; or*

(b) *furnish the operating ranges of the meter and the means to clearly and conspicuously display this information on or immediately adjacent to the device.*

The operating range shall specify the following:

(a) *the temperature range over which the meter may be used and still comply with the applicable requirements;*

(b) *the moisture range for each grain or seed for which the meter is to be used;*

(c) *the temperature range for each grain or seed for which the meter is to be used; and*

(d) *the maximum allowable difference in temperature between the meter and the sample for which an accurate moisture determination can be made.*

Examples of clearly indicating these conditions include an error indication, flashing the displayed moisture value, or blanking the display.
[Nonretroactive as of January 1, 1989]

(Amended 1986 and 1988)

S.1.6.3.   Value of Smallest Unit. – The value of the minimum indicated or recorded moisture indication shall not be greater than 0.1 %.

S.1.7.   Electric Power Supply.

S.1.7.1.   Power Supply, Voltage and Frequency.

(a) *A meter that operates using alternating current must perform within the tolerances defined in Section T.2. Tolerance Values over the line voltage range 100 V to 130 V, or 200 V to 250 V rms as designed, and over the frequency range of 59.5 Hz to 60.5 Hz.*

(b) *Battery-operated instruments shall not indicate or record values outside the applicable tolerance limits when battery power output is excessive or deficient.*
[Nonretroactive as of January 1, 1989]

S.1.7.2.   Power Interruption. – *A power interruption shall not cause an indicating or recording element to display or record any values outside the applicable tolerance limits.*
[Nonretroactive as of January 1, 1989]

(Added 1988)

S.1.8.   Level Indicating Means. – *A meter shall be equipped with a level indicator and leveling adjustments if its performance is changed by an amount greater than the applicable tolerance when the meter is moved from a level position to a position that is out of level in any upright direction by up to 5 % (approximately 3 degrees).*

The level-indicating means shall be readable without removing any meter parts requiring a tool.
[Nonretroactive as of January 1, 1989]

(Added 1988) (Amended 1994)

S.1.9.   Operating Temperature.

(a) *A meter shall not display or record any usable values until the operating temperature necessary for accurate determination has been attained, or the meter shall bear a conspicuous statement adjacent to the indication stating that the meter shall be turned on for a time period specified by the manufacturer prior to use.*

(b) *A meter shall meet the requirements of T.2. Tolerance Values when operated in the temperature range of 2 °C to 40 °C (35 °F to 104 °F) or within the range specified by the meter manufacturer.*

> (c) *If the manufacturer specifies a temperature range, the range shall be at least 10 °C (20 °F) and shall be marked on the device.*

[Nonretroactive as of January 1, 1989]

(Added 1988)

S.2. Design of Measuring Elements.

S.2.1. Design of Zero-Setting and Test Point Mechanisms. – If a grain moisture meter is equipped with a zero setting and/or test point mechanism(s), this (these) mechanism(s) shall be adjustable only with a tool outside and entirely separate from this mechanism or enclosed in a cabinet. This requirement shall not apply to manual operations that the operator must make (following operating instructions) in order to obtain a meter reading on a grain sample.

S.2.2. Provision for Sealing. – Provision shall be made for applying a security seal in a manner that requires the security seal to be broken before an adjustment can be made to any component of the grain moisture meter that is set by the manufacturer or authorized service representative and not intended to be adjusted by the user.

S.3. Accessory Equipment. – When the operating instructions for a moisture meter require accessory equipment separate from and external to the moisture meter, such equipment shall be appropriate and complete for the measurement.

S.3.1. Grain-Test Scale. – If the moisture meter requires the weighing of the grain sample, the weighing device shall meet the requirements of the General Code and those applicable portions of the Scales Code.

S.3.2. Thermometers or Other Temperature Sensing Equipment.

> (a) The temperature sensing equipment or thermometer shall be designed to be in direct contact with a grain sample in a closed container. It is acceptable to insert thermometer through a small hole in the lid of the container used to hold the grain sample.

> (b) A separate thermometer or other temperature sensing equipment shall have temperature divisions not greater than the temperature increments used by the manufacturer in the correction table.

(Amended 1988)

S.3.3. Conversion and Correction Tables. – Conversion and correction tables, charts, graphs, slide rules, or other apparatus to convert the conventional scale values read from a moisture meter to moisture content values, if such apparatus is required, shall be appropriate and correct for the moisture meter being used and shall be marked with the following information:

> (a) name and address or trademark of the manufacturer;

> (b) the type or design of the device with which it is intended to be used;

> (c) date of issue;

> (d) the kinds or classes of grain or seed for which the device is designed to measure moisture content;

> (e) the limitations of use, including but not confined to the moisture measurement range, grain or seed temperature, kind or class of grain or seed, moisture meter temperature, voltage and frequency ranges, electromagnetic interferences, and necessary accessory equipment; but

> (f) values exceeding any measurement range shall not be included.

(Added 1984)

S.3.4.   Operating Instructions and Use Limitations. – Operating instructions shall be furnished by the manufacturer with each device with all of the information required by paragraph S.3.3. Conversion and Correction Tables. Complete information concerning the accuracy, sensitivity, and use of accessory equipment (e.g., test weight per bushel equipment, thermometer, etc.) necessary in obtaining moisture content shall be included.

N. Notes

N.1.   Testing Procedures.

N.1.1.   Transfer Standards. – Official grain samples shall be used as the official transfer standards with moisture content values assigned by the reference methods. The reference methods shall be the oven drying methods as specified by the USDA GIPSA. Tolerances shall be applied to the average of at least three measurements on each official grain sample. Official grain samples shall be clean and naturally moist, but not tempered (i.e., water not added).
(Amended 1992)

N.1.2.   Minimum Test. – A minimum test of a grain moisture meter shall consist of tests with:

(a) samples (need not exceed three) of each grain or seed for which the device is used; and

(b) samples having at least two different moisture content values within the operating range of the device.
(Amended 1986 and 1989)

N.1.3.   Temperature Measuring Equipment. – The accuracy of accessory temperature measuring equipment shall be determined by comparison with a calibrated temperature sensor, such as a total immersion thermometer with 0.1 °C (0.2 °F) subdivisions, indicating over a range of from 0 °C to 40 °C (32 °F to 104 °F) with a maximum error of ± 0.1 °C (0.2 °F). Tests shall be conducted at two temperatures using liquid baths (e.g., ice water and room temperature water). The two temperatures selected shall not exceed the range of temperatures identified in the moisture meter operating instructions.
(Amended 1988)

T. Tolerances[1]

T.1.   To Underregistration and to Overregistration. – The tolerances hereinafter prescribed shall be applied to errors of underregistration and errors of overregistration.

T.2.   Tolerance Values. – Maintenance and acceptance tolerances shall be as shown in Table T.2. Acceptance and Maintenance Tolerances for Grain Moisture Meters. Tolerances are expressed as a fraction of the percent moisture content of the official grain sample, together with a minimum tolerance.

Table T.2. Acceptance and Maintenance Tolerances for Grain Moisture Meters		
Type of Grain or Seed	**Tolerance**	**Minimum Tolerance**
Corn, oats, rice, sorghum, sunflower	0.05 of the percent moisture content	0.8 % in moisture content
All other cereal grains and oil seeds	0.04 of the percent moisture content	0.7 % in moisture content

T.3.   For Test Weight Per Bushel Devices. – The maintenance and acceptance tolerances on separate test weight per bushel devices used to determine the test weight per bushel of grain samples for the purposes of making density

---

[1] These tolerances do not apply to tests in which grain moisture meters are the transfer standards.

corrections in moisture determination shall be 0.193 kg/hL or 0.15 lb/bu. The test methods used shall be those specified by the USDA GIPSA using a dockage-free sample of dry hard red winter wheat.

(Amended 1992 and 2003)

T.4. Thermometers or Other Temperature Sensing Equipment. – The tolerance for a separate thermometer or temperature sensing equipment used to determine the temperature of grain samples for the purpose of making temperature corrections in moisture determinations shall be ± 0.5 °C (1 °F).

(Added 1988)

UR.   User Requirements

UR.1.   Selection Requirements.

UR.1.1.   Value of the Smallest Unit on Primary Indicating and Recording Elements. – The value of the smallest unit on a moisture meter, whether the moisture meter reads directly in terms of moisture content, or when the conventional scale unit is converted or corrected to moisture content, shall be equal to or less than 0.1 %.

UR.1.2.   Environment. – Equipment shall be suitable for the environment in which it is used including, but not limited to, the effects of wind, weather, and RFI.

UR.2.   Installation Requirements. – The grain moisture meter shall be installed in an environment within the range of temperature and/or other environmental factors specified:

(a) in the operating manual; and

(b) on the conversion or correction tables if such tables are necessary for the operation of the device.

UR.3.   Use Requirements.

UR.3.1.   Operating Instructions. – The operating instructions for the use of the grain moisture meter shall be readily available to the user, service technician, and weights and measures official at the place of installation. It shall include a list of accessory equipment, conversion and correction charts if any are required to obtain moisture content values, and the kinds of grain or seed to be measured with the moisture meter.

(Amended 1988)

UR.3.2.   Other Devices not used for Commercial Measurement. – If there are other moisture meters on the premises not used for trade or determining other charges for services, these devices shall be clearly and conspicuously marked "Not for Use in Trade or Commerce."

UR.3.3.   Maintaining Integrity of Grain Samples. – Whenever there is a time lapse (temperature change) between taking the sample and testing the sample, means to prevent condensation of moisture or loss of moisture from grain samples shall be used. For example, a cold grain sample may be kept in a closed container in order to permit the cold grain to come to the operating temperature range of the meter before the grain moisture measurements are made.

UR.3.4.   Printed Tickets. – Printed tickets shall be free from any previous indication of moisture content or type of grain or seed selected.

UR.3.5.   Accessory Devices. – Accessory devices, if necessary in the determination of a moisture content value, shall be in close proximity to the moisture meter and allow immediate use.

UR.3.6.   Sampling. – A grain sample shall be obtained by following appropriate sampling methods and equipment. These include, but are not limited to, grain probes of appropriate length used at random locations in the bulk, the use of a pelican sampler, or other techniques and equipment giving equivalent results. The grain sample shall be taken such that it is representative of the lot.

**UR.3.7.	Location.** – See G-UR.3.3. Position of Equipment.

**UR.3.8.	Level Condition.** – If equipped with a level indicator, a meter shall be maintained in a level condition.
(Added 1988)

**UR.3.9.	Operating Limitation.** – Unless otherwise specified by the meter manufacturer, moisture determinations shall not be made when the difference in temperatures between the grain sample and the meter exceeds 10 °C (20 °F).
(Added 1988)

**UR.3.10.	Current Calibration Chart or Data.** – Grain moisture determinations shall be made using only the most recently published calibration charts or calibration data.
(Added 1988)

**UR.3.11.	Posting of Meter Operating Range.** – The operating range of the grain moisture meter shall be clearly and conspicuously posted in the place of business such that the information is readily visible from a reasonable customer position. The posted information shall include the following:

(a) The temperature range over which the meter may be used and still comply with the applicable requirements. If the temperature range varies for different grains or seed, the range shall be specified for each.

(b) The moisture range for each grain or seed for which the meter is to be used.

(c) The temperature range for each grain or seed for which the meter is to be used.

(d) The maximum allowable difference in temperature that may exist between the meter and the sample for which an accurate moisture determination can be made.

(Added 1988)

THIS PAGE INTENTIONALLY LEFT BLANK

Table of Contents

THIS PAGE INTENTIONALLY LEFT BLANK

Section 5.57. Near-Infrared Grain Analyzers

A. Application

A.1. General. – This code applies to near-infrared (NIR) grain analyzers; that is, devices used to indicate the constituent values (other than moisture content) of grain using near-infrared reflectance or transmittance technology. These instruments may analyze either whole grain or ground grain samples. The code consists of general requirements applicable to all NIR analyzers and specific requirements applicable only to certain types of NIR analyzers, grain types, or grain constituents. In addition to meeting the requirements of this code, a whole grain NIR analyzer that displays a measured moisture value must also comply with the requirements of the Grain Moisture Meters Code.

(Added 2001)

A.2. Exceptions. – This code does not apply to devices used for in-motion measurement of grain constituent values.

A.3. Calibrations. – The National Type Evaluation Program (NTEP) Certificate of Conformance (CC) shall indicate the native moisture basis of each calibration. The "native" moisture basis is the default moisture basis of the sealable constituent calibration (or constituent calibration pair when a non-displayed moisture calibration is also involved). If an NIR analyzer uses a self-generated moisture measurement internally but does NOT display or record a moisture value, the moisture calibration shall be considered to be a part of the constituent calibration. For such calibrations, the CC shall state: "Includes non-displayed moisture calibration." Changes to any part of such calibrations shall require changes to the CC.

(Added 2001)

A.4. Additional Code Requirements. – In addition to the requirements of this code, Near-Infrared Grain Analyzers shall meet the requirements of Section 1.10. General Code.

S. Specifications

S.1. Design of Indicating, Recording, and Measuring Elements.

S.1.1. Digital Indications and Recording Elements.

(a) *Analyzers shall be equipped with a digital indicating element.*

(b) *The minimum height for the digits used to display constituent values shall be 10 mm.*

(c) *Analyzers shall be equipped with a communication interface that permits interfacing with a recording element and transmitting the date, grain type or class, constituent values, the moisture basis for each constituent value (except moisture), and calibration version identification. If the analyzer converts constituent results to a manually entered moisture basis, the "native" concentration and the "native" moisture basis must appear on the printed ticket in addition to the converted results and the manually entered moisture basis.*

(Amended 2001 and 2003)

(d) *A digital indicating element shall not display, and a recording element shall not record, any constituent value before the end of the measurement cycle.*

(e) *Constituent content shall be recorded and displayed as percent of total mass at the specified moisture basis. The moisture basis shall also be recorded and displayed for each constituent content result (except moisture). If a whole grain analyzer that is calibrated to display results on an "as is" moisture basis does NOT display or record a moisture value, it must clearly indicate that results are expressed*

on an "as is" moisture basis. Ground grain analyzers must ALWAYS display and record a moisture measurement for "as is" content results (except moisture).
(Amended 2001 and 2003)

    (f) An analyzer shall not display or record any constituent value that is beyond the operating range of the device unless the constituent value representation includes a clear error indication (and recorded error message with the recorded representation).

    (g) If an NIR analyzer is used to determine a moisture value, either to determine the moisture of an "as is" constituent content measurement or to convert from one moisture basis to another, the moisture measurement must be concurrent with the measurement of other constituents.

    (h) If the analyzer incorporates a built-in printer, or if a printer is available as an accessory to the analyzer, the information appearing on the printout shall be arranged in a consistent and unambiguous manner.
    (Added 2003)
[Nonretroactive as of January 1, 2003]
(Added 2001)

S.1.2. Selecting and Recording Grain Class and Constituent. – Provision shall be made for selecting and recording the type or class or multi-class group of grain and the constituent(s) to be measured. The means to select the grain type or class or multi-class group and the constituent(s) shall be readily visible and the type or class or multi-class group of grain and the constituent(s) selected shall be clearly and definitely identified in letters (such as HRWW, HRSW, WHEAT, etc., or PROT, etc.). A symbol to identify the display of the type or class or multi-class group of grain and constituent(s) selected is permitted provided that it is clearly defined adjacent to the display. Minimum acceptable abbreviations are listed in Table S.1.2. Grain Types and Multi-Class Groups Considered for Type Evaluation and Calibration and Their Minimum Acceptable Abbreviations.
[Nonretroactive as of January 1, 2003]

If more than one calibration is included for a given grain type, the calibrations must be clearly distinguished from one another.
[Nonretroactive as of January 1, 2004]
(Amended 2003 and 2007)

Table S.1.2. Grain Types and Multi-Class Groups Considered for Type Evaluation and Calibration and Their Minimum Acceptable Abbreviations		
Grain Type	**Grain Class**	**Minimum Acceptable Abbreviation**
Barley	Two-Rowed Barley	TRB
	Six-Rowed Barley	SRB
	All-Class Barley*	BARLEY
Corn	---	CORN
Soybeans	---	SOYB
Wheat	All-Class Wheat*	WHEAT
	Durum Wheat	DURW
	Hard Red Spring Wheat	HRSW
	Hard Red Winter Wheat	HRWW
	Hard White Wheat	HDWW
	Soft Red Winter Wheat	SRWW
	Soft White Wheat	SWW
	Wheat Excluding Durum*	WHTEXDUR

[Note: Grain Types marked with an asterisk (*) are "Multi-Class Calibrations"]
[Nonretroactive as of January 1, 2003]
(Table Amended 2001 and 2007)

S.1.3.   Operating Range. *– An analyzer shall automatically and clearly indicate when the operating range of the device has been exceeded. The statement of the operating range shall be specified in the operator's manual and shall operate as follows:*

(a) *The ambient temperature range over which the analyzer may be used and still comply with the applicable requirements shall be specified. The minimum temperature range shall be 10 °C to 30 °C. No constituent value may be displayed when the temperature range is exceeded. An appropriate error message shall be displayed when the temperature of the analyzer is outside its specified operating range.*

(b) *The constituent range at the moisture basis specified in Table N.1.1. Constant Moisture Basis for Type Evaluation and Field Inspections shall be specified for each grain or seed for which the analyzer is to be used. A constituent value may be displayed when the constituent range is exceeded if accompanied by a clear indication that the constituent range has been exceeded.*
(Amended 2001)

(c) *For whole grain analyzers only, the temperature range shall be specified for each grain or seed for which an analyzer is to be used. The minimum temperature range for each grain shall be 10 °C to 30 °C. No constituent value may be displayed when the temperature range is exceeded. An appropriate error message shall be displayed when the temperature of the grain sample exceeds the temperature range for the grain. The requirements of this subsection (c) are not applicable to ground grain analyzers.*

(d) *For whole grain analyzers, the maximum allowable difference in temperature between the instrument environment (ambient temperature) and the sample for which an accurate constituent determination can be made shall be specified. The minimum temperature range shall cover at least 10 °C. No constituent value may be displayed when the maximum allowable temperature difference is exceeded. An appropriate error message shall be displayed when the difference between the ambient temperature and*

the sample temperature exceeds the specified difference. The requirements of this subsection (d) are not applicable to ground grain analyzers.
[Nonretroactive and effective as of January 1, 2003]

S.1.4. Operating Temperature.

(a) *An analyzer shall not display or record any usable values until the internal operating temperature necessary to meet tolerance requirements has been attained, or the analyzer shall bear a conspicuous statement adjacent to the indication stating that the analyzer shall be turned on for a time period specified by the manufacturer prior to use.*

(b) *If an instrument does not meet tolerance requirements because there is an upper internal operating temperature limit that could be exceeded when operating within the ambient temperature range specified by the manufacturer, then a means of sensing and indicating an over-temperature condition must be provided.*
[Nonretroactive as of January 1, 2003]

S.1.5. Value of Smallest Unit. – *The display shall permit constituent value determination to both 0.01 % and 0.1 % resolution. The 0.1 % resolution is for commercial transactions; the 0.01 % resolution is for calibration purposes only, not for commercial purposes.*
[Nonretroactive as of January 1, 2003]

S.2. Design of NIR Analyzers.

S.2.1. Minimum Sample Size. – *Analyzers shall be designed to measure constituent values of representative size grain samples. The minimum allowable sample size used in analysis shall be 20 g.*
[Nonretroactive as of January 1, 2003]

S.2.2. Electric Power Supply.

S.2.2.1. Power Supply, Voltage and Frequency. – *An analyzer that operates using alternating current must perform within tolerance requirements over the line voltage range 100 V to 130 V and over the frequency range of 59.5 Hz to 60.5 Hz.*
[Nonretroactive as of January 1, 2003]

S.2.2.2. Power Interruption. – *A power interruption shall not cause an indicating or recording element to display or record any values outside the applicable tolerance limits.*
[Nonretroactive as of January 1, 2003]

S.2.3. Level Indicating Means. – *Analyzers shall be equipped with a level indicator and leveling adjustments if its performance is changed by an amount greater than the tolerance requirement when the instrument is moved from a level position into a position that is out of level in any upright direction by up to 5 % (approximately three degrees). The level-indicating means shall be readable without removing any instrument parts requiring a tool.*
[Nonretroactive as of January 1, 2003]

S.2.4. Environmental Conditions. – *Instrument optics and electronics must be protected from exposure to dust by either sealing these areas or by protecting them with a dust filtration system suitable for the removal of air-borne grain dust.*
[Nonretroactive as of January 1, 2003]

S.2.5. Calibration Transfer and Verification.

S.2.5.1. Calibration Transfer. – *The instrument hardware/software design and calibration procedures shall permit calibration development and the mathematical transfer of calibrations between instruments of like models.*

Note: Only the manufacturer or the manufacturer's designated service agency may make calibration transfer or slope adjustments on near-infrared grain analyzers and, except for instrument failure and repair, only during a prescribed period of time during the year. This does not preclude the possibility of the operator installing the manufacturer-specified calibration constants or standardization parameters under the instructions of the manufacturer or the manufacturer's designated service agency. Nor does it preclude operator bias adjustments when made under the conditions specified in UR.2.8. Slope and Bias Adjustments.
[Nonretroactive and effective as of January 1, 2003]

(Note added 1995) (Amended 1995)

S.2.5.2.   Calibration Version. *– An instrument must be capable of displaying either calibration constants, a unique calibration name, or a unique calibration version number for use in verifying that the latest version of the calibration is being used to make constituent determinations, and that the appropriate instrument settings have been made for the calibration being used.*
[Nonretroactive as of January 1, 2003]

(Amended 2001)

S.2.5.3.   Calibration Corruption. *– If calibration constants are digitally stored in an electronically alterable form, the analyzer shall be designed to make automatic checks to detect corruption of calibration constants. An error message must be displayed if calibration constants have been electronically altered.*
[Nonretroactive as of January 1, 2003].

S.2.6.   Provision for Sealing. *– An event logger is required in the device; it must include an event counter (000 to 999), the parameter ID, the date and time of the change, and the new value of the parameter (for calibration changes consisting of multiple constants, the calibration version number may be used rather than the calibration constants.)*

*A printed copy of the information must be available through the device or through another on-site device. The event logger shall have a capacity to retain records equal to 25 times the number of sealable parameters in the device, but not more than 1000 records are required. (**Note:** Does not require 1000 changes to be stored for each parameter.)*
[Nonretroactive as of January 1, 2003]

(Amended 1997)

S.3.   Accessory Equipment. *– When the operating instructions for an NIR analyzer require accessory equipment separate from and external to the analyzer, such equipment shall be appropriate and complete for the measurement.*
[Nonretroactive as of January 1, 2003]

S.3.1. Grinders. *– The make and model of grinder used for ground grain NIR analyzers must be specified by the manufacturer and required as auxiliary equipment in the determination of constituent values for applicable grain types.*
[Nonretroactive as of January 1, 2003]

S.4.   Operating Instructions and Use Limitations. *– The manufacturer shall furnish operating instructions for the device and accessories that include complete information concerning the accuracy, sensitivity, and use of accessory equipment necessary in obtaining a constituent value. Operating instructions shall include the following information:*

(a)   name and address or trademark of the manufacturer;

(b)   the type or design of the device for which the operating instructions are intended to be used;

(c)   date of issue;

(d)   the kind or classes of grain or seed for which the device is designed to measure constituent values; and

(e) *the limitations of use, including but not limited to constituent range, grain or seed temperature, kind or class of grain or seed, instrument temperature, voltage and frequency ranges, electromagnetic interferences, and necessary accessory equipment.*

[Nonretroactive as of January 1, 2003]

N. Notes

N.1. Testing Procedures.

N.1.1.   Field Inspection. – Whole grain samples shall be used as the official field inspection standards. Five samples per grain type or class shall be used to check instrument performance. Each sample will be analyzed once. One of the samples will be analyzed an additional four times to test instrument repeatability. For ground grain instruments, the ground sample will be repacked four times. A new grind is not required. Test results must be converted to the standard moisture bases shown in Table N.1.1. Constant Moisture Basis for Type Evaluation and Field Inspection before applying the tolerances of Table T.2. Acceptance and Maintenance Tolerances for NIR Grain Analyzers. Test results on whole grain analyzers that produce results on an "as is" basis without displaying or recording a moisture value shall be converted to the standard moisture bases shown in Table N.1.1. Constant Moisture Basis for Type Evaluation and Field Inspection using sample moisture values determined with the facility's moisture meter (which must be certified for commercial use).

(Amended 2001)

Table N.1.1. *Constant Moisture Basis for Type Evaluation and Field Inspection*		
Grain Type or Class	*Constituents(s)*	*Moisture Basis*
Durum Wheat, Hard Red Spring Wheat, Hard Red Winter Wheat, Hard White Wheat, Soft Red Winter Wheat, Soft White Wheat	*Protein*	*12 %*
Soybeans	*Protein* *Oil*	*13 %*
Two-rowed Barley *Six-rowed Barley*	*Protein*	*0 % (dry basis)*
Corn	*Protein* *Oil* *Starch*	*0 % (dry basis)*

[Nonretroactive as of January 1, 2003]
(Table Added 2001)

Constituent values shall be assigned to test samples by the Grain Inspection, Packers and Stockyards Administration (GIPSA). Tolerances shall be applied to individual sample measurements, the average of individual measurements on each of the five test samples, and the maximum difference (range) in results for five analyses on one of the test samples.

(Amended 2001)

N.1.2.   Standard Reference Samples. – Reference samples used for field inspection purposes shall be clean and selected to reasonably represent the constituent range. These samples shall be selected such that the difference between constituent values obtained using the GIPSA standard reference method and an official GIPSA NIR grain analyzer does not exceed one-half of the acceptance tolerance shown in Table T.2. Acceptance and Maintenance Tolerances for NIR Grain Analyzers for individual test samples or 0.375 times the acceptance tolerance shown for the average of five samples.

(Amended 2001and 2003)

T. Tolerances

T.1.   To Underregistration and to Overregistration. – The tolerances hereinafter prescribed shall be applied to errors of underregistration and errors of overregistration and shall be based on constituent values expressed at the moisture bases shown in Table N.1.1. Constant Moisture Basis for Type Evaluation and Field Inspection.
(Amended 2001)

T.2.   Tolerance Values. – Acceptance and maintenance tolerances shall be equal. Tolerances for individual samples and the average for five samples are as shown in Table T.2. Acceptance and Maintenance Tolerances for NIR Grain Analyzers.

Table T.2. Acceptance and Maintenance Tolerances for NIR Grain Analyzers				
Type or Class of Grain	**Constituent**	**Individual Samples (percent)**	**Average for Five Samples (percent)**	**Range for Five Retests (percent)**
Durum Wheat, Hard Red Spring Wheat, Hard Red Winter Wheat, Hard White Wheat, Soft Red Winter Wheat, Soft White Wheat	protein	0.60	0.40	0.40
Soybeans	protein	0.80	0.60	0.60
	oil	0.70	0.50	0.50
Two-rowed Barley Six-rowed Barley	protein	0.70	0.50	0.50
Corn	protein	0.80	0.60	0.60
	oil	0.70	0.50	0.50
	starch	1.00	0.80	0.80

(Amended 2001)

UR.   User Requirements

UR.1.   Installation Requirements. – The NIR analyzer shall be installed in an environment within the range of temperature and/or other environmental factors specified in the operating manual.

UR.2.   User Requirements.

UR.2.1.   Operating Instructions. – The operating instructions for the NIR analyzer shall be readily available to the user, service technician, and weights and measures official at the place of installation. It shall include a list of accessory equipment if any are required to obtain constituent values, and the type or class of grain to be measured with the NIR analyzer. If an NIR analyzer has the capability, the user is permitted to select the moisture basis to be used on any measurement.
(Amended 2001)

UR.2.2.   Other Devices Not Used for Commercial Measurement. – If there are other NIR analyzers on the premises not used for trade or determining other charges for services, these devices shall be clearly and conspicuously marked "Not for Use in Trade or Commerce."

UR.2.3.   Printed Tickets.

    (a) Printed tickets shall be free from any previous indication of constituent or grain type selected. The printed ticket shall indicate constituent values and the moisture basis associated with each constituent value (except moisture). If the analyzer is calibrated to display results on an "as is" moisture basis and does NOT display or record a moisture value, the ticket must clearly indicate that results are expressed on an "as is" moisture basis.

    (Amended 2001)

    (b) The customer shall be given a printed ticket showing the date, grain type or class, constituent results, and calibration version identification. If the analyzer converts constituent results to a manually entered moisture basis, the "native" concentration and the "native" moisture basis must appear on the printed ticket in addition to the converted results and the manually entered moisture basis. If the manually entered moisture basis is intended to be the moisture value for an "as is" constituent concentration measurement, that moisture value must have been obtained on the same sample and must have been measured on a moisture meter certified for commercial use. The information presented on the ticket shall be arranged in a consistent and unambiguous manner. The ticket shall be generated by the near-infrared grain analyzer system.

[Nonretroactive as of January 1, 2003]

(Amended 2001)

UR.2.4.   Grinders. – Place grinders in a separate room from the NIR analyzer to avoid instrument contamination. If a separate room is not available, the grinder may be in the same room with the NIR analyzer provided the grinder is not placed within one meter of the air intake on the NIR.

UR.2.5.   Sampling. – Samples shall be obtained by following appropriate sampling methods and equipment. These include, but are not limited to grain probes of appropriate length used at random locations in the bulk, the use of a pelican sampler, or other techniques and equipment giving equivalent results. The sample shall be taken such that it is representative of the lot. If an NIR analyzer permits user entry of the moisture value for an "as is" constituent measurement, that moisture value must have been obtained on the same sample and must have been measured on a moisture meter certified for commercial use.

(Amended 2001)

UR.2.6.   Level Condition. – If equipped with a level indicator, an analyzer shall be maintained in a level condition.

UR.2.7.   Operating Limitation. – Constituent determinations shall not be made when the difference in temperatures between the grain sample and the instrument environment (ambient temperature) exceeds manufacturer recommendations.

UR.2.8.   Slope and Bias Adjustments. – Bias changes shall be made only on the basis of tests run on a current set of Standard Reference Samples (SRS) traceable to GIPSA Master Instruments.[1] A written explanation and record of all calibration changes, including those changes made by a manufacturer or the manufacturer's designated service agency, shall be maintained. The log shall indicate the date and magnitude of changes in bias and slope constants and the instrument serial number. A Calibration Adjustment Data Sheet for each log entry shall be available for inspection upon request by the field inspector. Data Sheets shall be retained by the user for a period of no less than 18 months following any calibration adjustment. The Data Sheet must show: date of test and adjustment, serial number of the instrument, calibration identification, the nature of the adjustment, the unique identification number and source of sample sets used, and, for each sample in the set, reference values, initial instrument results (except in the cases of instrument failure and repair), and instrument results after calibration adjustment or instrument repair.

(Amended 1995)

---

[1] Established error must be known.

Table of Contents

Section 5.58.   Multiple Dimension Measuring Devices

A. Application

A.1.   General. – This code applies to dimension and volume measuring devices used for determining the dimensions and/or volume of objects for the purpose of calculating freight, storage, or postal charges based on the dimensions and/or volume occupied by the object. A multiple dimension measuring device:

(a)   is generally used to measure hexahedron-shaped objects; and
(Added 2008)

(b)   may be used to measure irregularly-shaped objects.
(Added 2008)

(Amended 2008)

A.2.   Other Devices Designed to Make Multiple Measurement Automatically to Determine a Volume. – Insofar as they are clearly applicable, the provisions of this code apply also to devices designed to make multiple measurements automatically to determine a volume for other applications as defined by Section 1.10. General Code paragraph G-A.1. Commercial and Law-Enforcement Equipment.

A.3.   Additional Code Requirements. – In addition to the requirements of this code, Multiple Dimension Measuring Devices shall meet the requirements of Section 1.10. General Code.

A.4.   Exceptions. – This code does not apply to:

(a)   devices designed to indicate automatically (with or without value-computing capabilities) the length of fabric passed through the measuring elements (also see Section 5.50. for Fabric-Measuring Devices);

(b)   devices designed to indicate automatically the length of cordage, rope, wire, cable, or similar flexible material passed through the measuring elements (also see Section 5.51. for Wire- and Cordage-Measuring Devices); or

(c)   any linear measure, measure of length, or devices used to measure individual dimensions for the purpose of assessing a charge per unit of measurement of the individual dimension (also see Section 5.52. for Linear Measures).

A.5.   Type Evaluation. – The National Type Evaluation Program (NTEP) will accept for type evaluation only those devices that comply with all requirements of this code.

S. Specifications

S.1.   Design of Indicating and Recording Elements and of Recorded Representations.

S.1.1.   Zero or Ready Indication.

(a)   Provision shall be made to indicate or record either a zero or ready condition.

(b)   A zero or ready condition may be indicated by other than a continuous digital zero indication, provided that an effective automatic means is provided to inhibit a measuring operation when the device is in an out-of-zero or non-ready condition.

S.1.2.   Digital Indications. – Indicated and recorded values shall be presented digitally.

S.1.3.   **Negative Values.** – Except when in the tare mode, negative values shall not be indicated or recorded.

S.1.4.   **Dimensions Indication.** – If in normal operation the device indicates or records only volume, a testing mode shall be provided to indicate dimensions for all objects measured.

S.1.5.   **Value of Dimension/Volume Division Units.** – The value of a device division "d" expressed in a unit of dimension shall be presented in a decimal format. The value of "d" for each measurement axis shall be in the same unit of measure and expressed as:

(a) 1, 2, or 5;

(b) a decimal multiple or submultiple of 1, 2, or 5; or

(c) a binary submultiple of a specific U.S. customary unit of measure.

Examples: device divisions may be 0.01, 0.02, 0.05; 0.1, 0.2, or 0.5; 1, 2, or 5; 10, 20, 50, or 100; 0.5, 0.25, 0.125, 0.0625, etc.
(Amended 2016)

S.1.5.1.   **For Indirect Sales.** – In addition to the values specified in S.1.5. Value of Dimension/Volume Division Units, the value of the division may be 0.3 inch and 0.4 inch.

S.1.5.2.   **Devices Capable of Measuring Irregularly-Shaped Objects.** – For devices capable of measuring irregularly shaped objects, the value of the division size (d) shall be the same for the length axis (x) and the width axis (y) and may be different for the height axis (z), provided that electronic rotation of the object to determine the smallest hexahedron is calculated in only a two-dimension horizontal plane, retaining the stable side plane as the bottom of the hexahedron.
(Added 2008)

S.1.6.   **Customer Indications and Recorded Representations.** – Multiple dimension measuring devices or systems must provide information as specified in Table S.1.6. Required Information to be Provided by Multiple Dimension Measuring Systems. As a minimum, all devices or systems must be able to meet either column I or column II in Table S.1.6. Required Information to be Provided by Multiple Dimension Measuring Systems.
(Amended 2004)

Information	Column I[1] Provided by device	Column II[1] Provided by invoice or other means		Column III Provided by invoice or other means as specified in contractual agreement
Table S.1.6. **Required Information to be Provided by Multiple Dimension Measuring Systems**				
		Customer present	Customer not present	
1. Device identification[2]	D or P	P	P	P or A
2. Error message (when applicable)	D or P	P	N/A	N/A
3. Hexahedron dimensions[3]	D or P	P	P	P or A
4. Hexahedron volume (if used)[3]	D or P	P	P	P or A
5. Actual weight (if used)[3]	D or P	P	P	P or A
6. Tare (if used)[3]	D or P	N/A	N/A	N/A
7. Hexahedron measurement statement[4]	D or P or M	P	P	P or G

A = AVAILABLE UPON REQUEST BY CUSTOMER[5]

D = DISPLAYED

G = PUBLISHED GUIDELINES OR CONTRACTS

M = MARKED

N/A = NOT APPLICABLE

P = PRINTED or RECORDED IN A MEMORY DEVICE and AVAILABLE UPON REQUEST BY CUSTOMER[5]

Notes:

[1] As a minimum all devices or systems must be able to meet either column I or column II.

[2] This is only required in systems where more than one device or measuring element is being used.

[3] Some devices or systems may not utilize all of these values; however as a minimum either hexahedron dimensions or hexahedron volume must be displayed or printed.

[4] This is an explanation that the dimensions and/or volume shown are those of the smallest hexahedron in which the object that was measured may be enclosed rather than those of the object itself.

[5] The information "available upon request by customer" shall be retained by the party having issued the invoice for at least 30 calendar days after the date of invoicing.

(Amended 2004)

S.1.7.   Minimum Lengths. – Except for entries of tare, the minimum length to be measured by a device is 12 divisions. The manufacturer may specify a longer minimum length.

S.1.8.   Indications Below Minimum and Above Maximum. – When objects are smaller than the minimum dimensions identified in paragraph S.1.7. Minimum Lengths or larger than any of the maximum dimensions plus 9 d, and/or maximum volume marked on the device plus 9 d, or when a combination of dimensions for the object being measured exceeds the measurement capability of the device, the indicating or recording element shall either:

(a) not indicate or record any usable values; or

(b) identify the indicated or recorded representation with an error indication.
(Amended 2004)

S.1.9.   Operating Temperature. – An indicating or recording element shall not indicate nor record any usable values until the operating temperature necessary for accurate measuring and a stable zero reference or ready condition has been attained.

S.1.10.   Adjustable Components. – Adjustable components shall be held securely in adjustment and, except for a zeroing mechanism (when applicable), shall be located within the housing of the element.

S.1.11.   Provision for Sealing.

(a) A device shall be designed with provision(s) for applying a security seal that must be broken, or for using other approved means of providing security (e.g., data change audit trail available at the time of inspection), before any change that detrimentally affects the metrological integrity of the device can be made to any measuring element.

(b) Audit trails shall use the format set forth in Table S.1.11. Categories of Devices and Methods of Sealing for Multiple Dimension Measuring Systems.

Table S.1.11. Categories of Devices and Methods of Sealing for Multiple Dimension Measuring Systems	
Categories of Devices	**Methods of Sealing**
Category 1: No remote configuration.	Seal by physical seal or two event counters: one for calibration parameters and one for configuration parameters.
Category 2: Remote configuration capability, but access is controlled by physical hardware. Device shall clearly indicate that it is in the remote configuration mode and record such message if capable of printing in this mode.	The hardware enabling access for remote communication must be at the device and sealed using a physical seal or two event counters: one for calibration parameters and one for configuration parameters.
Category 3: Remote configuration capability access may be unlimited or controlled through a software switch (e.g., password).	An event logger is required in the device; it must include an event counter (000 to 999), the parameter ID, the date and time of the change, and the new value of the parameter. A printed copy of the information must be available through the device or through another on-site device. The event logger shall have a capacity to retain records equal to 10 times the number of sealable parameters in the device, but not more than 1000 records are required. (**Note:** Does not require 1000 changes to be stored for each parameter.)

S.2.   Design of Zero and Tare.

S.2.1.   Zero or Ready Adjustment. – A device shall be equipped with means by which the zero reference or ready condition can be adjusted, or the zero reference or ready condition shall be automatically maintained. The zero reference or ready control circuits shall be interlocked so that their use is prohibited during measurement operations.

S.2.2.   Tare. – The tare function shall operate only in a backward direction (that is, in a direction of under-registration) with respect to the zero reference or ready condition of the device. The value of the tare division or increment shall be equal to the division of its respective axis on the device. There shall be a clear indication that tare has been taken.

S.2.2.1.   Maximum Value of Tare for Multi-Interval (Variable Division-Value Devices. – A multi-interval device shall not accept any tare value greater than the maximum capacity of the lowest range of the axis for which the tare is being entered.

(Added 2016)

S.2.2.2.   Net Values, Mathematical Agreement. – All net values resulting from a device subtracting a tare entry from a gross value indication shall be indicated and recorded, if so equipped, to the nearest division of the measuring range in which the net value occurs. In instances where the tare value entered on a multi-interval device is in a lower partial measuring range (or segment) than the gross indication, the system shall either alter the tare entered or round the net result after subtraction of the tare in order to achieve correct mathematical agreement.

Consider a multi-interval device having two partial measuring ranges for the "x" axis:

- Partial measuring range 1: 0 to 100 inches in 0.2 inch increments

- Partial measuring range 2: 100 to 300 inches in 0.5 inch increments

The following examples clarify the two acceptable methods this device can use to achieve mathematical agreement when tare has been entered in a lower partial measuring range than the gross indication.
(Added 2016)

Acceptable Example 1. Altering of a Tare Entry to Achieve Accurate Net Indication			
Gross Indication of Item Being Measured	**Tare Entered**	**Value of Tare after Being Altered by the Device**	**Acceptable Net Indication**
154.5 in	41.2 in	41.0 in	113.5 in
154.5 in	41.4 in	41.5 in	113.0 in

(Added 2016)

Acceptable Example 2. Rounding of the Net Result (Following the Subtraction of Tare) to Achieve Accurate Net Indication			
Gross Indication of Item Being Measured	**Tare Entered**	**Net Result Before Rounding (Gross Indication minus Tare Entered)**	**Acceptable Net Indication Rounded to Nearest 0.5 Inch**
154.5 in	41.2 in	113.3 in	113.5 in
154.5 in	41.4 in	113.1 in	113.0 in

(Added 2016)

S.3.   Systems with Two or More Measuring Elements. – A multiple dimension measuring system with a single indicating or recording element, or a combination indicating-recording element, that is coupled to two or more measuring elements with independent measuring systems, shall be provided with means to prohibit the activation of any measuring element (or elements) not in use, and shall be provided with automatic means to indicate clearly and definitely which measuring element is in use.

Note: This requirement does not apply to individual devices that use multiple emitters/sensors within a device in combination to measure objects in the same measurement field.

(Amended 2004)

S.4. Marking Requirements. – (Also see G-S.1. Identification, G-S.4. Interchange or Reversal of Parts, G-S.5.2.5. Permanence, G-S.6. Marking Operational Controls, Indications, and Features, G-S.7. Lettering, G-UR.2.1.1. Visibility of Identification, and G-UR.3.1. Method of Operation.)

S.4.1. Multiple Dimension Measuring Devices, Main Elements, and Components of Measuring Devices. – Multiple dimension measuring devices, main elements of multiple dimension measuring devices when not contained in a single enclosure for the entire dimension/volume measuring device, and other components shall be marked as specified in Table S.4.1.a. and explained in the accompanying notes, Table S.4.1.b. Multiple Dimension Measuring Systems Notes for Table S.4.1.a.

Table S.4.1.a. Marking Requirements for Multiple Dimension Measuring Systems				
	Multiple Dimension Measuring Equipment			
To Be Marked With ⇓	**Multiple Dimension Measuring Device and Indicating Element in Same Housing**	**Indicating Element not Permanently Attached to Multiple Dimension Measuring Element**	**Multiple Dimension Measuring Element not Permanently Attached to the Indicating Element**	**Other Equipment (1)**
Manufacturer's ID	x	x	x	x
Model Designation	x	x	x	x
Serial Number and Prefix	x	x	x	x (2)
Certificate of Conformance Number (8)	x	x	x	x (8)
Minimum and Maximum Dimensions for Each Axis for Each Range in Each Axis (3)(9)	x	x	x	
Value of Measuring Division, d (for each axis and range) (9)	x	x	x	
Temperature Limits (4)(9)	x	x	x	
Minimum and Maximum speed (5)(9)	x	x	x	
Special Application (6)(9)	x	x	x	
Limitation of Use (7)(9)	x	x	x	

(Amended 2016)

| **Table S.4.1.b.** |
| **Multiple Dimension Measuring Systems Notes for Table S.4.1.a.** |

1. Necessary to the dimension and/or volume measuring system, but having no effect on the measuring value, e.g., auxiliary remote display, keyboard, etc.

2. Modules without "intelligence" on a modular system (e.g., printer, keyboard module, etc.) are not required to have serial numbers.

3. The minimum and maximum dimensions (using upper or lower case type) shall be marked. For example:
   Length: min _____   max _____
   Width: min _____   max _____
   Height: min _____   max _____

4. Required if the range is other than − 10 °C to 40 °C (14 °F to 104 °F).

5. Multiple dimension measuring devices, which require that the object or device be moved relative to one another, shall be marked with the minimum and maximum speeds at which the device is capable of making measurements that are within the applicable tolerances.

6. A device designed for a special application rather than general use shall be conspicuously marked with suitable words visible to the operator and the customer restricting its use to that application.

7. Materials, shapes, structures, combination of object dimensions, speed, spacing, minimum protrusion size, or object orientations that are inappropriate for the device or those that are appropriate.

8. Required only if a Certificate of Conformance has been issued for the equipment.

9. This marking information may be readily accessible via the display. Instructions for displaying the information shall be described in the NTEP CC.

(Amended 2004, 2008, and 2016)

S.4.2.   Location of Marking Information. – The required marking information shall be so located that it is readily observable without the necessity of the disassembly of a part requiring the use of any means separate from the device.

N. Notes

N.1. Test Procedures.

N.1.1.   General. – The device shall be tested using test standards and objects of known and stable dimensions.

N.1.2.   Position Test. – Measurements are made using different positions of the test object and consistent with the manufacturer's specified use for the device.

> **N.1.2.1.   Irregularly-Shaped Test Object Placement.** – Irregularly-shaped test objects must be measured while placed on a stable side. The rotation of the object to determine the smallest hexahedron should be calculated in a two-dimensional plane, retaining the stable side plane as the bottom of the hexahedron.
> (Added 2008)

N.1.3.   Disturbance Tests, Field Evaluation. – A disturbance test shall be conducted at a given installation when the presence of disturbances specified in T.6. has been verified and characterized if those conditions are considered "usual and customary."

N.1.4.   Test Object Size. – Test objects may vary in size from the smallest dimension to the largest dimension marked on the device, and for field verification examinations, shall be an integer multiple of "d."

N.1.4.1.   Test Objects. – Verification of devices may be conducted using appropriate test objects of various sizes and of stable dimensions. Test object dimensions must be known to an expanded uncertainty (coverage factor $k = 2$) of not more than one-third of the applicable device tolerance. The dimensions shall also be checked to the same uncertainty when used at the extreme values of the influence factors.

The dimension of all test objects shall be verified using a reference standard that is traceable to NIST (or equivalent national laboratory) and meet the tolerances expressed in NIST Handbook 44 Fundamental Considerations, paragraph 3.2. (i.e., one-third of the smallest tolerance applied to the device).
(Added 2004)

N.1.4.2.   Irregularly-Shaped Test Objects. – For irregularly-shaped test objects, at least one angle shall be obtuse and the smallest dimension for an axis shall be equal to or greater than the minimum dimension for that axis.
(Added 2008)
(Amended 2008 and 2012)

N.1.5.   Digital Zero Stability. – A zero indication change test shall be conducted on all devices which show a digital zero.   After the removal of any test object, the zero indication shall not change.   (Also see G-UR.4.2. Abnormal Performance.)

T. Tolerances

T.1.   Design. – The tolerance for a multiple dimension measuring device is a performance requirement independent of the design principle used.

T.2.   Tolerance Application.

T.2.1.   Type Evaluation. – For type evaluations, the tolerance values apply to tests within the influence factor limits of temperature and power supply voltage specified in T.5.1. Temperature and T.5.2. Power Supply Voltage.

T.2.2.   Subsequent Verification. – For subsequent verifications, the tolerance values apply regardless of the influence factors in effect at the time of the verification.   (Also see G-N.2. Testing with Nonassociated Equipment.)

T.2.3.   Multi-interval (Variable Division-Value) Devices. – When there exist two or more partial measuring ranges (or segments) specified for any of the "dimensioning" axes (length (x), width (y), or height (z)) and the division values corresponding to those partial measuring ranges (or segments) within the same "dimensioning" axis differ, the tolerance values shall be based on the value of the division of the range in use.
(Amended 2016)

T.2.4.   Mixed-Interval Devices. – For devices that measure to a different division value in at least one dimensioning axes and all axes are single range, the tolerance values shall be based on the value of the division of the axis in use.
(Added 2016)

T.3.   Tolerance Values. – The maintenance and acceptance tolerance values shall be ± 1 division.
(Amended 2004)

T.4.   Position Tests. – For a test standard measured several times in different positions by the device all indications shall be within applicable tolerances.

T.5.　Influence Factors. – The following factors are applicable to tests conducted under controlled conditions only.

T.5.1.　Temperature. – Devices shall satisfy the tolerance requirements under the following temperature conditions.

T.5.1.1.　Temperature Limits. – If not marked on the device, the temperature limits shall be − 10 °C to 40 °C (14 °F to 104 °F).

T.5.1.2.　Minimum Temperature Range. – If temperature limits are specified for the device, the range shall be at least 30 °C or 54 °F.

T.5.1.3.　Temperature Effect on Zero Indication. – The zero indication shall not vary by more than one division per 5 °C (9 °F) change in temperature.

T.5.2.　Power Supply Voltage.

T.5.2.1.　Alternating Current Power Supply. – Devices that operate using alternating current must perform within the conditions defined in paragraphs T.3. through T.6., inclusive, from − 15 % to + 10 % of the marked nominal line voltage(s) at 60 Hz, or the voltage range marked by the manufacturer, at 60 Hz.
(Added 2004)

T.5.2.2.　Direct Current Power Supply. – Devices that operate using direct current shall operate and perform within the applicable tolerance at any voltage level at which the device is capable of displaying metrological registrations.
(Added 2004)
(Amended 2004)

T.6.　Disturbances, Field Evaluation. – The following requirements apply to devices when subjected to disturbances which may normally exist in the surrounding environment. These disturbances include radio frequency interference (RFI), electromagnetic interference (EMI), acoustic changes, ambient light emissions, etc. The difference between the measurement indication with the disturbance and the measurement indication without the disturbance shall not exceed one division "d" or the equipment shall:

(a)　blank the indication;

(b)　provide an error message; or

(c)　the indication shall be so completely unstable that it could not be interpreted, or transmitted into memory or to a recording element, as a correct measurement value.

UR.　User Requirements

UR.1.　Selection Requirements. – Equipment shall be suitable for the service in which it is used with respect to elements of its design, including but not limited to, its maximum capacity, value of the division, minimum capacity, and computing capability.

UR.1.1.　Value of the Indicated and Recorded Division. – The value of the division recorded shall be the same as the division value indicated.

UR.2.　Installation Requirements.

UR.2.1.　Supports. – A device that is portable and is being used on a counter, table, or the floor shall be so positioned that it is firmly and securely supported.

UR.2.2.   Foundation, Supports, and Clearance. – The foundations and support of a device installed in a fixed location shall be such as to provide strength, rigidity, and permanence of all components, and clearance shall be provided around all live parts to the extent that no contacts may result when the measuring element is empty, nor throughout the performance range of the device such that the operation or performance of the device is adversely affected.

UR.2.3.   Protection from Environmental Factors. – The indicating and measuring elements of a device shall be adequately protected from environmental factors such as wind, weather, and RFI that may adversely affect the operation or performance of the device.

UR.3.   Use Requirements.

UR.3.1.   Minimum and Maximum Measuring Ranges. – A device shall not be used to measure objects smaller than the minimum or larger than the maximum dimensions marked on the device.

UR.3.2.   Special Designs. – A multiple dimension measuring device designed and marked for a special application shall not be used for other than its intended purpose.

UR.3.3.   Object Placement. – If the object being measured must be transported (e.g., shipped) on a stable side, that irregularly-shaped object must be measured while placed on that stable side. The electronic rotation of the object to determine the smallest hexahedron shall be calculated in a two-dimensional horizontal plane, retaining the stable side plane as the bottom of the hexahedron.
(Added 2008)
(Amended 2008)

UR.4.   Maintenance Requirements.

UR.4.1.   Zero or Ready Condition. – The zero-setting adjustment of a multiple dimension measuring device shall be maintained so that, with no object in or on the measuring element, the device shall indicate or record a zero or ready condition.

UR.4.2.   Level Condition. – If a multiple dimension measuring device is equipped with a level-condition indicator, the device shall be maintained in a level condition.

UR.4.3.   Device Modification. – The measuring capabilities of a device shall not be changed from the manufacturer's design unless the modification has been approved by the manufacturer and the weights and measures authority having jurisdiction over the device.

UR.5.   Customer Information Provided. – The user of a multiple dimension measuring device or system shall provide transaction information to the customer as specified in Table UR.5. Customer Information Provided.
(Added 2004)

Table UR.5. Customer Information Provided			
Information	**No Contractual Agreement**		**Contractual Agreement**
	Customer Present	**Customer not Present**	
1. Object identification	N/A	P	P or A
2. Billing method (scale or dimensional weight if used)	D or P	P	P or A
3. Billing rate or rate chart	D or P or A	P or G or A	P or A
4. Dimensional weight (if used)	P	P	P or A
5. Conversion factor (if dimensional weight is used)	D or P or A	P	P or G
6. Dimensional weight statement[1] (if dimensional weight is used)	D or P	P	P or G
7. Total price	P	P	P or A

A = Available upon Request by Customer[2]

D = Displayed

G = Published Guidelines or Contracts

M = Marked

N/A = Not Applicable

P = Printed

Notes:

[1] This is an explanation that the dimensional weight is not a true weight but is a calculated value obtained by applying a conversion factor to the hexahedron dimensions or volume of the object.

[2] The information "available upon request by customer" shall be retained by the party having issued the invoice for at least 30 calendar days after the date of invoicing.

(Added 2004)

THIS PAGE INTENTIONALLY LEFT BLANK

Table of Contents

THIS PAGE INTENTIONALLY LEFT BLANK

Section 5.59.   Electronic Livestock, Meat, and Poultry Evaluation Systems and/or Devices

The status of Section 5.59. Electronic Livestock, Meat, and Poultry Evaluation Systems and/or Devices was changed from "tentative" to "permanent" effective January 1, 2013.
(Added 2005) (Amended 2012)

A. Application

A.1.   General. – This code applies to electronic devices or systems for measuring the composition or quality constituents of live animals, livestock and poultry carcasses, and individual cuts of meat or a combination thereof for the purpose of determining value.

A.2.   Additional Code Requirements. – In addition to the requirements of this code, Electronic Livestock, Meat, and Poultry Evaluation Systems shall meet the requirements of Section 1.10. General Code.

A.3.   Exceptions. – This code does not apply to scales used to weigh live animals, livestock and poultry carcasses, and individual cuts of meat unless the scales are part of an integrated system designed to measure composition or quality constituents.   Scales used in integrated systems must also meet NIST Handbook 44, Section 2.20. Scale requirements.

S. Specifications

S.1.   Design and Manufacture. – All design and manufacturing specifications shall comply with American Society for Testing Materials (ASTM) International Standard F2342 Standard Specification for Design and Construction of Composition or Quality Constituent Measuring Devices or Systems.

N. Notes

N.1.   Method of Test. – Performance tests shall be conducted in accordance with ASTM Standard F2343 Test Method for Livestock, Meat, and Poultry Evaluation Devices.

N.2.   Testing Standards. – ASTM Standard F2343 requires device or system users to maintain accurate reference standards that meet the tolerance expressed in NIST Handbook 44 Fundamental Considerations, paragraph 3.2. Tolerances for Standards (i.e., one-third of the smallest tolerance applied).

N.3.   Verification. – Device or system users are required to verify and document the accuracy of a device or system on each production day as specified by ASTM Standard F2341 Standard Practice of User Requirements for Livestock, Meat, and Poultry Evaluation Devices or Systems.

N.3.1.   Official Tests. – Officials are encouraged to periodically witness the required "in house" verification of accuracy.   Officials may also conduct official tests using the on-site testing standards or other appropriate standards belonging to the jurisdiction with statutory authority over the device or system.

T. Tolerances

T.1. Tolerances on Individual Measurements. – Maintenance and acceptance tolerances on an individual measurement shall be as shown in Table T.1. Tolerances.

Table T.1. Tolerances	
Individual linear measurement of a single constituent	± 1 mm (0.039 in)
Measurement of area	± 1.6 cm^2 (0.25 in^2)
For measurements of other constituents	As specified in ASTM Standard F2343

UR.   User Requirements

UR.1.   Installation Requirements.

   UR.1.1.   Installation. – All devices and systems shall be installed in accordance with manufacturer's instructions.

UR.2.   Maintenance of Equipment.

   UR.2.1.   Maintenance. – All devices and systems shall be continually maintained in an accurate condition and in accordance with the manufacturer's instructions and ASTM Standard F2341.

UR.3.   Use Requirements.

   UR.3.1.   Limitation of Use. – All devices and systems shall be used to make measurements in a manner specified by the manufacturer.

UR.4.   Testing Standards. – The user of a commercial device shall make available to the official with statutory authority over the device testing standards that meet the tolerance expressed in Fundamental Considerations, paragraph 3.2. Tolerances for Standards (i.e., one-third of the smallest tolerance applied). The accuracy of the testing standards shall be verified annually or on a frequency as required by the official with statutory authority and shall be traceable to the appropriate SI standard.

Table of Contents

THIS PAGE INTENTIONALLY LEFT BLANK

Appendix A.   Fundamental Considerations Associated with the Enforcement of Handbook 44 Codes

1.   Uniformity of Requirements

1.1.   National Conference Codes. – Weights and measures jurisdictions are urged to promulgate and adhere to the National Conference codes, to the end that uniform requirements may be in force throughout the country. This action is recommended even though a particular jurisdiction does not wholly agree with every detail of the National Conference codes. Uniformity of specifications and tolerances is an important factor in the manufacture of commercial equipment. Deviations from standard designs to meet the special demands of individual weights and measures jurisdictions are expensive, and any increase in costs of manufacture is, of course, passed on to the purchaser of equipment. On the other hand, if designs can be standardized by the manufacturer to conform to a single set of technical requirements, production costs can be kept down, to the ultimate advantage of the general public. Moreover, it seems entirely logical that equipment that is suitable for commercial use in the "specification" states should be equally suitable for such use in other states.

Another consideration supporting the recommendation for uniformity of requirements among weights and measures jurisdictions is the cumulative and regenerative effect of the widespread enforcement of a single standard of design and performance. The enforcement effort in each jurisdiction can then reinforce the enforcement effort in all other jurisdictions. More effective regulatory control can be realized with less individual effort under a system of uniform requirements than under a system in which even minor deviations from standard practice are introduced by independent state action.

Since the National Conference codes represent the majority opinion of a large and representative group of experienced regulatory officials, and since these codes are recognized by equipment manufacturers as their basic guide in the design and construction of commercial weighing and measuring equipment, the acceptance and promulgation of these codes by each state are strongly recommended.

1.2.   Form of Promulgation. – A convenient and very effective form of promulgation already successfully used in a considerable number of states is promulgation by citation of National Institute of Standards and Technology Handbook 44. It is especially helpful when the citation is so made that, as amendments are adopted from time to time by the National Conference on Weights and Measures, these automatically go into effect in the state regulatory authority. For example, the following form of promulgation has been used successfully and is recommended for consideration:

> The specifications, tolerances, and other technical requirements for weighing and measuring devices as recommended by the National Conference on Weights and Measures and published in the National Institute of Standards and Technology Handbook 44, "Specifications, Tolerances, and Other Technical Requirements for Weighing and Measuring Devices," and supplements thereto or revisions thereof, shall apply to commercial weighing and measuring devices in the state.

In some states, it is preferred to base technical requirements upon specific action of the state legislature rather than upon an act of promulgation by a state officer. The advantages cited above may be obtained and may yet be surrounded by adequate safeguards to insure proper freedom of action by the state enforcing officer if the legislature adopts the National Conference requirements by language somewhat as follows:

> The specifications, tolerances, and other technical requirements for weighing and measuring devices as recommended by the National Conference on Weights and Measures shall be the specifications, tolerances, and other technical requirements for weighing and measuring devices of the state except insofar as specifically modified, amended, or rejected by a regulation issued by the state (insert title of enforcing officer).

2.   Tolerances for Commercial Equipment

2.1.   Acceptance and Maintenance Tolerances. – The official tolerances prescribed by a weights and measures jurisdiction for commercial equipment are the limits of inaccuracy officially permissible within that jurisdiction. It is recognized that errorless value or performance of mechanical equipment is unattainable. Tolerances are established, therefore, to fix the range of inaccuracy within which equipment will be officially approved for commercial use. In the case of classes of equipment on which the magnitude of the errors of value or performance may be expected to change as a result of use, two sets of tolerances are established: acceptance tolerances and maintenance tolerances.

Acceptance tolerances are applied to new or newly reconditioned or adjusted equipment, and are smaller than (usually one-half of) the maintenance tolerances. Maintenance tolerances thus provide an additional range of inaccuracy within which equipment will be approved on subsequent tests, permitting a limited amount of deterioration before the equipment will be officially rejected for inaccuracy and before reconditioning or adjustment will be required. In effect, there is assured a reasonable period of use for equipment after it is placed in service before reconditioning will be officially required. The foregoing comments do not apply, of course, when only a single set of tolerance values is established, as is the case with equipment such as glass milk bottles and graduates, which maintain their original accuracy regardless of use, and measure-containers, which are used only once.

2.2.   Theory of Tolerances. – Tolerance values are so fixed that the permissible errors are sufficiently small that there is no serious injury to either the buyer or the seller of commodities, yet not so small as to make manufacturing or maintenance costs of equipment disproportionately high. Obviously, the manufacturer must know what tolerances his equipment is required to meet, so that he can manufacture economically. His equipment must be good enough to satisfy commercial needs, but should not be subject to such stringent tolerance values as to make it unreasonably costly, complicated, or delicate.

2.3.   Tolerances and Adjustments. – Tolerances are primarily accuracy criteria for use by the regulatory official. However, when equipment is being adjusted for accuracy, either initially or following repair or official rejection, the objective should be to adjust as closely as practicable to zero error. Equipment owners should not take advantage of tolerances by deliberately adjusting their equipment to have a value, or to give performance, at or close to the tolerance limit. Nor should the repair or service personnel bring equipment merely within tolerance range when it is possible to adjust closer to zero error.[1]

3.   Testing Apparatus

3.1.   Adequacy.[2] – Tests can be made properly only if, among other things, adequate testing apparatus is available. Testing apparatus may be considered adequate only when it is properly designed for its intended use, when it is so constructed that it will retain its characteristics for a reasonable period under conditions of normal use, when it is available in denominations appropriate for a proper determination of the value or performance of the commercial equipment under test, and when it is accurately calibrated.

3.2.   Tolerances for Standards. – Except for work of relatively high precision, it is recommended that the accuracy of standards used in testing commercial weighing and measuring equipment be established and maintained so that the use of corrections is not necessary. When the standard is used without correction, its combined error and uncertainty must be less than one-third of the applicable device tolerance.

Device testing is complicated to some degree when corrections to standards are applied. When using a correction for a standard, the uncertainty associated with the corrected value must be less than one-third of the applicable device

---

[1] See General Code, Section 1.10.; User Requirement G-UR.4.3. Use of Adjustments.

[2] Recommendations regarding the specifications and tolerances for suitable field standards may be obtained from the Office of Weights and Measures of the National Institute of Standards and Technology. Standards will meet the specifications of the National Institute of Standards and Technology Handbook 105-Series standards (or other suitable and designated standards). This section shall not preclude the use of additional field standards and/or equipment, as approved by the Director, for uniform evaluation of device performance.

tolerance. The reason for this requirement is to give the device being tested as nearly as practicable the full benefit of its own tolerance.

3.3.   Accuracy of Standards. – Prior to the official use of testing apparatus, its accuracy should invariably be verified. Field standards should be calibrated as often as circumstances require. By their nature, metal volumetric field standards are more susceptible to damage in handling than are standards of some other types. A field standard should be calibrated whenever damage is known or suspected to have occurred or significant repairs have been made. In addition, field standards, particularly volumetric standards, should be calibrated with sufficient frequency to affirm their continued accuracy, so that the official may always be in an unassailable position with respect to the accuracy of his testing apparatus. Secondary field standards, such as special fabric testing tapes, should be verified much more frequently than such basic standards as steel tapes or volumetric provers to demonstrate their constancy of value or performance.

Accurate and dependable results cannot be obtained with faulty or inadequate field standards. If either the service person or official is poorly equipped, their results cannot be expected to check consistently. Disagreements can be avoided and the servicing of commercial equipment can be expedited and improved if service persons and officials give equal attention to the adequacy and maintenance of their testing apparatus.

4.   Inspection of Commercial Equipment

4.1.   Inspection Versus Testing. – A distinction may be made between the inspection and the testing of commercial equipment that should be useful in differentiating between the two principal groups of official requirements; i.e., specifications and performance requirements. Although the term inspection is frequently loosely used to include everything that the official has to do in connection with commercial equipment, it is useful to limit the scope of that term primarily to examinations made to determine compliance with design, maintenance, and user requirements. The term testing may then be limited to those operations carried out to determine the accuracy of value or performance of the equipment under examination by comparison with the actual physical standards of the official. These two terms will be used herein in the limited senses defined.

4.2.   Necessity for Inspection. – It is not enough merely to determine that the errors of equipment do not exceed the appropriate tolerances. Specification and user requirements are as important as tolerance requirements and should be enforced. Inspection is particularly important, and should be carried out with unusual thoroughness whenever the official examines a type of equipment not previously encountered.

This is the way the official learns whether or not the design and construction of the device conform to the specification requirements. But even a device of a type with which the official is thoroughly familiar and that he has previously found to meet specification requirements should not be accepted entirely on faith. Some part may have become damaged, or some detail of design may have been changed by the manufacturer, or the owner or operator may have removed an essential element or made an objectionable addition. Such conditions may be learned only by inspection. Some degree of inspection is therefore an essential part of the official examination of every piece of weighing or measuring equipment.

4.3.   Specification Requirements. – A thorough knowledge by the official of the specification requirements is a prerequisite to competent inspection of equipment. The inexperienced official should have his specifications before him when making an inspection, and should check the requirements one by one against the equipment itself. Otherwise some important requirement may be overlooked. As experience is gained, the official will become progressively less dependent on the handbook, until finally observance of faulty conditions becomes almost automatic and the time and effort required to do the inspecting are reduced to a minimum. The printed specifications, however, should always be available for reference to refresh the official's memory or to be displayed to support his decisions, and they are an essential item of his kit.

Specification requirements for a particular class of equipment are not all to be found in the separate code for that class. The requirements of the General Code apply, in general, to all classes of equipment, and these must always be considered in combination with the requirements of the appropriate separate code to arrive at the total of the requirements applicable to a piece of commercial equipment.

4.4.   General Considerations. – The simpler the commercial device, the fewer are the specification requirements affecting it, and the more easily and quickly can adequate inspection be made. As mechanical complexity increases, however, inspection becomes increasingly important and more time consuming, because the opportunities for the existence of faulty conditions are multiplied. It is on the relatively complex device, too, that the official must be on the alert to discover any modification that may have been made by an operator that might adversely affect the proper functioning of the device.

It is essential for the officials to familiarize themselves with the design and operating characteristics of the devices that he inspects and tests. Such knowledge can be obtained from the catalogs and advertising literature of device manufacturers, from trained service persons and plant engineers, from observation of the operations performed by service persons when reconditioning equipment in the field, and from a study of the devices themselves.

Inspection should include any auxiliary equipment and general conditions external to the device that may affect its performance characteristics. In order to prolong the life of the equipment and forestall rejection, inspection should also include observation of the general maintenance of the device and of the proper functioning of all required elements. The official should look for worn or weakened mechanical parts, leaks in volumetric equipment, or elements in need of cleaning.

4.5.   Misuse of Equipment. – Inspection, coupled with judicious inquiry, will sometimes disclose that equipment is being improperly used, either through ignorance of the proper method of operation or because some other method is preferred by the operator. Equipment should be operated only in the manner that is obviously indicated by its construction or that is indicated by instructions on the equipment, and operation in any other manner should be prohibited.

4.6.   Recommendations. – A comprehensive knowledge of each installation will enable the official to make constructive recommendations to the equipment owner regarding proper maintenance of his weighing and measuring devices and the suitability of his equipment for the purposes for which it is being used or for which it is proposed that it be used. Such recommendations are always in order and may be very helpful to an owner. The official will, of course, carefully avoid partiality toward or against equipment of specific makes, and will confine his recommendations to points upon which he is qualified, by knowledge and experience, to make suggestions of practical merit.

4.7.   Accurate and Correct Equipment. – Finally, the weights and measures official is reminded that commercial equipment may be accurate without being correct. A piece of equipment is accurate when its performance or value (that is, its indications, its deliveries, its recorded representations, or its capacity or actual value, etc., as determined by tests made with suitable standards) conforms to the standard within the applicable tolerances and other performance requirements. Equipment that fails so to conform is inaccurate. A piece of equipment is correct when, in addition to being accurate, it meets all applicable specification requirements. Equipment that fails to meet any of the requirements for correct equipment is incorrect. Only equipment that is correct should be sealed and approved for commercial use.[3]

5.   Correction of Commercial Equipment

5.1.   Adjustable Elements. – Many types of weighing and measuring instruments are not susceptible to adjustment for accuracy by means of adjustable elements. Linear measures, liquid measures, graduates, measure-containers, milk and lubricating-oil bottles, farm milk tanks, dry measures, and some of the more simple types of scales are in this category. Other types (for example, taximeters and odometers and some metering devices) may be adjusted in the field, but only by changing certain parts such as gears in gear trains.

Some types, of which fabric-measuring devices and cordage-measuring devices are examples, are not intended to be adjusted in the field and require reconditioning in shop or factory if inaccurate. Liquid-measuring devices and most scales are equipped with adjustable elements, and some vehicle-tank compartments have adjustable indicators. Field adjustments may readily be made on such equipment. In the discussion that follows, the principles pointed out and the recommendations made apply to adjustments on any commercial equipment, by whatever means accomplished.

---

[3] See Section 1.10. General Code and Appendix D. Definitions.

5.2.   When Corrections Should Be Made. – One of the primary duties of a weights and measures official is to determine whether equipment is suitable for commercial use. If a device conforms to all legal requirements, the official "marks" or "seals" it to indicate approval. If it does not conform to all official requirements, the official is required to take action to ensure that the device is corrected within a reasonable period of time. Devices with performance errors that could result in serious economic injury to either party in a transaction should be prohibited from use immediately and not allowed to be returned to service until necessary corrections have been made. The official should consider the most appropriate action, based on all available information and economic factors.

Some officials contend that it is justifiable for the official to make minor corrections and adjustments if there is no service agency nearby or if the owner or operator depends on this single device and would be "out of business" if the use of the device were prohibited until repairs could be made. Before adjustments are made at the request of the owner or the owner's representative, the official should be confident that the problem is not due to faulty installation or a defective part, and that the adjustment will correct the problem. The official should never undertake major repairs, or even minor corrections, if services of commercial agencies are readily available. The official should always be mindful of conflicts of interest before attempting to perform any services other than normal device examination and testing duties.

(Amended 1995)

5.3.   Gauging. – In the majority of cases, when the weights and measures official tests commercial equipment, he is verifying the accuracy of a value or the accuracy of the performance as previously established either by himself or by someone else. There are times, however, when the test of the official is the initial test on the basis of which the calibration of the device is first determined or its performance first established. The most common example of such gauging is in connection with vehicle tanks the compartments of which are used as measures. Frequently the official makes the first determination on the capacities of the compartments of a vehicle tank, and his test results are used to determine the proper settings of the compartment indicators for the exact compartment capacities desired. Adjustments of the position of an indicator under these circumstances are clearly not the kind of adjustments discussed in the preceding paragraph.

6. Rejection of Commercial Equipment

6.1.   Rejection and Condemnation. – The Uniform Weights and Measures Law contains a provision stating that the director shall reject and order to be corrected such physical weights and measures or devices found to be incorrect. Weights and measures and devices that have been rejected, may be seized if not corrected within a reasonable time or if used or disposed of in a manner not specifically authorized. The director shall remove from service and may seize weights and measures found to be incorrect that are not capable of being made correct.

These broad powers should be used by the official with discretion. The director should always keep in mind the property rights of an equipment owner, and cooperate in working out arrangements whereby an owner can realize at least something from equipment that has been rejected. In cases of doubt, the official should initially reject rather than condemn outright. Destruction and confiscation of equipment are harsh procedures. Power to seize and destroy is necessary for adequate control of extreme situations, but seizure and destruction should be resorted to only when clearly justified.

On the other hand, rejection is clearly inappropriate for many items of measuring equipment. This is true for most linear measures, many liquid and dry measures, graduates, measure-containers, milk bottles, lubricating-oil bottles, and some scales. When such equipment is "incorrect," it is either impractical or impossible to adjust or repair it, and the official has no alternative to outright condemnation. When only a few such items are involved, immediate destruction or confiscation is probably the best procedure. If a considerable number of items are involved (as, for example, a stock of measures in the hands of a dealer or a large shipment of bottles), return of these to the manufacturer for credit or replacement should ordinarily be permitted provided that the official is assured that they will not get into commercial use. In rare instances, confiscation and destruction are justified as a method of control when less harsh methods have failed.

In the case of incorrect mechanisms such as fabric-measuring devices, taximeters, liquid-measuring devices, and most scales, repair of the equipment is usually possible, so rejection is the customary procedure. Seizure may occasionally

be justified, but in the large majority of instances this should be unnecessary. Even in the case of worn-out equipment, some salvage is usually possible, and this should be permitted under proper controls.
(Amended 1995)

7.   Tagging of Equipment

7.1.   Rejected and Condemned. – It will ordinarily be practicable to tag or mark as rejected each item of equipment found to be incorrect and considered susceptible of proper reconditioning. However, it can be considered justifiable not to mark as rejected incorrect devices capable of meeting acceptable performance requirements if they are to be allowed to remain in service for a reasonable time until minor problems are corrected since marks of rejection may tend to be misleading about a device's ability to produce accurate measurements during the correction period. The tagging of equipment as condemned, or with a similar label to indicate that it is permanently out of service, is not recommended if there is any other way in which the equipment can definitely be put out of service. Equipment that cannot successfully be repaired should be dismantled, removed from the premises, or confiscated by the official rather than merely being tagged as "condemned."
(Amended 1995)

7.2.   Nonsealed and Noncommercial. – Rejection is not appropriate if measuring equipment cannot be tested by the official at the time of his regular visit–for example, when there is no gasoline in the supply tank of a gasoline-dispensing device. Some officials affix to such equipment a nonsealed tag stating that the device has not been tested and sealed and that it must not be used commercially until it has been officially tested and approved. This is recommended whenever considerable time will elapse before the device can be tested.

Where the official finds in the same establishment, equipment that is in commercial use and also equipment suitable for commercial use that is not presently in service, but which may be put into service at some future time, he may treat the latter equipment in any of the following ways:

(a) Test and approve the same as commercial equipment in use.

(b) Refrain from testing it and remove it from the premises to preclude its use for commercial purposes.

(c) Mark the equipment nonsealed.

Where the official finds commercial equipment and noncommercial equipment installed or used in close proximity, he may treat the noncommercial equipment in any of the following ways:

(a) Test and approve the same as commercial equipment.

(b) Physically separate the two groups of equipment so that misuse of the noncommercial equipment will be prevented.

(c) Tag it to show that it has not been officially tested and is not to be used commercially.

8.   Records of Equipment

8.1.   The official will be well advised to keep careful records of equipment that is rejected, so that he may follow up to insure that the necessary repairs have been made. As soon as practicable following completion of repairs, the equipment should be retested. Complete records should also be kept of equipment that has been tagged as nonsealed or noncommercial. Such records may be invaluable should it subsequently become necessary to take disciplinary steps because of improper use of such equipment.

9.   Sealing of Equipment

9.1.   Types of Seals and Their Locations. – Most weights and measures jurisdictions require that all equipment officially approved for commercial use (with certain exceptions to be pointed out later) be suitably marked or sealed to show approval. This is done primarily for the benefit of the public to show that such equipment has been officially examined and approved. The seal of approval should be as conspicuous as circumstances permit and should be of such a character and so applied that it will be reasonably permanent. Uniformity of position of the seal on similar types of equipment is also desirable as a further aid to the public.

The official will need more than one form of seal to meet the requirements of different kinds of equipment. Good quality, weather-resistant, water-adhesive, or pressure-sensitive seals or decalcomania seals are recommended for fabric-measuring devices, liquid-measuring devices, taximeters, and most scales, because of their permanence and good appearance. Steel stamps are most suitable for liquid and dry measures, for some types of linear measures, and for weights. An etched seal, applied with suitable etching ink, is excellent for steel tapes, and greatly preferable to a seal applied with a steel stamp. The only practicable seal for a graduate is one marked with a diamond or carbide pencil, or one etched with glass-marking ink. For a vehicle tank, the official may wish to devise a relatively large seal, perhaps of metal, with provision for stamping data relative to compartment capacities, the whole to be welded or otherwise permanently attached to the shell of the tank. In general, the lead-and-wire seal is not suitable as an approval seal.

9.2. Exceptions. – Commercial equipment such as measure-containers, milk bottles, and lubricating-oil bottles are not tested individually because of the time element involved. Because manufacturing processes for these items are closely controlled, an essentially uniform product is produced by each manufacturer. The official normally tests samples of these items prior to their sale within his jurisdiction and subsequently makes spot checks by testing samples selected at random from new stocks.

Another exception to the general rule for sealing approved equipment is found in certain very small weights whose size precludes satisfactory stamping with a steel die.

10. Rounding Off Numerical Values

10.1. Definition. – To round off or round a numerical value is to change the value of recorded digits to some other value considered more desirable for the purpose at hand by dropping or changing certain figures. For example, if a computed, observed, or accumulated value is 4738, this can be rounded off to the nearest thousand, hundred, or ten, as desired. Such rounded-off values would be, respectively, 5000, 4700, and 4740. Similarly, a value such as 47.382 can be rounded off to two decimal places, to one decimal place, or to the units place. The rounded-off figures in this example would be, respectively, 47.38, 47.4, and 47.

10.2. General Rules. – The general rules for rounding off may be stated briefly as follows:

(a) When the figure next beyond the last figure or place to be retained is less than 5, the figure in the last place retained is to be kept unchanged. When rounding off 4738 to the nearest hundred, it is noted that the figure 3 (next beyond the last figure to be retained) is less than 5. Thus the rounded-off value would be 4700. Likewise, 47.382 rounded to two decimal places becomes 47.38.

(b) When the figure next beyond the last figure or place to be retained is greater than 5, the figure in the last place retained is to be increased by 1. When rounding off 4738 to the nearest thousand, it is noted that the figure 7 (next beyond the last figure to be retained) is greater than 5. Thus the rounded-off value would be 5000. Likewise, 47.382 rounded to one decimal place becomes 47.4.

(c) When the figure next beyond the last figure to be retained is 5 followed by any figures other than zero(s), treat as in (b) above; that is, the figure in the last place retained is to be increased by 1. When rounding off 4501 to the nearest thousand, 1 is added to the thousands figure and the result becomes 5000.

(d) When the figure next beyond the last figure to be retained is 5 and there are no figures, or only zeros, beyond this 5, the figure in the last place to be retained is to be left unchanged if it is even (0, 2, 4, 6, or 8) and is to be increased by 1 if it is odd (1, 3, 5, 7, or 9). This is the odd and even rule, and may be stated as follows: "If odd, then add." Thus, rounding off to the first decimal place, 47.25 would become 47.2 and 47.15 would become 47.2. Also, rounded to the nearest thousand, 4500 would become 4000 and 1500 would become 2000.

It is important to remember that, when there are two or more figures to the right of the place where the last significant figure of the final result is to be, the entire series of such figures must be rounded off in one step and not in two or more successive rounding steps. [Expressed differently, when two or more such figures are involved, these are not to be rounded off individually, but are to be rounded off as a group.] Thus, when rounding off 47.3499 to the first decimal place, the result becomes 47.3. In arriving at this result, the figures "499" are treated as a group. Since the 4 next beyond the last figure to be retained is less than 5, the "499" is dropped (see subparagraph (a) above). It would be incorrect to round off these figures successively to the left so that 47.3499 would become 47.350 and then 47.35 and then 47.4.

10.3. Rules for Reading of Indications. – An important aspect of rounding off values is the application of these rules to the reading of indications of an indicator-and-graduated-scale combination (where the majority of the indications may be expected to lie somewhere between two graduations) if it is desired to read or record values only to the nearest graduation. Consider a vertical graduated scale and an indicator. Obviously, if the indicator is between two graduations but is closer to one graduation than it is to the other adjacent graduation, the value of the closer graduation is the one to be read or recorded.

In the case where, as nearly as can be determined, the indicator is midway between two graduations, the odd-and-even rule is invoked, and the value to be read or recorded is that of the graduation whose value is even. For example, if the indicator lies exactly midway between two graduations having values of 471 and 472, respectively, the indication should be read or recorded as 472, this being an even value. If midway between graduations having values of 474 and 475, the even value 474 should be read or recorded. Similarly, if the two graduations involved had values of 470 and 475, the even value of 470 should be read or recorded.

A special case not covered by the foregoing paragraph is that of a graduated scale in which successive graduations are numbered by twos, all graduations thus having even values; for example, 470, 472, 474, etc. When, in this case, an indication lies midway between two graduations, the recommended procedure is to depart from the practice of reading or recording only to the value of the nearest graduation and to read or record the intermediate odd value. For example, an indication midway between 470 and 472 should be read as 471.

10.4. Rules for Common Fractions. – When applying the rounding-off rules to common fractions, the principles are to be applied to the numerators of the fractions that have, if necessary, been reduced to a common denominator. The principle of "5s" is changed to the one-half principle; that is, add if more than one-half, drop if less than one-half, and apply the odd-and even rule if exactly one-half.

For example, a series of values might be $1^1/_{32}$, $1^2/_{32}$, $1^3/_{32}$, $1^4/_{32}$, $1^5/_{32}$, $1^6/_{32}$, $1^7/_{32}$, $1^8/_{32}$, $1^9/_{32}$. Assume that these values are to be rounded off to the nearest eighth ($^4/_{32}$). Then,

$1^1/_{32}$ becomes 1. ($^1/_{32}$ is less than half of $^4/_{32}$ and accordingly is dropped.)

$1^2/_{32}$ becomes 1. ($^2/_{32}$ is exactly one-half of $^4/_{32}$; it is dropped because it is rounded (down) to the "even" eighth, which in this instance is $^0/_8$.)

$1^3/_{32}$ becomes $1^4/_{32}$ or $1^1/_8$. ($^3/_{32}$ is more than half of $^4/_{32}$, and accordingly is rounded "up" to $^4/_{32}$ or $^1/_8$.)

$1^4/_{32}$ remains unchanged, being an exact eighth ($1^1/_8$).

$1^5/_{32}$ becomes $1^4/_{32}$ or $1^1/_8$. ($^5/_{32}$ is $^1/_{32}$ more than an exact $^1/_8$; $^1/_{32}$ is less than half of $^4/_{32}$ and accordingly is dropped.)

$1^6/_{32}$ becomes $1^2/_8$ or $1^1/_4$. ($^6/_{32}$ is $^2/_{32}$ more than an exact $^1/_8$; $^2/_{32}$ is exactly one-half of $^4/_{32}$, and the final fraction is rounded (up) to the "even" eighth, which in this instance is $^2/_8$.)

$1^7/_{32}$ becomes $1^2/_8$ or $1^1/_4$. ($^7/_{32}$ is $^3/_{32}$ more than an exact $^1/_8$; $^3/_{32}$ is more than one-half of $^4/_{32}$ and accordingly the final fraction is rounded (up) to $^2/_8$ or $^1/_4$.)

$1^8/_{32}$ remains unchanged, being an exact eighth ($1^2/_8$ or $1^1/_4$.)

1$\frac{9}{32}$ becomes 1$\frac{2}{8}$ or 1$\frac{1}{4}$. ($\frac{9}{32}$ is $\frac{1}{32}$ more than an exact $\frac{1}{8}$; $\frac{1}{32}$ is less than half of $\frac{4}{32}$ and accordingly is dropped.)

THIS PAGE INTENTIONALLY LEFT BLANK

Table of Contents

THIS PAGE INTENTIONALLY LEFT BLANK

Appendix B.   Units and Systems of Measurement Their Origin, Development, and Present Status

1. Introduction

The National Institute of Standards and Technology (NIST) (formerly the National Bureau of Standards) was established by Act of Congress in 1901 to serve as a national scientific laboratory in the physical sciences, and to provide fundamental measurement standards for science and industry. In carrying out these related functions the Institute conducts research and development in many fields of physics, mathematics, chemistry, and engineering. At the time of its founding, the Institute had custody of two primary standards – the meter bar for length and the kilogram cylinder for mass. With the phenomenal growth of science and technology over the past century, the Institute has become a major research institution concerned not only with everyday weights and measures, but also with hundreds of other scientific and engineering standards that are necessary to the industrial progress of the nation. Nevertheless, the country still looks to NIST for information on the units of measurement, particularly their definitions and equivalents.

The subject of measurement systems and units can be treated from several different standpoints. Scientists and engineers are interested in the methods by which precision measurements are made. State weights and measures officials are concerned with laws and regulations that assure equity in the marketplace, protect public health and safety, and with methods for verifying commercial weighing and measuring devices. But a vastly larger group of people is interested in some general knowledge of the origin and development of measurement systems, of the present status of units and standards, and of miscellaneous facts that will be useful in everyday life. This material has been prepared to supply that information on measurement systems and units that experience has shown to be the common subject of inquiry.

2.   Units and Systems of Measurement

The expression "weights and measures" is often used to refer to measurements of length, mass, and capacity or volume, thus excluding such quantities as electrical and time measurements and thermometry. This section on units and measurement systems presents some fundamental information to clarify the concepts of this subject and to eliminate erroneous and misleading use of terms.

It is essential that the distinction between the terms "units" and "standards" be established and kept in mind.

A unit is a special quantity in terms of which other quantities are expressed. In general, a unit is fixed by definition and is independent of such physical conditions as temperature. Examples: the meter, the liter, the gram, the yard, the pound, the gallon.

A standard is a physical realization or representation of a unit. In general, it is not entirely independent of physical conditions, and it is a representation of the unit only under specified conditions. For example, a meter standard has a length of one meter when at some definite temperature and supported in a certain manner. If supported in a different manner, it might have to be at a different temperature to have a length of one meter.

2.1.   Origin and Early History of Units and Standards.

2.1.1.   General Survey of Early History of Measurement Systems. – Weights and measures were among the earliest tools invented by man. Primitive societies needed rudimentary measures for many tasks: constructing dwellings of an appropriate size and shape, fashioning clothing, or bartering food or raw materials.

Man understandably turned first to parts of the body and the natural surroundings for measuring instruments. Early Babylonian and Egyptian records and the Bible indicate that length was first measured with the forearm, hand, or finger and that time was measured by the periods of the sun, moon, and other heavenly bodies. When it was necessary to compare the capacities of containers such as gourds or clay or metal vessels, they were filled with plant seeds which were then counted to measure the volumes. When means for weighing were invented, seeds and stones served as standards. For instance, the "carat," still used as a unit for gems, was derived from the carob seed.

Our present knowledge of early weights and measures comes from many sources. Archaeologists have recovered some rather early standards and preserved them in museums. The comparison of the dimensions of buildings with the descriptions of contemporary writers is another source of information. An interesting example of this is the comparison of the dimensions of the Greek Parthenon with the description given by Plutarch from which a fairly accurate idea of the size of the Attic foot is obtained. In some cases, we have only plausible theories and we must sometimes select the interpretation to be given to the evidence.

For example, does the fact that the length of the double-cubit of early Babylonia was equal (within two parts per thousand) to the length of the seconds pendulum at Babylon suggest a scientific knowledge of the pendulum at a very early date, or do we merely have a curious coincidence? By studying the evidence given by all available sources, and by correlating the relevant facts, we obtain some idea of the origin and development of the units. We find that they have changed more or less gradually with the passing of time in a complex manner because of a great variety of modifying influences. We find the units modified and grouped into measurement systems: the Babylonian system, the Egyptian system, the Phileterian system of the Ptolemaic age, the Olympic system of Greece, the Roman system, and the British system, to mention only a few.

2.1.2.   Origin and Development of Some Common Customary Units. – The origin and development of units of measurement has been investigated in considerable detail and a number of books have been written on the subject. It is only possible to give here, somewhat sketchily, the story about a few units.

Units of length: The <u>cubit</u> was the first recorded unit used by ancient peoples to measure length. There were several cubits of different magnitudes that were used. The common cubit was the length of the forearm from the elbow to the tip of the middle finger. It was divided into the span of the hand (one-half cubit), the palm or width of the hand (one sixth), and the digit or width of a finger (one twenty-fourth). The Royal or Sacred Cubit, which was 7 palms or 28 digits long, was used in constructing buildings and monuments and in surveying. The <u>inch</u>, <u>foot</u>, and <u>yard</u> evolved from these units through a complicated transformation not yet fully understood. Some believe they evolved from cubic measures; others believe they were simple proportions or multiples of the cubit. In any case, the Greeks and Romans inherited the foot from the Egyptians. The Roman foot was divided into both 12 unciae (inches) and 16 digits. The Romans also introduced the mile of 1000 paces or double steps, the pace being equal to five Roman feet. The Roman mile of 5000 feet was introduced into England during the occupation. Queen Elizabeth, who reigned from 1558 to 1603, changed, by statute, the mile to 5280 feet or 8 furlongs, a furlong being 40 rods of 5½ yards each.

The introduction of the <u>yard</u> as a unit of length came later, but its origin is not definitely known. Some believe the origin was the double cubit, others believe that it originated from cubic measure. Whatever its origin, the early yard was divided by the binary method into 2, 4, 8, and 16 parts called the half-yard, span, finger, and nail. The association of the yard with the "gird" or circumference of a person's waist or with the distance from the tip of the nose to the end of the thumb of Henry I are probably standardizing actions, since several yards were in use in Great Britain.

The <u>point</u>, which is a unit for measuring print type, is recent. It originated with Pierre Simon Fournier in 1737. It was modified and developed by the Didot brothers, Francois Ambroise and Pierre Francois, in 1755. The point was first used in the United States in 1878 by a Chicago type foundry (Marder, Luse, and Company). Since 1886, a point has been exactly 0.351 459 8 millimeters, or about $1/72$ inch.

Units of mass: The <u>grain</u> was the earliest unit of mass and is the smallest unit in the apothecary, avoirdupois, Tower, and Troy systems. The early unit was a grain of wheat or barleycorn used to weigh the precious metals silver and gold. Larger units preserved in stone standards were developed that were used as both units of mass and of monetary currency. The <u>pound</u> was derived from the mina used by ancient civilizations. A smaller unit was the shekel, and a larger unit was the talent. The magnitude of these units varied from place to place. The Babylonians and Sumerians had a system in which there were 60 shekels in a mina and 60 minas in a talent. The Roman talent consisted of 100 libra (pound) which were smaller in magnitude than the mina. The Troy pound used in England and the United States for monetary purposes, like the Roman pound, was divided into 12 ounces, but the Roman uncia (ounce) was smaller. The carat is a unit for measuring gemstones that had its origin in the carob seed, which later was standardized at $1/444$ ounce and then 0.2 gram.

Goods of commerce were originally traded by number or volume. When weighing of goods began, units of mass based on a volume of grain or water were developed. For example, the talent in some places was approximately equal to the mass of one cubic foot of water. Was this a coincidence or by design? The diverse magnitudes of units having the same name, which still appear today in our dry and liquid measures, could have arisen from the various commodities traded. The larger avoirdupois pound for goods of commerce might have been based on volume of water, which has a higher bulk density than grain. For example, the Egyptian hon was a volume unit about 11 % larger than a cubic palm and corresponded to one mina of water. It was almost identical in volume to the present U.S. pint.

The stone, quarter, hundredweight, and ton were larger units of mass used in Great Britain. Today only the stone continues in customary use for measuring personal body weight. The present stone is 14 pounds, but an earlier unit appears to have been 16 pounds. The other units were multiples of 2, 8, and 160 times the stone, or 28, 112, and 2240 pounds, respectively. The hundredweight was approximately equal to two talents. In the United States the ton of 2240 pounds is called the "long ton." The "short ton" is equal to 2000 pounds.

Units of time and angle: We can trace the division of the circle into 360 degrees and the day into hours, minutes, and seconds to the Babylonians who had a sexagesimal system of numbers. The 360 degrees may have been related to a year of 360 days.

2.2. The Metric System.

2.2.1. Definition, Origin, and Development.
– Metric systems of units have evolved since the adoption of the first well-defined system in France in 1791. During this evolution the use of these systems spread throughout the world, first to the non-English-speaking countries, and more recently to the English-speaking countries. The first metric system was based on the centimeter, gram, and second (cgs) and these units were particularly convenient in science and technology. Later metric systems were based on the meter, kilogram, and second (mks) to improve the value of the units for practical applications. The present metric system is the International System of Units (SI). It is also based on the meter, kilogram and second as well as additional base units for temperature, electric current, luminous intensity, and amount of substance. The International System of Units is referred to as the modern metric system.

The adoption of the metric system in France was slow, but its desirability as an international system was recognized by geodesists and others. On May 20, 1875, an international treaty known as the International Metric Convention or the Treaty of the Meter was signed by seventeen countries including the United States. This treaty established the following organizations to conduct international activities relating to a uniform system for measurements:

(1) The General Conference on Weights and Measures (French initials: CGPM), an intergovernmental conference of official delegates of member nations and the supreme authority for all actions;

(2) The International Committee of Weights and Measures (French initials: CIPM), consisting of selected scientists and metrologists, which prepares and executes the decisions of the CGPM and is responsible for the supervision of the International Bureau of Weights and Measures;

(3) The International Bureau of Weights and Measures (French initials: BIPM), a permanent laboratory and world center of scientific metrology, the activities of which include the establishment of the basic standards and scales of the principal physical quantities and maintenance of the international prototype standards.

The National Institute of Standards and Technology provides official United States representation in these organizations. The CGPM, the CIPM, and the BIPM have been major factors in the continuing refinement of the metric system on a scientific basis and in the evolution of the International System of Units.

Multiples and submultiples of metric units are related by powers of ten. This relationship is compatible with the decimal system of numbers and it contributes greatly to the convenience of metric units.

2.2.2.   International System of Units. – At the end of World War II, a number of different systems of measurement still existed throughout the world. Some of these systems were variations of the metric system, and others were based on the customary U.S. customary system of the English-speaking countries. It was recognized that additional steps were needed to promote a worldwide measurement system. As a result the 9th GCPM, in 1948, asked the ICPM to conduct an international study of the measurement needs of the scientific, technical, and educational communities. Based on the findings of this study, the 10th General Conference in 1954 decided that an international system should be derived from six base units to provide for the measurement of temperature and optical radiation in addition to mechanical and electromagnetic quantities. The six base units recommended were the meter, kilogram, second, ampere, Kelvin degree (later renamed the kelvin), and the candela.

In 1960, the 11th General Conference of Weights and Measures named the system based on the six base quantities the International System of Units, abbreviated SI from the French name: Le Système International d'Unités. The SI metric system is now either obligatory or permissible throughout the world.

2.2.3.   Units and Standards of the Metric System. – In the early metric system there were two fundamental or base units, the meter and the kilogram, for length and mass. The other units of length and mass, and all units of area, volume, and compound units such as density were derived from these two fundamental units.

The meter was originally intended to be one ten-millionth part of a meridional quadrant of the earth. The Meter of the Archives, the platinum length standard which was the standard for most of the 19th century, at first was supposed to be exactly this fractional part of the quadrant. More refined measurements over the earth's surface showed that this supposition was not correct. In 1889, a new international metric standard of length, the International Prototype Meter, a graduated line standard of platinum-iridium, was selected from a group of bars because precise measurements found it to have the same length as the Meter of the Archives. The meter was then defined as the distance, under specified conditions, between the lines on the International Prototype Meter without reference to any measurements of the earth or to the Meter of the Archives, which it superseded. Advances in science and technology have made it possible to improve the definition of the meter and reduce the uncertainties associated with artifacts. From 1960 to 1983, the meter was defined as the length equal to 1 650 763.73 wavelengths in a vacuum of the radiation corresponding to the transition between the specified energy levels of the krypton 86 atom. Since 1983 the meter has been defined as the length of the path traveled by light in a vacuum during an interval of $1/299\,792\,458$ of a second.

The kilogram, originally defined as the mass of one cubic decimeter of water at the temperature of maximum density, was known as the Kilogram of the Archives. It was replaced after the International Metric Convention in 1875 by the International Prototype Kilogram which became the unit of mass without reference to the mass of a cubic decimeter of water or to the Kilogram of the Archives. Each country that subscribed to the International Metric Convention was assigned one or more copies of the international standards; these are known as National Prototype Meters and Kilograms.

The liter is a unit of capacity or volume. In 1964, the 12th GCPM redefined the liter as being one cubic decimeter. By its previous definition – the volume occupied, under standard conditions, by a quantity of pure water having a mass of one kilogram – the liter was larger than the cubic decimeter by 28 parts per 1 000 000. Except for determinations of high precision, this difference is so small as to be of no consequence.

The modern metric system (SI) includes two classes of units:

   (a)   base units for length, mass, time, temperature, electric current, luminous intensity, and amount of substance; and

   (b)   derived units for all other quantities (e.g., work, force, power) expressed in terms of the seven base units.

For details, see NIST Special Publication 330 (2008), The International System of Units (SI) and NIST Special Publication 811 (2008), Guide for the Use of the International System of Units.

2.2.4.   International Bureau of Weights and Measures. – The International Bureau of Weights and Measures (BIPM) was established at Sèvres, a suburb of Paris, France, by the International Metric Convention of May 20, 1875. The BIPM maintains the International Prototype Kilogram, many secondary standards, and equipment for comparing standards and making precision measurements. The Bureau, funded by assessment of the signatory governments, is truly international. In recent years the scope of the work at the Bureau has been considerably broadened. It now carries on researches in the fields of electricity, photometry and radiometry, ionizing radiations, and time and frequency besides its work in mass, length, and thermometry.

2.2.5.   Status of the Metric System in the United States. – The use of the metric system in this country was legalized by Act of Congress in 1866, but was not made obligatory then or since. Following the signing of the Convention of the Meter in 1875, the United States acquired national prototype standards for the meter and the kilogram. U.S. Prototype Kilogram No. 20 continues to be the primary standard for mass in the United States. It is recalibrated from time to time at the BIPM. The prototype meter has been replaced by modern stabilized lasers following the most recent definition of the meter.

From 1893 until 1959, the yard was defined as equal exactly to $^{3600}/_{3937}$ meter. In 1959, a small change was made in the definition of the yard to resolve discrepancies both in this country and abroad. Since 1959, we define the yard as equal exactly to 0.9144 meter; the new yard is shorter than the old yard by exactly two parts in a million. At the same time, it was decided that any data expressed in feet derived from geodetic surveys within the United States would continue to bear the relationship as defined in 1893 (one foot equals $^{1200}/_{3937}$ meter). We call this foot the U.S. Survey Foot, while the foot defined in 1959 is called the International Foot. Measurements expressed in U.S. statute miles, survey feet, rods, chains, links, or the squares thereof, and acres should be converted to the corresponding metric values by using pre-1959 conversion factors if more than five significant figure accuracy is required.

Since 1970, actions have been taken to encourage the use of metric units of measurement in the United States. A brief summary of actions by Congress is provided below as reported in the Federal Register Notice dated July 28, 1998.

Section 403 of Public Law 93-380, the Education Amendment of 1974, states that it is the policy of the United States to encourage educational agencies and institutions to prepare students to use the metric system of measurement as part of the regular education program. Under both this act and the Metric Conversion Act of 1975, the "metric system of measurement" is defined as the International System of Units as established in 1960 by the General Conference on Weights and Measures and interpreted or modified for the United States by the Secretary of Commerce (Section 4(4)- Public Law 94-168; Section 403(a)(3)- Public Law 93-380). The Secretary has delegated authority under these subsections to the Director of the National Institute of Standards and Technology.

Section 5164 of Public Law 100-418, the Omnibus Trade and Competitiveness Act of 1988, amends Public Law 94-168, The Metric Conversion Act of 1975. In particular, Section 3, The Metric Conversion Act is amended to read as follows:

"Sec. 3. It is therefore the declared policy of the United States–

(1)   to designate the metric system of measurement as the preferred system of weights and measures for United States trade and commerce;

(2)   to require that each federal agency, by a date certain and to the extent economically feasible by the end of the fiscal year 1992, use the metric system of measurement in its procurements, grants, and other business-related activities, except to the extent that such use is impractical or is likely to cause

significant inefficiencies or loss of markets to U.S. firms, such as when foreign competitors are producing competing products in non-metric units;

(3)   to seek ways to increase understanding of the metric system of measurement through educational information and guidance and in government publications; and

(4)   to permit the continued use of traditional systems of weights and measures in nonbusiness activities."

The Code of Federal Regulations makes the use of metric units mandatory for agencies of the federal government. (Federal Register, Vol. 56, No. 23, page 160, January 2, 1991.)

2.3.   British and United States Systems of Measurement. – In the past, the customary system of weights and measures in the British Commonwealth countries and that in the United States were very similar; however, the SI metric system is now the official system of units in the United Kingdom, while the customary units are still predominantly used in the United States. Because references to the units of the old British customary system are still found, the following discussion describes the differences between the U.S. and British customary systems of units.

After 1959, the U.S. and the British inches were defined identically for scientific work and were identical in commercial usage. A similar situation existed for the U.S. and the British pounds, and many relationships, such as 12 inches = 1 foot, 3 feet = 1 yard, and 1760 yards = 1 international mile, were the same in both countries; but there were some very important differences.

In the first place, the U.S. customary bushel and the U.S. gallon, and their subdivisions differed from the corresponding British Imperial units. Also the British ton is 2240 pounds, whereas the ton generally used in the United States is the short ton of 2000 pounds. The American colonists adopted the English wine gallon of 231 cubic inches. The English of that period used this wine gallon and they also had another gallon, the ale gallon of 282 cubic inches. In 1824, the British abandoned these two gallons when they adopted the British Imperial gallon, which they defined as the volume of 10 pounds of water, at a temperature of 62 °F, which, by calculation, is equivalent to 277.42 cubic inches. At the same time, they redefined the bushel as 8 gallons.

In the customary British system, the units of dry measure are the same as those of liquid measure. In the United States these two are not the same; the gallon and its subdivisions are used in the measurement of liquids and the bushel, with its subdivisions, is used in the measurement of certain dry commodities. The U.S. gallon is divided into four liquid quarts and the U.S. bushel into 32 dry quarts. All the units of capacity or volume mentioned thus far are larger in the customary British system than in the U.S. system. But the British fluid ounce is smaller than the U.S. fluid ounce, because the British quart is divided into 40 fluid ounces whereas the U.S. quart is divided into 32 fluid ounces.

From this we see that in the customary British system an avoirdupois ounce of water at 62 °F has a volume of one fluid ounce, because 10 pounds is equivalent to 160 avoirdupois ounces, and 1 gallon is equivalent to 4 quarts, or 160 fluid ounces. This convenient relation does not exist in the U.S. system because a U.S. gallon of water at 62 °F weighs about 8⅓ pounds, or 133⅓ avoirdupois ounces, and the U.S. gallon is equivalent to 4 x 32, or 128 fluid ounces.

1 U.S. fluid ounce	= 1.041 British fluid ounces
1 British fluid ounce	= 0.961 U.S. fluid ounce
1 U.S. gallon	= 0.833 British Imperial gallon
1 British Imperial gallon	= 1.201 U.S. gallons

Among other differences between the customary British and the United States measurement systems, we should note that they abolished the use of the troy pound in England January 6, 1879; they retained only the troy ounce and its subdivisions, whereas the troy pound is still legal in the United States, although it is not now greatly used. We can mention again the common use, for body weight, in England of the stone of 14 pounds, this being a unit now unused in the United States, although its influence was shown in the practice until World War II of selling flour by the barrel of 196 pounds (14 stone). In the apothecary system of liquid measure the British add a unit, the fluid scruple, equal to one third of a fluid drachm (spelled <u>dram</u> in the United States) between their minim and their fluid drachm. In the United States, the general practice now is to sell dry commodities, such as fruits and vegetables, by their mass.

2.4.   Subdivision of Units. – In general, units are subdivided by one of three methods: (a) decimal, into tenths; (b) duodecimal, into twelfths; or (c) binary, into halves (twos). Usually the subdivision is continued by using the same method. Each method has its advantages for certain purposes, and it cannot properly be said that any one method is "best" unless the use to which the unit and its subdivisions are to be put is known.

For example, if we are concerned only with measurements of length to moderate precision, it is convenient to measure and to express these lengths in feet, inches, and binary fractions of an inch, thus 9 feet, $4^3/8$ inches. However, if these lengths are to be subsequently used to calculate area or volume, that method of subdivision at once becomes extremely inconvenient.   For that reason, civil engineers, who are concerned with areas of land, volumes of cuts, fills, excavations, etc., instead of dividing the foot into inches and binary subdivisions of the inch, divide it decimally; that is, into tenths, hundredths, and thousandths of a foot.

The method of subdivision of a unit is thus largely made based on convenience to the user. The fact that units have commonly been subdivided into certain subunits for centuries does not preclude their also having another mode of subdivision in some frequently used cases where convenience indicates the value of such other method. Thus, while we usually subdivide the gallon into quarts and pints, most gasoline-measuring pumps, of the price-computing type, are graduated to show tenths, hundredths, or thousandths of a gallon.

Although the mile has for centuries been divided into rods, yards, feet, and inches, the odometer part of an automobile speedometer shows tenths of a mile. Although we divide our dollar into 100 parts, we habitually use and speak of halves and quarters. An illustration of rather complex subdividing is found on the scales used by draftsmen. These scales are of two types: (a) architects, which are commonly graduated with scales in which $3/32$, $3/16$, $1/8$, $1/4$, $3/8$, $1/2$, $3/4$, 1, $1^{1}/2$, and 3 inches, respectively, represent 1 foot full scale, and also having a scale graduated in the usual manner to $1/16$ inch; and (b) engineers, which are commonly subdivided to 10, 20, 30, 40, 50, and 60 parts to the inch.

The dictum of convenience applies not only to subdivisions of a unit but also to multiples of a unit. Land elevations above sea level are given in feet although the height may be several miles; the height of aircraft above sea level as given by an altimeter is likewise given in feet, no matter how high it may be.

On the other hand, machinists, toolmakers, gauge makers, scientists, and others who are engaged in precision measurements of relatively small distances, even though concerned with measurements of length only, find it convenient to use the inch, instead of the tenth of a foot, but to divide the inch decimally to tenths, hundredths, thousandths, etc., even down to millionths of an inch. Verniers, micrometers, and other precision measuring instruments are usually graduated in this manner. Machinist scales are commonly graduated decimally along one edge and are also graduated along another edge to binary fractions as small as $1/64$ inch. The scales with binary fractions are used only for relatively rough measurements.

It is seldom convenient or advisable to use binary subdivisions of the inch that are smaller than $1/64$. In fact, $1/32$-, $1/16$-, or $1/8$-inch subdivisions are usually preferable for use on a scale to be read with the unaided eye.

2.5.   Arithmetical Systems of Numbers. – The subdivision of units of measurement is closely associated with arithmetical systems of numbers. The systems of units used in this country for commercial and scientific work, having many origins as has already been shown, naturally show traces of the various number systems associated with their origins and developments. Thus, (a) the binary subdivision has come down to us from the Hindus, (b) the duodecimal system of fractions from the Romans, (c) the decimal system from the Chinese and Egyptians, some developments having been made by the Hindus, and (d) the sexagesimal system (division by 60) now illustrated in the subdivision of units of angle and of time, from the ancient Babylonians. The use of decimal numbers in measurements is becoming the standard practice.

3.   Standards of Length, Mass, and Capacity or Volume

3.1.   Standards of Length. – The meter, which is defined in terms of the speed of light in a vacuum, is the unit on which all length measurements are based.

The yard is defined[1] as follows:

    1 yard = 0.914 4 meter, and

    1 inch = 25.4 millimeters exactly.

3.1.1.   **Calibration of Length Standards.** – NIST calibrates standards of length including meter bars, yard bars, miscellaneous precision line standards, steel tapes, invar geodetic tapes, precision gauge blocks, micrometers, and limit gauges. It also measures the linear dimensions of miscellaneous apparatus such as penetration needles, cement sieves, and hemacytometer chambers. In general, NIST accepts for calibration only apparatus of such material, design, and construction as to ensure accuracy and permanence sufficient to justify calibration by the Institute. NIST performs calibrations in accordance with fee schedules, copies of which may be obtained from NIST.

NIST does not calibrate carpenters' rules, machinist scales, draftsman scales, and the like. Such apparatus, if they require calibration, should be submitted to state or local weights and measures officials.

3.2.   **Standards of Mass.** – The primary standard of mass for this country is United States Prototype Kilogram 20, which is a platinum-iridium cylinder kept at NIST. We know the value of this mass standard in terms of the International Prototype Kilogram, a platinum-iridium standard which is kept at the International Bureau of Weights and Measures.

In Colonial Times the British standards were considered the primary standards of the United States. Later, the U.S. avoirdupois pound was defined in terms of the Troy Pound of the Mint, which is a brass standard kept at the United States Mint in Philadelphia. In 1911, the Troy Pound of the Mint was superseded, for coinage purposes, by the Troy Pound of the Institute.

The avoirdupois pound is defined in terms of the kilogram by the relation:

    1 avoirdupois pound = 0.453 592 37 kilogram.[2]

These changes in definition have not made any appreciable change in the value of the pound.

The grain is $1/7000$ of the avoirdupois pound and is identical in the avoirdupois, troy, and apothecary systems. The troy ounce and the apothecary ounce differ from the avoirdupois ounce but are equal to each other, and equal to 480 grains. The avoirdupois ounce is equal to 437.5 grains.

3.2.1.   **Mass and Weight.** – The mass of a body is a measure of its inertial property or how much matter it contains. The weight of a body is a measure of the force exerted on it by gravity or the force needed to support it. Gravity on earth gives a body a downward acceleration of about 9.8 m/s^2. (In common parlance, weight is often used as a synonym for mass in weights and measures.) The incorrect use of weight in place of mass should be phased out, and the term mass used when mass is meant.

Standards of mass are ordinarily calibrated by comparison to a reference standard of mass. If two objects are compared on a balance and give the same balance indication, they have the same "mass" (excluding the effect of air buoyancy). The forces of gravity on the two objects are balanced. Even though the value of the acceleration of gravity, g, is different from location to location, because the two objects of equal mass in the same location (where both masses are acted upon by the same g) will be affected in the same manner and by the same amount by any change in the value of g, the two objects will balance each other under any value of g.

However, on a spring balance the mass of a body is not balanced against the mass of another body. Instead, the gravitational force on the body is balanced by the restoring force of a spring. Therefore, if a very sensitive spring balance is used, the indicated mass of the body would be found to change if the spring balance and the

---

[1] See Federal Register for July 1, 1959. Also see next-to-last paragraph of 2.2.5.
[2] See Federal Register for July 1, 1959.

body were moved from one locality to another locality with a different acceleration of gravity. But a spring balance is usually used in one locality and is adjusted or calibrated to indicate mass at that locality.

3.2.2.   Effect of Air Buoyancy. – Another point that must be taken into account in the calibration and use of standards of mass is the buoyancy or lifting effect of the air. A body immersed in any fluid is buoyed up by a force equal to the force of gravity on the displaced fluid. Two bodies of equal mass, if placed one on each pan of an equal-arm balance, will balance each other in a vacuum. A comparison in a vacuum against a known mass standard gives "true mass." If compared in air, however, they will not balance each other unless they are of equal volume. If of unequal volume, the larger body will displace the greater volume of air and will be buoyed up by a greater force than will the smaller body, and the larger body will appear to be of less mass than the smaller body.

The greater the difference in volume, and the greater the density of the air in which we make the comparison weighing, the greater will be the apparent difference in mass. For that reason, in assigning a precise numerical value of mass to a standard, it is necessary to base this value on definite values for the air density and the density of the mass standard of reference.

The apparent mass of an object is equal to the mass of just enough reference material of a specified density (at 20 °C) that will produce a balance reading equal to that produced by the object if the measurements are done in air with a density of 1.2 mg/cm^3 at 20 °C. The original basis for reporting apparent mass is apparent mass versus brass. The apparent mass versus a density of 8.0 g/cm^3 is the more recent definition, and is used extensively throughout the world. The use of apparent mass versus 8.0 g/cm^3 is encouraged over apparent mass versus brass. The difference in these apparent mass systems is insignificant in most commercial weighing applications.

A full discussion of this topic is given in NIST Monograph 133, Mass and Mass Values, by Paul E. Pontius [for sale by the National Technical Information Service, 5285 Port Royal Road, Springfield, VA   22161 (COM 7450309)].

3.2.3.   Calibrations of Standards of Mass. – Standards of mass regularly used in ordinary trade should be tested by state or local weights and measures officials. NIST calibrates mass standards submitted, but it does not manufacture or sell them. Information regarding the mass calibration service of NIST and the regulations governing the submission of standards of mass to NIST for calibration are contained in NIST Special Publication 250, Calibration and Related Measurement Services of NIST, latest edition.

3.3.   Standards of Capacity. – Units of capacity or volume, being derived units, are in this country defined in terms of linear units. Laboratory standards have been constructed and are maintained at NIST. These have validity only by calibration with reference either directly or indirectly to the linear standards. Similarly, NIST has made and distributed standards of capacity to the several states. Other standards of capacity have been verified by calibration for a variety of uses in science, technology, and commerce.

3.3.1.   Calibrations of Standards of Capacity. – NIST makes calibrations on capacity or volume standards that are in the customary units of trade; that is, the gallon, its multiples, and submultiples, or in metric units. Further, NIST calibrates precision-grade volumetric glassware which is normally in metric units. NIST makes calibrations in accordance with fee schedules, copies of which may be obtained from NIST.

3.4.   Maintenance and Preservation of Fundamental Standard of Mass. – It is a statutory responsibility of NIST to maintain and preserve the national standard of mass at NIST and to realize all the other base units. The U.S. Prototype Kilogram maintained at NIST is fully protected by an alarm system. All measurements made with this standard are conducted in special air-conditioned laboratories to which the standard is taken a sufficiently long time before the observations to ensure that the standard will be in a state of equilibrium under standard conditions when the measurements or comparisons are made. Hence, it is not necessary to maintain the standard at standard conditions, but care is taken to prevent large changes of temperature. More important is the care to prevent any damage to the standard because of careless handling.

4.   Specialized Use of the Terms "Ton" and "Tonnage"

As weighing and measuring are important factors in our everyday lives, it is quite natural that questions arise about the use of various units and terms and about the magnitude of quantities involved. For example, the words "ton" and "tonnage" are used in widely different senses, and a great deal of confusion has arisen regarding the application of these terms.

The ton is used as a unit of measure in two distinct senses: (1) as a unit of mass, and (2) as a unit of capacity or volume.

In the first sense, the term has the following meanings:

(a)   The short, or net ton of 2000 pounds.

(b)   The long, gross, or shipper's ton of 2240 pounds.

(c)   The metric ton of 1000 kilograms, or 2204.6 pounds.

In the second sense (capacity), it is usually restricted to uses relating to ships and has the following meaning:

(a)   The register ton of 100 cubic feet.

(b)   The measurement ton of 40 cubic feet.

(c)   The English water ton of 224 British Imperial gallons.

In the United States and Canada the ton (mass) most commonly used is the short ton. In Great Britain, it is the long ton, and in countries using the metric system, it is the metric ton. The register ton and the measurement ton are capacity or volume units used in expressing the tonnage of ships. The English water ton is used, chiefly in Great Britain, in statistics dealing with petroleum products.

There have been many other uses of the term ton such as the timber ton of 40 cubic feet and the wheat ton of 20 bushels, but their uses have been local and the meanings have not been consistent from one place to another.
Properly, the word "tonnage" is used as a noun only in respect to the capacity or volume and dimensions of ships, and to the amount of the ship's cargo. There are two distinct kinds of tonnage; namely, vessel tonnage and cargo tonnage and each of these is used in various meanings. The several kinds of vessel tonnage are as follows:

Gross tonnage, or gross register tonnage, is the total cubical capacity or volume of a ship expressed in register tons of 100 cubic feet, or 2.83 cubic meters, less such space as hatchways, bakeries, galleys, etc., as are exempted from measurement by different governments. There is some lack of uniformity in the gross tonnages as given by different nations due to lack of agreement on the spaces that are to be exempted. Official merchant marine statistics of most countries are published in terms of the gross register tonnage. Press references to ship tonnage are usually to the gross tonnage.

The net tonnage, or net register tonnage, is the gross tonnage less the different spaces specified by maritime nations in their measurement rules and laws. The spaces deducted are those totally unavailable for carrying cargo, such as the engine room, coal bunkers, crew quarters, chart and instrument room, etc. The net tonnage is used in computing how much cargo that can be loaded on a ship. It is used as the basis for wharfage and other similar charges.

The register under-deck tonnage is the cubical capacity of a ship under her tonnage deck expressed in register tons. In a vessel having more than one deck, the tonnage deck is the second from the keel.

There are several variations of displacement tonnage.

The <u>dead weight tonnage</u> is the difference between the "loaded" and "light" <u>displacement tonnages</u> of a vessel. It is expressed in terms of the long ton of 2240 pounds, or the metric ton of 2204.6 pounds, and is the weight of fuel, passengers, and cargo that a vessel can carry when loaded to its maximum draft.

The second variety of tonnage, <u>cargo tonnage</u>, refers to the weight of the particular items making up the cargo. In overseas traffic it is usually expressed in long tons of 2240 pounds or metric tons of 2204.6 pounds. The short ton is only occasionally used. Therefore, the <u>cargo tonnage</u> is very distinct from <u>vessel tonnage</u>.

THIS PAGE INTENTIONALLY LEFT BLANK

Table of Contents

THIS PAGE INTENTIONALLY LEFT BLANK

Appendix C. General Tables of Units of Measurement

These tables have been prepared for the benefit of those requiring tables of units for occasional ready reference. In Section 4 of this Appendix, the tables are carried out to a large number of decimal places and exact values are indicated by underlining. In most of the other tables, only a limited number of decimal places are given, therefore making the tables better adapted to the average user.

1. Tables of Metric Units of Measurement

In the metric system of measurement, designations of multiples and subdivisions of any unit may be arrived at by combining with the name of the unit the prefixes deka, hecto, and kilo meaning, respectively, 10, 100, and 1000, and deci, centi, and milli, meaning, respectively, one-tenth, one-hundredth, and one-thousandth. In some of the following metric tables, some such multiples and subdivisions have not been included for the reason that these have little, if any, currency in actual usage.

In certain cases, particularly in scientific usage, it becomes convenient to provide for multiples larger than 1000 and for subdivisions smaller than one-thousandth. Accordingly, the following prefixes have been introduced and these are now generally recognized:

yotta,	(Y)	meaning 10^{24}	deci,	(d),	meaning 10^{-1}
zetta,	(Z),	meaning 10^{21}	centi,	(c),	meaning 10^{-2}
exa,	(E),	meaning 10^{18}	milli,	(m),	meaning 10^{-3}
peta,	(P),	meaning 10^{15}	micro,	(μ),	meaning 10^{-6}
tera,	(T),	meaning 10^{12}	nano,	(n),	meaning 10^{-9}
giga,	(G),	meaning 10^{9}	pico,	(p),	meaning 10^{-12}
mega,	(M),	meaning 10^{6}	femto,	(f),	meaning 10^{-15}
kilo,	(k),	meaning 10^{3}	atto,	(a),	meaning 10^{-18}
hecto,	(h),	meaning 10^{2}	zepto,	(z),	meaning 10^{-21}
deka,	(da),	meaning 10^{1}	yocto,	(y),	meaning 10^{-24}

Thus a kilometer is 1000 meters and a millimeter is 0.001 meter.

Units of Length

10 millimeters (mm)	= 1 centimeter (cm)
10 centimeters	= 1 decimeter (dm) = 100 millimeters
10 decimeters	= 1 meter (m) = 1000 millimeters
10 meters	= 1 dekameter (dam)
10 dekameters	= 1 hectometer (hm) = 100 meters
10 hectometers	= 1 kilometer (km) = 1000 meters

Units of Area

100 square millimeters (mm²)	= 1 square centimeter (cm²)
100 square centimeters	= 1 square decimeter (dm²)
100 square decimeters	= 1 square meter (m²)
100 square meters	= 1 square dekameter (dam²) = 1 are
100 square dekameters	= 1 square hectometer (hm²) = 1 hectare (ha)
100 square hectometers	= 1 square kilometer (km²)

Units of Liquid Volume

10 milliliters (mL)	= 1 centiliter (cL)
10 centiliters	= 1 deciliter (dL) = 100 milliliters
10 deciliters	= 1 liter[1] = 1000 milliliters
10 liters	= 1 dekaliter (daL)
10 dekaliters	= 1 hectoliter (hL) = 100 liters
10 hectoliters	= 1 kiloliter (kL) = 1000 liters

Units of Volume

1000 cubic millimeters (mm^3)	= 1 cubic centimeter (cm^3)
1000 cubic centimeters	= 1 cubic decimeter (dm^3)
	= 1 000 000 cubic millimeters
1000 cubic decimeters	= 1 cubic meter (m^3)
	= 1 000 000 cubic centimeters
	= 1 000 000 000 cubic millimeters

Units of Mass

10 milligrams (mg)	= 1 centigram (cg)
10 centigrams	= 1 decigram (dg) = 100 milligrams
10 decigrams	= 1 gram (g) = 1000 milligrams
10 grams	= 1 dekagram (dag)
10 dekagrams	= 1 hectogram (hg) = 100 grams
10 hectograms	= 1 kilogram (kg) = 1000 grams
1000 kilograms	= 1 megagram (Mg) or 1 metric ton (t)

2. Tables of U.S. Customary Units of Measurement[2]

In these tables where <u>foot</u> or <u>mile</u> is underlined, it is survey foot or U.S. statute mile rather than international foot or mile that is meant.

Units of Length

12 inches (in)	= 1 foot (ft)
3 feet	= 1 yard (yd)
16½ <u>feet</u>	= 1 rod (rd), pole, or perch
40 rods	= 1 furlong (fur) = 660 <u>feet</u>
8 furlongs	= 1 U.S. statute mile (mi) = 5280 <u>feet</u>
1852 meters (m)	= 6076.115 49 feet (approximately)
	= 1 international nautical mile

---

[1] By action of the 12th General Conference on Weights and Measures (1964), the liter is a special name for the cubic decimeter.

[2] This section lists units of measurement that have traditionally been used in the United States. In keeping with the Omnibus Trade and Competitiveness Act of 1988, the ultimate objective is to make the International System of Units the primary measurement system used in the United States.

Units of Area[3]

144 square inches (in^2)	= 1 square foot (ft^2)
9 square feet	= 1 square yard (yd^2)
	= 1296 square inches
272¼ square feet	= 1 square rod (rd^2)
160 square rods	= 1 acre = 43 560 square feet
640 acres	= 1 square mile (mi^2)
1 mile square	= 1 section of land
6 miles square	= 1 township
	= 36 sections = 36 square miles

Units of Volume[3]

1728 cubic inches (in^3)	= 1 cubic foot (ft^3)
27 cubic feet	= 1 cubic yard (yd^3)

Gunter's or Surveyors Chain Units of Measurement

0.66 foot (ft)	= 1 link (li)
100 links	= 1 chain (ch)
	= 4 rods = 66 feet
80 chains	= 1 U.S. statute mile (mi)
	= 320 rods = 5280 feet

Units of Liquid Volume[4]

4 gills (gi)	= 1 pint (pt) = 28.875 cubic inches (in^3)
2 pints	= 1 quart (qt) = 57.75 cubic inches
4 quarts	= 1 gallon (gal) = 231 cubic inches
	= 8 pints = 32 gills

Apothecaries Units of Liquid Volume

60 minims	= 1 fluid dram (fl dr or ƒ ℨ)
	= 0.225 6 cubic inch (in^3)
8 fluid drams	= 1 fluid ounce (fl oz or ƒ ℥)
	= 1.804 7 cubic inches
16 fluid ounces	= 1 pint (pt)
	= 28.875 cubic inches
	= 128 fluid drams
2 pints	= 1 quart (qt) = 57.75 cubic inches
	= 32 fluid ounces = 256 fluid drams
4 quarts	= 1 gallon (gal) = 231 cubic inches
	= 128 fluid ounces = 1024 fluid drams

---

[3] Squares and cubes of customary but not of metric units are sometimes expressed by the use of abbreviations rather than symbols. For example, sq ft means square foot, and cu ft means cubic foot.

[4] When necessary to distinguish the liquid pint or quart from the dry pint or quart, the word "liquid" or the abbreviation "liq" should be used in combination with the name or abbreviation of the liquid unit.

Units of Dry Volume[5]

2 pints (pt)	= 1 quart (qt) = 67.200 6 cubic inches (in³)
8 quarts	= 1 peck (pk) = 537.605 cubic inches
	= 16 pints
4 pecks	= 1 bushel (bu) = 2150.42 cubic inches
	= 32 quarts

Avoirdupois Units of Mass[6]

[The "grain" is the same in avoirdupois, troy, and apothecaries units of mass.]

1 µlb	= 0.000 001 pound (lb)
27¹¹/₃₂ grains (gr)	= 1 dram (dr)
16 drams	= 1 ounce (oz)
	= 437½ grains
16 ounces	= 1 pound (lb)
	= 256 drams
	= 7000 grains
100 pounds	= 1 hundredweight (cwt)[7]
20 hundredweights	= 1 ton (tn)[8]
	= 2000 pounds[7]

In "gross" or "long" measure, the following values are recognized:

112 pounds (lb)	= 1 gross or long hundredweight (cwt)[7]
20 gross or long hundredweights	= 1 gross or long ton
	= 2240 pounds[7]

Troy Units of Mass

[The "grain" is the same in avoirdupois, troy, and apothecaries units of mass.]

24 grains (gr)	= 1 pennyweight (dwt)
20 pennyweights	= 1 ounce troy (oz t) = 480 grains
12 ounces troy	= 1 pound troy (lb t)
	= 240 pennyweights = 5760 grains

---

[5] When necessary to distinguish dry pint or quart from the liquid pint or quart, the word "dry" should be used in combination with the name or abbreviation of the dry unit.

[6] When necessary to distinguish the avoirdupois dram from the apothecaries dram, or to distinguish the avoirdupois dram or ounce from the fluid dram or ounce, or to distinguish the avoirdupois ounce or pound from the troy or apothecaries ounce or pound, the word "avoirdupois" or the abbreviation "avdp" should be used in combination with the name or abbreviation of the avoirdupois unit.

[7] When the terms "hundredweight" and "ton" are used unmodified, they are commonly understood to mean the 100-pound hundredweight and the 2000-pound ton, respectively; these units may be designated "net" or "short" when necessary to distinguish them from the corresponding units in gross or long measure.

[8] As of January 1, 2014, "tn" is the required abbreviation for "short ton." Devices manufactured between January 1, 2008, and December 31, 2013, may use an abbreviation other than "tn" to specify "short ton."

Apothecaries Units of Mass

[The "grain" is the same in avoirdupois, troy, and apothecaries units of mass.]

20 grains (gr)	= 1 scruple (s ap or ℈)
3 scruples	= 1 dram apothecaries (dr ap or ℨ)
	= 60 grains
8 drams apothecaries	= 1 ounce apothecaries (oz ap or ℥)
	= 24 scruples = 480 grains
12 ounces apothecaries	= 1 pound apothecaries (lb ap)
	= 96 drams apothecaries
	= 288 scruples = 5760 grains

3. Notes on British Units of Measurement

In Great Britain, the yard, the avoirdupois pound, the troy pound, and the apothecaries pound are identical with the units of the same names used in the United States. The tables of British linear measure, troy mass, and apothecaries mass are the same as the corresponding United States tables, except for the British spelling "drachm" in the table of apothecaries mass. The table of British avoirdupois mass is the same as the United States table up to 1 pound; above that point the table reads:

14 pounds	= 1 stone
2 stones	= 1 quarter = 28 pounds
4 quarters	= 1 hundredweight = 112 pounds
20 hundredweight	= 1 ton = 2240 pounds

The present British gallon and bushel – known as the "Imperial gallon" and "Imperial bushel" – are, respectively, about 20 % and 3 % larger than the United States gallon and bushel. The Imperial gallon is defined as the volume of 10 avoirdupois pounds of water under specified conditions, and the Imperial bushel is defined as 8 Imperial gallons. Also, the subdivision of the Imperial gallon as presented in the table of British apothecaries fluid measure differs in two important respects from the corresponding United States subdivision, in that the Imperial gallon is divided into 160 fluid ounces (whereas the United States gallon is divided into 128 fluid ounces), and a "fluid scruple" is included. The full table of British measures of capacity (which are used alike for liquid and for dry commodities) is as follows:

4 gills	= 1 pint
2 pints	= 1 quart
4 quarts	= 1 gallon
2 gallons	= 1 peck
8 gallons (4 pecks)	= 1 bushel
8 bushels	= 1 quarter

The full table of British apothecaries measure is as follows:

20 minims	= 1 fluid scruple
3 fluid scruples	= 1 fluid drachm
	= 60 minims
8 fluid drachms	= 1 fluid ounce
20 fluid ounces	= 1 pint
8 pints	= 1 gallon (160 fluid ounces)

4. Tables of Units of Measurement
(all underlined figures are exact)

Units of Length - International Measure[9]

Units		Inches	Feet	Yards	Miles	Centimeters	Meters
1 inch	=	1	0.083 333 33	0.027 777 78	0.000 015 782 83	2.54	0.025 4
1 foot	=	12	1	0.333 333 3	0.000 189 393 9	30.48	0.304 8
1 yard	=	36	3	1	0.000 568 181 8	91.44	0.914 4
1 mile	=	63 360	5 280	1 760	1	160 934.4	1609.344
1 centimeter	=	0.393 700 8	0.032 808 40	0.010 936 13	0.000 006 213 712	1	0.01
1 meter	=	39.370 08	3.280 840	1.093 613	0.000 621 371 2	100	1

Units of Length - Survey Measure[9]

Units		Links	Feet	Rods	Chains	Miles	Meters
1 link	=	1	0.66	0.04	0.01	0.000 125	0.201 168 4
1 foot	=	1.515 152	1	0.060 606 06	0.015 151 52	0.000 189 393 9	0.304 800 6
1 rod	=	25	16.5	1	0.25	0.003 125	5.029 210
1 chain	=	100	66	4	1	0.0125	20.116 84
1 mile	=	8 000	5 280	320	80	1	1609.347
1 meter	=	4.970 960	3.280 833	0.198 838 4	0.049 709 60	0.000 621 369 9	1

Units of Area - International Measure[10]
(all underlined figures are exact)

Units		Square Inches	Square Feet	Square Yards
1 square inch	=	1	0.006 944 444	0.000 771 604 9
1 square foot	=	144	1	0.111 111 1
1 square yard	=	1 296	9	1
1 square mile	=	4 014 489 600	27 878 400	3 097 600
1 square centimeter	=	0.155 000 3	0.001 076 391	0.000 119 599 0
1 square meter	=	1550.003	10.763 91	1.195 990

---

[9] One international foot = 0.999 998 survey foot (exactly)
One international mile = 0.999 998 survey mile (exactly)

[10] One square survey foot = 1.000 004 square international feet
One square survey mile = 1.000 004 square international miles

Note: 1 survey foot	= $^{1200}/_{3937}$ meter (exactly)	
1 international foot	= 12 x 0.0254 meter (exactly)	
1 international foot	= 0.0254 x 39.37 survey foot (exactly)	

Units		Square Miles	Square Centimeters	Square Meters
1 square inch	=	0.000 000 000 249 097 7	6.451 6	0.000 645 16
1 square foot	=	0.000 000 035 870 06	929.030 4	0.092 903 04
1 square yard	=	0.000 000 322 830 6	8361.273 6	0.836 127 36
1 square mile	=	1	25 899 881 103.36	2 589 988.110 336
1 square centimeter	=	0.000 000 000 038 610 22	1	0.0001
1 square meter	=	0.000 000 386 102 2	10 000	1

Units of Area - Survey Measure[10, 11]

Units		Square Feet	Square Rods	Square Chains	Acres
1 square foot	=	1	0.003 673 095	0.000 229 568 4	0.000 022 956 84
1 square rod	=	272.25	1	0.062 5	0.006 25
1 square chain	=	4 356	16	1	0.1
1 acre	=	43 560	160	10	1
1 square mile	=	27 878 400	102 400	6 400	640
1 square meter	=	10.763 87	0.039 536 70	0.002 471 044	0.000 247 104 4
1 hectare	=	107 638.7	395.367 0	24.710 44	2.471 044

Units		Square Miles	Square Meters	Hectares
1 square foot	=	0.000 000 035 870 06	0.092 903 41	0.000 009 290 341
1 square rod	=	0.000 009 765 625	25.292 95	0.002 529 295
1 square chain	=	0.000 156 25	404.687 3	0.040 468 73
1 acre	=	0.001 562 5	4 046.873	0.404 687 3
1 square mile	=	1	2 589 998	258.999 8
1 square meter	=	0.000 000 386 100 6	1	0.000 1
1 hectare	=	0.003 861 006	10 000	1

---

[10] One square survey foot   = 1.000 004 square international feet
     One square survey mile   = 1.000 004 square international miles

[11] One international foot   = 0.999 998 survey foot (exactly)
     One international mile   = 0.999 998 survey mile (exactly)

Units of Volume
(all underlined figures are exact)

Units		Cubic Inches	Cubic Feet	Cubic Yards
1 cubic inch	=	1	0.000 578 703 7	0.000 021 433 47
1 cubic foot	=	1 728	1	0.037 037 04
1 cubic yard	=	46 656	27	1
1 cubic centimeter	=	0.061 023 74	0.000 035 314 67	0.000 001 307 951
1 cubic decimeter	=	61.023 74	0.035 314 67	0.001 307 951
1 cubic meter	=	61 023.74	35.314 67	1.307 951

Units		Milliliters (Cubic Centimeters)	Liters (Cubic Decimeters)	Cubic Meters
1 cubic inch	=	16.387 064	0.016 387 064	0.000 016 387 064
1 cubic foot	=	28 316.846 592	28.316 846 592	0.028 316 846 592
1 cubic yard	=	764 554.857 984	764.554 857 984	0.764 554 857 984
1 cubic centimeter	=	1	0.001	0.000 001
1 cubic decimeter	=	1 000	1	0.001
1 cubic meter	=	1 000 000	1000	1

Units of Capacity or Volume - Dry Volume Measure

Units		Dry Pints	Dry Quarts	Pecks	Bushels
1 dry pint	=	1	0.5	0.062 5	0.015 625
1 dry quart	=	2	1	0.125	0.031 25
1 peck	=	16	8	1	0.25
1 bushel	=	64	32	4	1
1 cubic inch	=	0.029 761 6	0.014 880 8	0.001 860 10	0.000 465 025
1 cubic foot	=	51.428 09	25.714 05	3.214 256	0.803 563 95
1 liter	=	1.816 166	0.908 083 0	0.113 510 4	0.028 377 59
1 cubic meter	=	1 816.166	908.083 0	113.510 4	28.377 59

Units		Cubic Inches	Cubic Feet	Liters	Cubic Meters
1 dry pint	=	<u>33.600 312 5</u>	0.019 444 63	0.550 610 5	0.000 550 610 5
1 dry quart	=	<u>67.200 625</u>	0.038 889 25	1.101 221	0.001 101 221
1 peck	=	<u>537.605</u>	0.311 114	8.809 768	0.008 809 768
1 bushel	=	<u>2 150.42</u>	1.244 456	<u>35.239 070 166 88</u>	<u>0.035 239 070 166 88</u>
1 cubic inch	=	<u>1</u>	0.000 578 703 7	<u>0.016 387 064</u>	<u>0.000 016 387 064</u>
1 cubic foot	=	<u>1728</u>	<u>1</u>	28.316 846 592	<u>0.028 316 846 592</u>
1 liter	=	61.023 74	0.035 314 67	<u>1</u>	<u>0.001</u>
1 cubic meter	=	61 023.74	35.314 67	<u>1000</u>	<u>1</u>

Units of Capacity or Volume - Liquid Volume Measure
(All underlined figures are exact)

Units		Minims	Fluid Drams	Fluid Ounces	Gills
1 minim	=	<u>1</u>	0.016 666 67	0.002 083 333	0.000 520 833 3
1 fluid dram	=	<u>60</u>	<u>1</u>	<u>0.125</u>	<u>0.031 25</u>
1 fluid ounce	=	<u>480</u>	<u>8</u>	<u>1</u>	0.25
1 gill	=	<u>1 920</u>	<u>32</u>	<u>4</u>	<u>1</u>
1 liquid pint	=	<u>7 680</u>	<u>128</u>	<u>16</u>	<u>4</u>
1 liquid quart	=	<u>15 360</u>	<u>256</u>	<u>32</u>	<u>8</u>
1 gallon	=	<u>61 440</u>	<u>1024</u>	<u>128</u>	<u>32</u>
1 cubic inch	=	265.974 0	4.432 900	0.554 112 6	0.138 528 1
1 cubic foot	=	459 603.1	7660.052	957.506 5	239.376 6
1 milliliter	=	16.230 73	0.270 512 2	0.033 814 02	0.008 453 506
1 liter	=	16 230.73	270.512 2	33.814 02	8.453 506

Units		Liquid Pints	Liquid Quarts	Gallons	Cubic Inches
1 minim	=	0.000 130 208 3	0.000 065 104 17	0.000 016 276 04	0.003 759 766
1 fluid dram	=	<u>0.007 812 5</u>	<u>0.003 906 25</u>	<u>0.000 976 562 5</u>	0.225 585 94
1 fluid ounce	=	<u>0.062 5</u>	<u>0.031 25</u>	<u>0.007 812 5</u>	<u>1.804 687 5</u>
1 gill	=	0.25	<u>0.125</u>	<u>0.031 25</u>	<u>7.218 75</u>
1 liquid pint	=	<u>1</u>	<u>0.5</u>	<u>0.125</u>	<u>28.875</u>
1 liquid quart	=	<u>2</u>	<u>1</u>	<u>0.25</u>	<u>57.75</u>
1 gallon	=	<u>8</u>	<u>4</u>	<u>1</u>	<u>231</u>
1 cubic inch	=	0.034 632 03	0.017 316 02	0.004 329 004	<u>1</u>
1 cubic foot	=	59.844 16	29.922 08	7.480 519	<u>1 728</u>
1 milliliter	=	0.002 113 376	0.001 056 688	0.000 264 172 1	0.061 023 74
1 liter	=	2.113 376	1.056 688	0.264 172 1	61.023 74

Units		Cubic Feet	Milliliters	Liters
1 minim	=	0.000 002 175 790	0.061 611 52	0.000 061 611 52
1 fluid dram	=	0.000 130 547 4	3.696 691	0.003 696 691
1 fluid ounce	=	0.001 044 379	29.573 53	0.029 573 53
1 gill	=	0.004 177 517	118.294 1	0.118 294 1
1 liquid pint	=	0.016 710 07	473.176 5	0.473 176 5
1 liquid quart	=	0.033 420 14	946.352 9	0.946 352 9
1 gallon	=	0.133 680 6	<u>3785.411 784</u>	<u>3.785 411 784</u>
1 cubic inch	=	0.000 578 703 7	16.387 06	0.016 387 06
1 cubic foot	=	<u>1</u>	28 316.85	28.316 85
1 milliliter	=	0.000 035 314 67	<u>1</u>	<u>0.001</u>
1 liter	=	0.035 314 67	<u>1 000</u>	<u>1</u>

Units of Mass Not Less Than Avoirdupois Ounces
(all underlined figures are exact)

Units		Avoirdupois Ounces	Avoirdupois Pounds	Short Hundred-weights	Short Tons
1 avoirdupois ounce	=	<u>1</u>	<u>0.0625</u>	<u>0.000 625</u>	<u>0.000 031 25</u>
1 avoirdupois pound	=	<u>16</u>	<u>1</u>	<u>0.01</u>	<u>0.000 5</u>
1 short hundredweight	=	<u>1 600</u>	<u>100</u>	<u>1</u>	<u>0.05</u>
1 short ton	=	<u>32 000</u>	<u>2 000</u>	<u>20</u>	<u>1</u>
1 long ton	=	<u>35 840</u>	<u>2 240</u>	<u>22.4</u>	<u>1.12</u>
1 kilogram	=	35.273 96	2.204 623	0.022 046 23	0.001 102 311
1 metric ton	=	35 273.96	2204.623	22.046 23	1.102 311

Units		Long Tons	Kilograms	Metric Tons
1 avoirdupois ounce	=	0.000 027 901 79	<u>0.028 349 523 125</u>	<u>0.000 028 349 523 125</u>
1 avoirdupois pound	=	0.000 446 428 6	<u>0.453 592 37</u>	<u>0.000 453 592 37</u>
1 short hundredweight	=	0.044 642 86	<u>45.359 237</u>	<u>0.045 359 237</u>
1 short ton	=	0.892 857 1	<u>907.184 74</u>	<u>0.907 184 74</u>
1 long ton	=	<u>1</u>	1016.046 908 8	1.016 046 908 8
1 kilogram	=	0.000 984 206 5	<u>1</u>	<u>0.001</u>
1 metric ton	=	0.984 206 5	<u>1 000</u>	<u>1</u>

Units of Mass Not Greater Than Pounds and Kilograms
(all underlined figures are exact)

Units		Grains	Apothecaries Scruples	Pennyweights	Avoirdupois Drams
1 grain	=	1	0.05	0.041 666 67	0.036 571 43
1 apoth. scruple	=	20	1	0.833 333 3	0.731 428 6
1 pennyweight	=	24	1.2	1	0.877 714 3
1 avdp. dram	=	27.343 75	1.367 187 5	1.139 323	1
1 apoth. dram	=	60	3	2.5	2.194 286
1 avdp. ounce	=	437.5	21.875	18.229 17	16
1 apoth. or troy oz.	=	480	24	20	17.554 29
1 apoth. or troy pound	=	5 760	288	240	210.651 4
1 avdp. pound	=	7 000	350	291.666 7	256
1 milligram	=	0.015 432 36	0.000 771 617 9	0.000 643 014 9	0.000 564 383 4
1 gram	=	15.432 36	0.771 617 9	0.643 014 9	0.564 383 4
1 kilogram	=	15432.36	771.617 9	643.014 9	564.383 4

Units		Apothecaries Drams	Avoirdupois Ounces	Apothecaries or Troy Ounces	Apothecaries or Troy Pounds
1 grain	=	0.016 666 67	0.002 285 714	0.002 083 333	0.000 173 611 1
1 apoth. scruple	=	0.333 333 3	0.045 714 29	0.041 666 67	0.003 472 222
1 pennyweight	=	0.4	0.054 857 14	0.05	0.004 166 667
1 avdp. dram	=	0.455 729 2	0.062 5	0.56 966 15	0.004 747 179
1 apoth. dram	=	1	0.137 142 9	0.125	0.010 416 67
1 avdp. ounce	=	7.291 667	1	0.911 458 3	0.075 954 86
1 apoth. or troy ounce	=	8	1.097 143	1	0.083 333 333
1 apoth. or troy pound	=	96	13.165 71	12	1
1 avdp. pound	=	116.666 7	16	14.583 33	1.215 278
1 milligram	=	0.000 257 206 0	0.000 035 273 96	0.000 032 150 75	0.000 002 679 229
1 gram	=	0.257 206 0	0.035 273 96	0.032 150 75	0.002 679 229
1 kilogram	=	257.206 0	35.273 96	32.150 75	2.679 229

Units		Avoirdupois Pounds	Milligrams	Grams	Kilograms
1 grain	=	0.000 142 857 1	64.798 91	0.064 798 91	0.000 064 798 91
1 apoth. scruple	=	0.002 857 143	1 295.978 2	1.295 978 2	0.001 295 978 2
1 pennyweight	=	0.003 428 571	1 555.173 84	1.555 173 84	0.001 555 173 84
1 avdp. dram	=	0.003 906 25	1 771.845 195 312 5	1.771 845 195 312 5	0.001 771 845 195 312 5
1 apoth. dram	=	0.008 571 429	3 887.934 6	3.887 934 6	0.003 887 934 6
1 avdp. ounce	=	0.062 5	28 349.523 125	28.349 523 125	0.028 349 523 125
1 apoth. or troy ounce	=	0.068 571 43	31 103.476 8	31.103 476 8	0.031 103 476 8
1 apoth. or troy pound	=	0.822 857 1	373 241.721 6	373.241 721 6	0.373 241 721 6
1 avdp. pound	=	1	453 592.37	453.592 37	0.453 592 37
1 milligram	=	0.000 002 204 623	1	0.001	0.000 001
1 gram	=	0.002 204 623	1 000	1	0.001
1 kilogram	=	2.204 623	1 000 000	1 000	1

5. Tables of Equivalents

In these tables it is necessary to differentiate between the "international foot" and the "survey foot." Therefore, the survey foot is underlined.

When the name of a unit is enclosed in brackets (thus, [1 hand] . . .), this indicates (1) that the unit is not in general current use in the United States, or (2) that the unit is believed to be based on "custom and usage" rather than on formal authoritative definition.

Equivalents involving decimals are, in most instances, rounded off to the third decimal place except where they are exact, in which cases these exact equivalents are so designated. The equivalents of the imprecise units "tablespoon" and "teaspoon" are rounded to the nearest milliliter.

Units of Length	
angstrom (Å)[12]	0.1 nanometer (exactly) 0.000 1 micrometer (exactly) 0.000 000 1 millimeter (exactly) 0.000 000 004 inch
1 cable's length	120 fathoms (exactly) 720 feet (exactly) 219 meters
1 centimeter (cm)	0.393 7 inch

---

[12] The angstrom is basically defined as 10^{-10} meter.

Units of Length	
1 chain (ch) (Gunter's or surveyors)	66 feet (exactly) 20.116 8 meters
1 decimeter (dm)	3.937 inches
1 dekameter (dam)	32.808 feet
1 fathom	6 feet (exactly) 1.828 8 meters
1 foot (ft)	0.304 8 meter (exactly)
1 furlong (fur)	10 chains (surveyors) (exactly) 660 feet (exactly) ⅛ U.S. statute mile (exactly) 201.168 meters
[1 hand]	4 inches
1 inch (in)	2.54 centimeters (exactly)
1 kilometer (km)	0.621 mile
1 league (land)	3 U.S. statute miles (exactly) 4.828 kilometers
1 link (li) (Gunter's or surveyors)	0.66 foot (exactly) 0.201 168 meter
1 meter (m)	39.37 inches 1.094 yards
1 micrometer	0.001 millimeter (exactly) 0.000 039 37 inch
1 mil	0.001 inch (exactly) 0.025 4 millimeter (exactly)
1 mile (mi) (U.S. statute)[13]	5280 feet survey (exactly) 1.609 kilometers
1 mile (mi) (international)	5280 feet international (exactly)
1 mile (mi) (international nautical)[14]	1.852 kilometers (exactly) 1.151 survey miles
1 millimeter (mm)	0.039 37 inch 0.001 meter (exactly)
1 nanometer (nm)	0.000 000 039 37 inch
1 Point (typography)	0.013 837 inch (exactly) 1/72 inch (approximately) 0.351 millimeter
1 rod (rd), pole, or perch	16½ feet (exactly) 5.029 2 meters

[13] The term "statute mile" originated with Queen Elizabeth I who changed the definition of the mile from the Roman mile of 5000 feet to the statute mile of 5280 feet. The international mile and the U.S. statute mile differ by about 3 millimeters although both are defined as being equal to 5280 feet. The international mile is based on the international foot (0.3048 meter) whereas the U.S. statute mile is based on the survey foot (1200/3937 meter).

[14] The international nautical mile of 1852 meters (6076.115 49 feet) was adopted effective July 1, 1954, for use in the United States. The value formerly used in the United States was 6080.20 feet = 1 nautical (geographical or sea) mile.

Units of Length	
1 yard (yd)	0.914 4 meter (exactly)

Units of Area	
1 acre[15]	43 560 square <u>feet</u> (exactly) 0.405 hectare
1 are	119.599 square yards 0.025 acre
1 hectare	2.471 acres
[1 square (building)]	100 square feet
1 square centimeter (cm²)	0.155 square inch
1 square decimeter (dm²)	15.500 square inches
1 square foot (ft²)	929.030 square centimeters
1 square inch (in²)	6.451 6 square centimeters (exactly)
1 square kilometer (km²)	247.104 acres 0.386 square mile
1 square meter (m²)	1.196 square yards 10.764 square feet
1 square mile (mi²)	258.999 hectares
1 square millimeter (mm²)	0.002 square inch
1 square rod (rd²), sq pole, or sq perch	25.293 square meters
1 square yard (yd²)	0.836 square meter

Units of Capacity or Volume	
1 barrel (bbl), liquid	31 to 42 gallons[16]
1 barrel (bbl), standard for fruits, vegetables, and other dry commodities, except cranberries	7056 cubic inches 105 dry quarts 3.281 bushels, struck measure
1 barrel (bbl), standard, cranberry	5826 cubic inches 86⁴⁵⁄₆₄ dry quarts 2.709 bushels, struck measure
1 bushel (bu) (U.S.) struck measure	2150.42 cubic inches (exactly) 35.238 liters
[1 bushel, heaped (U.S.)]	2747.715 cubic inches

---

[15] The question is often asked as to the length of a side of an acre of ground. An acre is a unit of area containing 43 560 square <u>feet</u>. It is not necessarily square, or even rectangular. But, if it is square, then the length of a side is equal to $\sqrt{43560\,\text{ft}^2} = 208.710\,\text{ft}$ (not exact).

[16] There are a variety of "barrels" established by law or usage. For example, federal taxes on fermented liquors are based on a barrel of 31 gallons; many state laws fix the "barrel for liquids" as 31½ gallons; one state fixes a 36-gallon barrel for cistern measurement; federal law recognizes a 40-gallon barrel for "proof spirits;" by custom, 42 gallons comprise a barrel of crude oil or petroleum products for statistical purposes, and this equivalent is recognized "for liquids" by four states.

Units of Capacity or Volume	
	1.278 bushels, struck measure[17]
[1 bushel (bu) (British Imperial) (struck measure)]	1.032 U.S. bushels, struck measure 2219.36 cubic inches
1 cord (cd) (firewood)	128 cubic feet (exactly)
1 cubic centimeter (cm³)	0.061 cubic inch
1 cubic decimeter (dm³)	61.024 cubic inches
1 cubic foot (ft³)	7.481 gallons 28.316 cubic decimeters
1 cubic inch (in³)	0.554 fluid ounce 4.433 fluid drams 16.387 cubic centimeters
1 cubic meter (m³)	1.308 cubic yards
1 cubic yard (yd³)	0.765 cubic meter
1 cup, measuring	8 fluid ounces (exactly) 237 milliliters ½ liquid pint (exactly)
1 dekaliter (daL)	2.642 gallons 1.135 pecks
1 dram, fluid (or liquid) (fl dr) or ƒ 3) (U.S.)	⅛ fluid ounce (exactly) 0.226 cubic inch 3.697 milliliters 1.041 British fluid drachms
[1 drachm, fluid (fl dr) (British)]	0.961 U.S. fluid dram 0.217 cubic inch 3.552 milliliters
1 gallon (gal) (U.S.)	231 cubic inches (exactly) 3.785 liters 0.833 British gallon 128 U.S. fluid ounces (exactly)
[1 gallon (gal) (British Imperial)]	277.42 cubic inches 1.201 U.S. gallons 4.546 liters 160 British fluid ounces (exactly)
1 gill (gi)	7.219 cubic inches 4 fluid ounces (exactly) 0.118 liter
1 hectoliter (hL)	26.418 gallons 2.838 bushels
1 liter (1 cubic decimeter exactly)	1.057 liquid quarts 0.908 dry quart 61.024 cubic inches
1 milliliter (mL)	0.271 fluid dram 16.231 minims 0.061 cubic inch

---

[17] Frequently recognized as 1¼ bushels, struck measure.

Units of Capacity or Volume	
1 ounce, fluid (or liquid) (fl oz) or $f\ \overline{3}$) (U.S.)	1.805 cubic inches 29.573 milliliters 1.041 British fluid ounces
[1 ounce, fluid (fl oz) (British)]	0.961 U.S. fluid ounce 1.734 cubic inches 28.412 milliliters
1 peck (pk)	8.810 liters
1 pint (pt), dry	33.600 cubic inches 0.551 liter
1 pint (pt), liquid	28.875 cubic inches exactly 0.473 liter
1 quart (qt), dry (U.S.)	67.201 cubic inches 1.101 liters 0.969 British quart
1 quart (qt), liquid (U.S.)	57.75 cubic inches (exactly) 0.946 liter 0.833 British quart
[1 quart (qt) (British)]	69.354 cubic inches 1.032 U.S. dry quarts 1.201 U.S. liquid quarts
1 tablespoon, measuring	3 teaspoons (exactly) 15 milliliters 4 fluid drams ½ fluid ounce (exactly)
1 teaspoon, measuring	⅓ tablespoon (exactly) 5 milliliters 1⅓ fluid drams[18]
1 water ton (English)	270.91 U.S. gallons 224 British Imperial gallons (exactly)

Units of Mass	
1 assay ton (AT)[19]	29.167 grams
1 carat (c)	200 milligrams (exactly) 3.086 grains
1 dram apothecaries (dr ap or 3)	60 grains (exactly) 3.888 grams
1 dram avoirdupois (dr avdp)	$27^{11}/32$ (= 27.344) grains 1.772 grams
1 gamma (γ)	1 microgram (exactly)

[18] The equivalent "1 teaspoon = 1⅓ fluid drams" has been found by the Bureau to correspond more closely with the actual capacities of "measuring" and silver teaspoons than the equivalent "1 teaspoon = 1 fluid dram," which is given by a number of dictionaries.

[19] Used in assaying. The assay ton bears the same relation to the milligram that a ton of 2000 pounds avoirdupois bears to the ounce troy; hence the mass in milligrams of precious metal obtained from one assay ton of ore gives directly the number of troy ounces to the net ton.

Units of Mass	
1 grain	64.798 91 milligrams (exactly)
1 gram (g)	15.432 grains 0.035 ounce, avoirdupois
1 hundredweight, gross or long[20] (gross cwt)	112 pounds (exactly) 50.802 kilograms
1 hundredweight, gross or short (cwt or net cwt)	100 pounds (exactly) 45.359 kilograms
1 kilogram (kg)	2.205 pounds
1 milligram (mg)	0.015 grain
1 ounce, avoirdupois (oz avdp)	437.5 grains (exactly) 0.911 troy or apothecaries ounce 28.350 grams
1 ounce, troy or apothecaries (oz t or oz ap or ℥)	480 grains (exactly) 1.097 avoirdupois ounces 31.103 grams
1 pennyweight (dwt)	1.555 grams
1 point	0.01 carat 2 milligrams
1 pound, avoirdupois (lb avdp)	7000 grains (exactly) 1.215 troy or apothecaries pounds 453.592 37 grams (exactly)
1 micropound (μlb) [the Greek letter mu in combination with the letters lb]	0.000 001 pound (exactly)
1 pound, troy or apothecaries (lb t or lb ap)	5760 grains (exactly) 0.823 avoirdupois pound 373.242 grams
1 scruple (s ap or ℈)	20 grains (exactly) 1.296 grams
1 ton, gross or long[21]	2240 pounds (exactly) 1.12 net tons (exactly) 1.016 metric tons
1 ton, metric (t)	2204.623 pounds 0.984 gross ton 1.102 net tons
1 ton, net or short (tn)[21]	2000 pounds (exactly) 0.893 gross ton 0.907 metric ton

---

[20] The gross or long ton and hundredweight are used commercially in the United States to only a very limited extent, usually in restricted industrial fields. The units are the same as the British "ton" and the "hundredweights."

[21] As of January 1, 2014, "tn" is the required abbreviation for "short ton." Devices manufactured between January 1, 2008, and December 31, 2013, may use an abbreviation other than "tn" to specify "short ton."

THIS PAGE INTENTIONALLY LEFT BLANK

Table of Contents

Appendix D.   Definitions

The specific code to which the definition applies is shown in [brackets] at the end of the definition. Definitions for the General Code [1.10] apply to all codes in Handbook 44.

A

absolute value. – The absolute value of a number is the magnitude of that number without considering the positive or negative sign. [2.20]

acceptance test. – The first official test of a farm milk tank, at a particular location, in which the tank is accepted as correct. This test applies to newly constructed tanks, relocated used tanks, and recalibrated tanks. [4.42]

accurate. – A piece of equipment is "accurate" when its performance or value – that is, its indications, its deliveries, its recorded representations, or its capacity or actual value, etc., as determined by tests made with suitable standards - conforms to the standard within the applicable tolerances and other performance requirements. Equipment that fails so to conform is "inaccurate." (Also see "correct.") [Appendix A]

all-class. – A description of a multi-class calibration that includes all the classes of a grain type. [5.56(a), 5.57]
(Added 2007)

analog or digital recorder. – An element used with a belt-conveyor scale that continuously records the rate-of-flow of bulk material over the scale (formerly referred to as a chart recorder). [2.21]
(Amended 1989)

analog type. – A system of indication or recording in which values are presented as a series of graduations in combination with an indicator, or in which the most sensitive element of an indicating system moves continuously during the operation of the device. [1.10]

animal scale. – A scale designed for weighing single heads of livestock. [2.20]
(Amended 1987)

apparent mass versus 8.0 g/cm³. – The apparent mass of an object versus 8.0 g/cm³ is the mass of material of density 8.0 g/cm³ that produces exactly the same balance reading as the object when the comparison is made in air with a density of 1.2 mg/cm³ at 20 °C. [3.37]

approval seal. – A label, tag, stamped or etched impression, or the like, indicating official approval of a device. (Also see "security seal.") [1.10]

assumed atmospheric pressure. – The average atmospheric pressure agreed to exist at the meter at various ranges of elevation, irrespective of variations in atmospheric pressure from time to time. [3.33]

audit trail. – An electronic count and/or information record of the changes to the values of the calibration or configuration parameters of a device. [1.10, 2.20, 2.21, 2.24, 3.30, 3.37, 5.56(a)]
(Added 1993)

automatic bulk weighing system. – A weighing system adapted to the automatic weighing of bulk commodities in successive drafts of predetermined amounts, automatically recording the no-load and loaded weight values and accumulating the net weight of each draft. [2.22]

automatic checkweigher. – An automatic weighing system that does not require the intervention of an operator during the weighing process and used to subdivide items of different weights into one or more subgroups, such as identifying packages that have acceptable or unacceptable fill levels according to the value of the difference between their weight and a pre-determined set point. These systems may be used to fill standard packages for compliance with net weight requirements. [2.24]

(Amended 2004)

automatic gravimetric filling machine (instrument). – A filling machine or instrument that fills containers or packages with predetermined and virtually constant mass of product from bulk by automatic weighing, and which comprises essentially an automatic feeding device or devices associated with one or more weighing unit and the appropriate discharge devices. [2.24]

(Added 2004)

automatic-indicating scale. – One on which the weights of applied loads of various magnitudes are automatically indicated throughout all or a portion of the weighing range of the scale. (A scale that automatically weighs out commodity in predetermined drafts, such as an automatic hopper scale, a packaging scale, and the like, is not an "automatic-indicating" scale.) [2.20. 2.22]

automatic temperature or density compensation. – The use of integrated or ancillary equipment to obtain from the output of a volumetric meter an equivalent mass, or an equivalent liquid volume at the assigned reference temperature below and a pressure of 14.696 lb/in^2 absolute.

> Cryogenic liquids: 21 °C (70 °F) [3.34]
> Hydrocarbon gas vapor: 15 °C (60 °F) [3.33]
> Liquid carbon dioxide: 21 °C (70 °F) [3.38]
> Liquefied petroleum gas (LPG) and Anhydrous ammonia: 15 °C (60 °F) [3.32]
> Petroleum liquid fuels and lubricants: 15 °C (60 °F) [3.30]

automatic weighing system (AWS). – An automatic weighing system is a weighing device that, in combination with other hardware and/or software components, automatically weighs discrete items and that does not require the intervention of an operator during the weighing process. Examples include, but are not limited to, weigh-labelers and checkweighers. [2.24]

(Amended 2004)

automatic zero-setting mechanism (AZSM). – See "automatic zero-setting mechanism" under "zero-setting mechanism." [2.22]

(Amended 2010)

automatic zero-setting mechanism (belt-conveyor scale). – A zero setting device that operates automatically without intervention of the operator after the belt has been running empty. [2.21]

(Added 2002)

automatic zero-tracking (AZT) mechanism. – Automatic means provided to maintain the zero balance indication, within specified limits, without the intervention of an operator. [2.20, 2.22, 2.24]

(Amended 2010)

auxiliary indicator. – Any indicator other than the master weight totalizer that indicates the weight of material determined by the scale. [2.21]

axle-load scale. – A scale permanently installed in a fixed location, having a load-receiving element specially adapted to determine the combined load of all wheels (1) on a single axle or (2) on a tandem axle of a highway vehicle. [2.20]

B

badge. – A metal plate affixed to the meter by the manufacturer showing the manufacturer's name, serial number and model number of the meter, and its rated capacity. [3.33]

balance, zero-load. – See "zero-load balance." [2.20]

balance indicator. – A combination of elements, one or both of which will oscillate with respect to the other, for indicating the balance condition of a nonautomatic indicating scale. The combination may consist of two indicating edges, lines, or points, or a single edge, line, or point and a graduated scale. [2.20]

balancing mechanism. – A mechanism (including a balance ball) that is designed for adjusting a scale to an accurate zero-load balance condition. [2.20]

base pressure. – The absolute pressure used in defining the gas measurement unit to be used, and is the gauge pressure at the meter plus an agreed atmospheric pressure. [3.33]

basic distance rate. – The charge for distance for all intervals except the initial interval. [5.54]

basic time rate. – The charge for time for all intervals except the initial interval. [5.54]

basic tolerances. – Basic tolerances are those tolerances on underregistration and on overregistration, or in excess and in deficiency, that are established by a particular code for a particular device under all normal tests, whether maintenance or acceptance. Basic tolerances include minimum tolerance values when these are specified. Special tolerances, identified as such and pertaining to special tests, are not basic tolerances. [1.10]

batching meter. – A device used for the purpose of measuring quantities of water to be used in a batching operation. [3.36]

beam. – See "weighbeam." [2.20]

beam scale. – One on which the weights of loads of various magnitudes are indicated solely by means of one or more weighbeam bars either alone or in combination with counterpoise weights. [2.20]

bell prover. – A calibrated cylindrical metal tank of the annular type with a scale thereon that, in the downward travel in a surrounding tank containing a sealing medium, displaces air through the meter being proved or calibrated. [3.33]

belt-conveyor. – An endless moving belt for transporting material from place to place. [2.21]

belt-conveyor scale. – A device that employs a weighing element in contact with a belt to sense the weight of the material being conveyed and the speed (travel) of the material, and integrates these values to produce total delivered weight. [2.21]

belt-conveyor scale systems area. – The scale system area refers to the scale suspension, weigh idlers attached to the scale suspension, 5 approach (−) idlers, and 5 retreat (+) idlers. [2.21]
(Added 2001)

belt load. – The weight of the material carried by the conveyor belt, expressed in terms of weight units per unit of length (e.g., pounds per foot, kilograms per meter). Also called "belt loading." [2.21]
(Added 2013)

belt revolution. – The amount of conveyor belt movement or travel that is equivalent to the total length of the conveyor belt. Also referred to as "belt circuit." [2.21]
(Added 2013)

billed weight. – The weight used in the computation of the freight, postal, or storage charge, whether actual weight or dimensional weight. [5.58]

binary submultiples. – Fractional parts obtained by successively dividing by the number two. Thus, one-half, one-fourth, one-eighth, one-sixteenth, and so on, are binary submultiples. [1.10]

built-for-purpose device. – Any main device or element which was manufactured with the intent that it be used as, or part of, a weighing or measuring device or system. [1.10]
(Added 2003)

C

calibration parameter. – Any adjustable parameter that can affect measurement or performance accuracy and, due to its nature, needs to be updated on an ongoing basis to maintain device accuracy, e.g., span adjustments, linearization factors, and coarse zero adjustments. [2.20, 2.21, 2.24, 3.30, 3.31, 3.32, 3.34, 3.35, 3.37, 3.38, 3.39, 5.56(a), 5.58]
(Added 1993) (Amended 2016)

carbon dioxide liquid-measuring device. – A system including a mechanism or machine of (a) the meter or (b) a weighing type of device mounted on a vehicle designed to measure and deliver liquid carbon dioxide. Means may be provided to indicate automatically, for one of a series of unit prices, the total money value of the quantity measured. [3.38]

car-wash timer. – A timer used in conjunction with a coin-operated device to measure the time during which car-wash water, cleaning solutions, or waxing solutions are dispensed. [5.55]

center-reading tank. – One so designed that the gauge rod or surface gauge, when properly positioned for use, will be approximately in the vertical axis of the tank, centrally positioned with respect to the tank walls. [4.43]

cereal grain and oil seeds. – Agricultural commodities including, but not limited to, corn, wheat, oats, barley, flax, rice, sorghum, soybeans, peanuts, dry beans, safflower, sunflower, fescue seed, etc. [5.56(a), 5.56(b)]

chart recorder. – See analog or digital recorder.
(Amended 1989)

check rate. – A rate of flow usually 20 % of the capacity rate. [3.33]

checkweighing scale. – One used to verify predetermined weight within prescribed limits. [2.24]

class of grain. – Hard Red Winter Wheat as distinguished from Hard Red Spring Wheat as distinguished from Soft Red Winter Wheat, etc. [5.56(a), 5.56(b), 5.57]

clear interval between graduations. – The distance between adjacent edges of successive graduations in a series of graduations. If the graduations are "staggered," the interval shall be measured, if necessary, between a graduation and an extension of the adjacent graduation. (Also see "minimum clear interval.") [1.10]

cleared. – A taximeter is "cleared" when it is inoperative with respect to all fare indication, when no indication of fare or extras is shown and when all parts are in those positions in which they are designed to be when the vehicle on which the taximeter is installed is not engaged by a passenger. [5.54]

cold-tire pressure. – The pressure in a tire at ambient temperature. [5.53, 5.54]

commercial equipment. – See "equipment."
(Added 2008)

computing scale. – One that indicates the money values of amounts of commodity weighed, at predetermined unit prices, throughout all or part of the weighing range of the scale. [2.20]

computing type or computing type device. – A device designed to indicate, in addition to weight or measure, the total money value of product weighed or measured, for one of a series of unit prices. [1.10]

concave curve. – A change in the angle of inclination of a belt conveyor where the center of the curve is above the conveyor. [2.21]

concentrated load capacity (CLC) (also referred to as Dual Tandem Axle Capacity[DTAC]). – A capacity rating of a vehicle or axle-load scale, specified by the manufacturer, defining the maximum load applied by a group of two axles with a centerline spaced four feet apart and an axle width of eight feet for which the weighbridge is designed. The concentrated load capacity rating is for both test and use. [2.20]
(Added 1988) (Amended 1991, 1994, and 2003)

configuration parameter. – Any adjustable or selectable parameter for a device feature that can affect the accuracy of a transaction or can significantly increase the potential for fraudulent use of the device and, due to its nature, needs to be updated only during device installation or upon replacement of a component, e.g., division value (increment), sensor range, and units of measurement. [2.20, 2.21, 2.24, 3.30, 3.37, 5.56(a)]
(Added 1993)

consecutive-car test train. – A train consisting of cars weighed on a reference scale, then coupled consecutively and run over the coupled-in-motion railway track scale under test. [2.20]
(Added 1990)

construction materials hopper scale. – A scale adapted to weighing construction materials such as sand, gravel, cement, and hot oil. [2.20]

contract sale. – A sale where a written agreement exists, prior to the point of sale, in which both buyer and seller have accepted pricing conditions of the sale. Examples include, but are not limited to: e-commerce, club sales, or pre-purchase agreements. Any devices used in the determination of quantity must comply with NIST Handbook 44. [3.30, 3.32, 3.37]
(Added 1993) (Amended 2002)

conventional scale. – If the use of conversion tables is necessary to obtain a moisture content value, the moisture meter indicating scale is called "conventional scale." The values indicated by the scale are dimensionless. [5.56(b)]

conversion table. – Any table, graph, slide rule, or other external device used to determine the moisture content from the value indicated by the moisture meter. [5.56(b)]

convex curve. – A change in the angle of inclination of a belt conveyor where the center of the curve is below the conveyor. [2.21]

conveyor stringers. – Support members for the conveyor on which the scale and idlers are mounted. [2.21]

correct. – A piece of equipment is "correct" when, in addition to being accurate, it meets all applicable specification requirements. Equipment that fails to meet any of the requirements for correct equipment is "incorrect." (Also see "accurate.") [Appendix A]

correction table. – Any table, graph, slide rule, or other external device used to determine the moisture content from the value indicated by the moisture meter when the indicated value is altered by a parameter not automatically corrected for in the moisture meter (for example, temperature or test weight). [5.56(b)]

counterbalance weight(s). – One intended for application near the butt of a weighbeam for zero-load balancing purposes. [2.20]

counterpoise weight(s). – A slotted or "hanger" weight intended for application near the tip of the weighbeam of a scale having a multiple greater than one. [2.20]

coupled-in-motion railroad weighing system. – A device and related installation characteristics consisting of (1) the associated approach trackage, (2) the scale (i.e., the weighing element, the load-receiving element, and the indicating element with its software), and (3) the exit trackage, which permit the weighing of railroad cars coupled in motion. [2.20, 2.23]

(Added 1992)

crane scale. – One with a nominal capacity of 5000 pounds or more designed to weigh loads while they are suspended freely from an overhead, track-mounted crane. [2.20]

cryogenic liquid-measuring device. – A system including a liquid-measuring element designed to measure and deliver cryogenic liquids in the liquid state. [3.34]

(Amended 1986 and 2003)

cryogenic liquids. – Fluids whose normal boiling point is below 120 kelvin (− 243 °F). [3.34]

cubic foot, gas. – The amount of a cryogenic liquid in the gaseous state at a temperature of 70 °F and under a pressure of 14.696 lb/in² absolute that occupies one cubic foot (1 ft³). (See NTP.) [3.34]

D

"d," dimension division value. – The smallest increment that the device displays for any axis and length of object in that axis. [5.58]

d, value scale division. – See "scale division, value of (d)." [2.20, 2.22]

D_{max} (maximum load of the measuring range). – Largest value of a quantity (mass) which is applied to a load cell during test or use. This value shall not be greater than E_{max}. [2.20]

(Added 2005)

D_{min} (minimum load of the measuring range). – Smallest value of a quantity (mass) which is applied to a load cell during test or use. This value shall not be less than E_{min}. [2.20]

(Added 2006)

dairy-product-test scale. – A scale used in determining the moisture content of butter and/or cheese or in determining the butterfat content of milk, cream, or butter. [2.20]

decimal submultiples. – Parts obtained by successively dividing by the number 10. Thus 0.1, 0.01, 0.001, and so on are decimal submultiples. [1.10]

decreasing-load test. – A test for automatic-indicating scales only, wherein the performance of the scale is tested as the load is reduced. [2.20, 2.22]

(Amended 1987)

deficiency. – See "excess and deficiency." [1.10]

diesel gallon equivalent (DGE). – Diesel gallon equivalent (DGE) means 6.384 pounds of compressed natural gas or 6.059 pounds of liquefied natural gas. [3.37]

(Added 2016)

digital type. – A system of indication or recording of the selector type or one that advances intermittently in which all values are presented digitally, or in numbers. In a digital indicating or recording element, or in digital representation, there are no graduations. [1.10]

dimensional weight (or dim, weight). – A value computed by dividing the object's volume by a conversion factor; it may be used for the calculation of charges when the value is greater than the actual weight. [5.58]
(Added 2004)

direct sale. – A sale in which both parties in the transaction are present when the quantity is being determined. An unattended automated or customer-operated weighing or measuring system is considered to represent the device/business owner in transactions involving an unattended device. [1.10]
(Amended 1993)

discharge hose. – A flexible hose connected to the discharge outlet of a measuring device or its discharge line. [3.30, 3.31, 3.32, 3.34, 3.37, 3.38]
(Added 1987)

discharge line. – A rigid pipe connected to the outlet of a measuring device. [3.30, 3.31, 3.32, 3.34, 3.37]
(Added 1987)

discrimination (of an automatic-indicating scale). – The value of the test load on the load-receiving element of the scale that will produce a specified minimum change of the indicated or recorded value on the scale. [2.20, 2.22]

dispenser. – See motor-fuel device. [3.30, 3.37]

distributed-car test train. – A train consisting of cars weighed first on a reference scale, cars coupled consecutively in groups at different locations within the train, then run over the coupled-in-motion railway track scale under test. The groups are typically placed at the front, middle, and rear of the train. [2.20]
(Added 1990)

dry hose. – A discharge hose intended to be completely drained at the end of each delivery of product. (Also see "dry-hose type.") [3.30, 3.31]
(Amended 2002)

dry-hose type. – A type of device in which it is intended that the discharge hose be completely drained following the mechanical operations involved in each delivery. (Also see "dry hose.") [3.30, 3.31, 3.34, 3.35]

dynamic monorail weighing system. – A weighing system which employs hardware or software to compensate for dynamic effects from the load or the system that do not exist in static weighing, in order to provide a stable indication. Dynamic factors may include shock or impact loading, system vibrations, oscillations, etc., and can occur even when the load is not moving across the load-receiving element. [2.20]
(Added 1999)

E

e, value of verification scale division. – See "verification scale division, value of (e)." [2.20]

E_{max} **(maximum capacity).** – Largest value of a quantity (mass) which may be applied to a load cell without exceeding the mpe. [2.20]
(Added 2005)

E_{min} **(minimum dead load).** – Smallest value of a quantity (mass) which may be applied to a load cell during test or use without exceeding the mpe. [2.20]
(Added 2006)

e_{min} **(minimum verification scale division).** – The smallest scale division for which a weighing element complies with the applicable requirements. [2.20, 2.21, 2.24]
(Added 1997)

electronic link. – An electronic connection between the weighing/load-receiving or other sensing element and indicating element where one recognizes the other and neither can be replaced without calibration. [2.20]

(Added 2001)

element. – A portion of a weighing or measuring device or system which performs a specific function and can be separated, evaluated separately, and is subject to specified full or partial error limits.

(Added 2002)

equal-arm scale. – A scale having only a single lever with equal arms (that is, with a multiple of one), equipped with two similar or dissimilar load-receiving elements (pan, plate, platter, scoop, or the like), one intended to receive material being weighed and the other intended to receive weights. There may or may not be a weighbeam. [2.20]

equipment, commercial. – Weights, measures, and weighing and measuring devices, instruments, elements, and systems or portion thereof, used or employed in establishing the measurement or in computing any basic charge or payment for services rendered on the basis of weight or measure. As used in this definition, measurement includes the determination of size, quantity, value, extent, area, composition (limited to meat and poultry), constituent value (for grain), or measurement of quantities, things, produce, or articles for distribution or consumption, purchased, offered, or submitted for sale, hire, or award. [1.10, 2.20, 2.21, 2.22, 2.24, 3.30, 3.31, 3.32, 3.33, 3.34, 3.35, 3.38, 4.40, 5.51, 5.56.(a), 5.56.(b), 5.57, 5.58, 5.59]

(Added 2008)

event counter. – A non-resettable counter that increments once each time the mode that permits changes to sealable parameters is entered and one or more changes are made to sealable calibration or configuration parameters of a device. [2.20, 2.21, 3.30, 3.37, 5.54, 5.56(a), 5.56(b), 5.57]

(Added 1993)

event logger. – A form of audit trail containing a series of records where each record contains the number from the event counter corresponding to the change to a sealable parameter, the identification of the parameter that was changed, the time and date when the parameter was changed, and the new value of the parameter. [2.20, 2.21, 3.30, 3.37, 5.54, 5.56(a), 5.56(b), 5.57]

(Added 1993)

excess and deficiency. – When an instrument or device is of such a character that it has a value of its own that can be determined, its error is said to be "in excess" or "in deficiency," depending upon whether its actual value is, respectively, greater or less than its nominal value. (Also see "nominal.") Examples of instruments having errors "in excess" are: a linear measure that is too long; a liquid measure that is too large; and a weight that is "heavy." Examples of instruments having errors "in deficiency" are: a lubricating-oil bottle that is too small; a vehicle tank compartment that is too small; and a weight that is "light." [1.10]

extras. – Charges to be paid by a passenger in addition to the fare, including any charge at a flat rate for the transportation of passengers in excess of a stated number and any charge for the transportation of baggage. [5.54]

F

face. – That side of a taximeter on which passenger charges are indicated. [5.54]

face. – That portion of a computing-type pump or dispenser which displays the actual computation of price per unit, delivered quantity, and total sale price. In the case of some electronic displays, this may not be an integral part of the pump or dispenser. [3.30]

(Added 1987)

fare. – That portion of the charge for the hire of a vehicle that is automatically calculated by a taximeter through the operation of the distance and/or time mechanism. [5.54]

farm milk tank. – A unit for measuring milk or other fluid dairy product, comprising a combination of (1) a stationary or portable tank, whether or not equipped with means for cooling its contents, (2) means for reading the level of liquid in the tank, such as a removable gauge rod or a surface gauge, and (3) a chart for converting level-of-liquid readings to volume; or such a unit in which readings are made on a gauge rod or surface gauge directly in terms of volume. Each compartment of a subdivided tank shall, for purposes of this code, be construed to be a "farm milk tank." [4.43]

feeding mechanism. – The means for depositing material to be weighed on the belt conveyor. [2.21]

ft³/h. – Cubic feet per hour. [3.33]

fifth wheel. – A commercially-available distance-measuring device which, after calibration, is recommended for use as a field transfer standard for testing the accuracy of taximeters and odometers on rented vehicles. [5.53, 5.54]

fifth-wheel test. – A distance test similar to a road test, except that the distance traveled by the vehicle under test is determined by a mechanism known as a "fifth wheel" that is attached to the vehicle and that independently measures and indicates the distance. [5.53, 5.54]

flat rate. – A rate selection that when applied results in the indication of a fixed (non-incrementing) amount for passenger charges. This rate shall be included on the statement of established rates that is required to be posted in the vehicle. [5.54.]
(Added 2016)

fractional bar. – A weighbeam bar of relatively small capacity for obtaining indications intermediate between notches or graduations on a main or tare bar. [2.20]

G

gasoline gallon equivalent (GGE). – Gasoline gallon equivalent (GGE) means 5.660 pounds of compressed natural gas. [3.37]
(Added 1994) (Amended 2016)

gauge pressure. – The difference between the pressure at the meter and the atmospheric pressure (psi). [3.33]

gauge rod. – A graduated, "dip-stick" type of measuring rod designed to be partially immersed in the liquid and to be read at the point where the liquid surface crosses the rod. [4.42]

gauging. – The process of determining and assigning volumetric values to specific graduations on the gauge or gauge rod that serve as the basis for the tank volume chart. [4.42]

graduated interval. – The distance from the center of one graduation to the center of the next graduation in a series of graduations. (Also see "value of minimum graduated interval.") [1.10]

graduation. – A defining line or one of the lines defining the subdivisions of a graduated series. The term includes such special forms as raised or indented or scored reference "lines" and special characters such as dots. (Also see "main graduation" and "subordinate graduation.") [1.10]

grain class. – Different grains within the same grain type. For example, there are six classes for the grain type "wheat:" Durum Wheat, Hard Red Spring Wheat, Hard Red Winter Wheat, Soft Red Winter Wheat, Hard White Wheat, and Soft White Wheat. [5.56(a), 5.57]
(Added 2007)

grain hopper scale. – One adapted to the weighing of individual loads of varying amounts of grain. [2.20]

grain moisture meter. – Any device indicating either directly or through conversion tables and/or correction tables the moisture content of cereal grains and oil seeds. Also termed "moisture meter." [5.56(a), 5.56(b)]

grain sample. – That portion of grain or seed taken from a bulk quantity of grain or seed to be bought or sold and used to determine the moisture content of the bulk. [5.56(a), 5.56(b)]

grain-test scale. – A scale adapted to weighing grain samples used in determining moisture content, dockage, weight per unit volume, etc. [2.20]

grain type. – See "kind of grain." [5.56(a), 5.57]
(Added 2007)

gravity discharge. – A type of device designed for discharge by gravity. [3.30, 3.31]

H

head pulley. – The pulley at the discharge end of the belt conveyor. The power drive to drive the belt is generally applied to the head pulley. [2.21]

hexahedron. – A geometric solid (i.e., box) with six rectangular or square plane surfaces. [5.58]
(Added 2008)

hired. – A taximeter is "hired" when it is operative with respect to all applicable indications of fare or extras. The indications of fare include time and distance where applicable unless qualified by another indication of "Time Not Recording" or an equivalent expression. [5.54]

hopper scale. – A scale designed for weighing bulk commodities whose load-receiving element is a tank, box, or hopper mounted on a weighing element. (Also see "automatic hopper scale," "grain hopper scale," and "construction materials hopper scale.") [2.20]

I

idlers or idler rollers. – Freely turning cylinders mounted on a frame to support the conveyor belt. For a flat belt, the idlers consist of one or more horizontal cylinders transverse to the direction of belt travel. For a troughed belt, the idlers consist of one or more horizontal cylinders and one or more cylinders at an angle to the horizontal to lift the sides of the belt to form a trough. [2.21]

idler space. – The center-to-center distance between idler rollers measured parallel to the belt. [2.21]

increasing-load test. – The normal basic performance test for a scale in which observations are made as increments of test load are successively added to the load-receiving element of the scale. [2.20, 2.22]

increment. – The value of the smallest change in value that can be indicated or recorded by a digital device in normal operation. [1.10]

index of an indicator. – The particular portion of an indicator that is directly utilized in making a reading. [1.10]

indicating element. – An element incorporated in a weighing or measuring device by means of which its performance relative to quantity or money value is "read" from the device itself as, for example, an index-and-graduated-scale combination, a weighbeam-and-poise combination, a digital indicator, and the like. (Also see "primary indicating or recording element.") [1.10]

indicator, balance. – See "balance indicator." [2.20]

initial distance or time interval. – The interval corresponding to the initial money drop. [5.54]

initial zero-setting mechanism. – See "initial zero-setting mechanism" under "zero-setting mechanism." [2.20] (Added 1990)

in-service light indicator. – A light used to indicate that a timing device is in operation. [5.55]

integrator. – A device used with a belt-conveyor scale that combines conveyor belt load (e.g., lb/ft) and belt travel (e.g., feet) to produce a total weight of material passing over the belt-conveyor scale. An integrator may be a separate, detached mechanism or may be a component within a totalizing device. (Also see "master weight totalizer.") [2.21] (Added 2013)

interval, clear, between graduations. – See "clear interval between graduations." [1.10]

interval, graduated. – See "graduated interval." [1.10]

irregularly-shaped object. – Any object that is not a hexahedron shape. [5.58] (Added 2008)

J

jewelers' scale. – One adapted to weighing gems and precious metals. [2.20]

K

kind of grain. – Corn as distinguished from soybeans as distinguished from wheat, etc. [5.56(a), 5.56(b)]

L

label. – A printed ticket, to be attached to a package, produced by a printer that is a part of a prepackaging scale or that is an auxiliary device. [2.20]

large-delivery device. – Devices used primarily for single deliveries greater than 200 gallons, 2000 pounds, 20 000 cubic feet, 2000 liters, or 2000 kilograms. [3.34, 3.38]

laundry-drier timer. – A timer used in conjunction with a coin-operated device to measure the period of time that a laundry drier is in operation. [5.55]

liquefied petroleum gas. – A petroleum product composed predominantly of any of the following hydrocarbons or mixtures thereof: propane, propylene, butanes (normal butane or isobutane), and butylenes. [3.31, 3.32, 3.33, 3.34, 3.37]

liquefied petroleum gas liquid-measuring device. – A system including a mechanism or machine of the meter type designed to measure and deliver liquefied petroleum gas in the liquid state by a definite quantity, whether installed in a permanent location or mounted on a vehicle. Means may or may not be provided to indicate automatically, for one of a series of unit prices, the total money value of the liquid measured. [3.33] (Amended 1987)

liquefied petroleum gas vapor-measuring device. – A system including a mechanism or device of the meter type, equipped with a totalizing index, designed to measure and deliver liquefied petroleum gas in the vapor state by definite volumes, and generally installed in a permanent location. The meters are similar in construction and operation to the conventional natural- and manufactured-gas meters. [3.33]

liquid fuel. – Any liquid used for fuel purposes, that is, as a fuel, including motor-fuel. [3.30, 3.31]

liquid-fuel device. – A device designed for the measurement and delivery of liquid fuels. [3.30]

liquid-measuring device. – A mechanism or machine designed to measure and deliver liquid by definite volume. Means may or may not be provided to indicate automatically, for one of a series of unit prices, the total money value of the liquid measured, or to make deliveries corresponding to specific money values at a definite unit price. [3.30]

liquid volume correction factor. – A correction factor used to adjust the liquid volume of a cryogenic product at the time of measurement to the liquid volume at NBP. [3.34]

livestock scale. – A scale equipped with stock racks and gates and adapted to weighing livestock standing on the scale platform. [2.20]
(Amended 1989)

load cell. – A device, whether electric, hydraulic, or pneumatic, that produces a signal (change in output) proportional to the load applied. [2.20, 2.21, 2.23]

load cell verification interval (v). – The load cell interval, expressed in units of mass, used in the test of the load cell for accuracy classification. [2.20, 2.21]
(Added 1996)

loading point. – A location on a conveyor where the material is received by the belt. The location of the discharge from a hopper, chute, or pre-feed device used to supply material to a conveyor. [2.21]
(Amended 2013)

load-receiving element. – That element of a scale that is designed to receive the load to be weighed; for example, platform, deck, rail, hopper, platter, plate, scoop. [2.20, 2.21, 2.23]

low-flame test. – A test simulating extremely low-flow rates such as caused by pilot lights. [3.33]

lubricant device. – A device designed for the measurement and delivery of liquid lubricants, including, but not limited to, heavy gear lubricants and automatic transmission fluids (automotive). [3.30]

M

m³/h. – Cubic meters per hour. [3.33]

main bar. – A principal weighbeam bar, usually of relatively large capacity as compared with other bars of the same weighbeam. (On an automatic-indicating scale equipped with a weighbeam, the main weighbeam bar is frequently called the "capacity bar.") [2.20]

main graduation. – A graduation defining the primary or principal subdivisions of a graduated series. (Also see "graduation.") [1.10]

main-weighbeam elements. – The combination of a main bar and its fractional bar, or a main bar alone if no fractional bar is associated with it. [2.20]

manual zero-setting mechanism. – See "manual zero-setting mechanism" under "zero-setting mechanism." [2.20]

manufactured device. – Any commercial weighing or measuring device shipped as new from the original equipment manufacturer. [1.10]
(Amended 2001)

mass flow meter. – A device that measures the mass of a product flowing through the system. The mass measurement may be determined directly from the effects of mass on the sensing unit or may be inferred by measuring the properties of the product, such as the volume, density, temperature, or pressure, and displaying the quantity in mass units. [3.37]

master meter test method. – A method of testing milk tanks that utilizes an approved master meter system for measuring test liquid removed from or introduced into the tank. [4.42]

master weight totalizer. – A primary indicating element used with a belt-conveyor scale that incorporates the function of an integrator to indicate the totalized weight of material passed over the scale. (Also see "integrator.") [2.21]
(Amended 2013)

material test. – The test of a belt-conveyor scale using material (preferably that for which the device is normally used) that has been weighed to an accuracy of 0.1 %. [2.21]
(Amended 1989)

maximum capacity. – The largest load that may be accurately weighed. [2.20, 2.24]
(Added 1999)

maximum cargo load. – The maximum cargo load for trucks is the difference between the manufacturer's rated gross vehicle weight and the actual weight of the vehicle having no cargo load. [5.53]

measurement field. – A region of space or the measurement pattern produced by the measuring instrument in which objects are placed or passed through, either singly or in groups, when being measured by a single device. [5.58]

measuring element. – That portion of a complete multiple dimension measuring device that does not include the indicating element. [5.58]

meter register. – An observation index for the cumulative reading of the gas flow through the meter. In addition there are one or two proving circles in which one revolution of the test hand represents ½, 1, 2, 5, or 10 cubic feet, or 0.025, 0.05, 0.1, 0.2, or 0.25 cubic meter, depending on meter size. If two proving circles are present, the circle representing the smallest volume per revolution is referred to as the "leak-test circle." [3.33]

metrological integrity (of a device). – The design, features, operation, installation, or use of a device that facilitates (1) the accuracy and validity of a measurement or transaction, (2) compliance of the device with weights and measures requirements, or (3) the suitability of the device for a given application. [1.10, 2.20]
(Added 1993)

minimum capacity. – The smallest load that may be accurately weighed. The weighing results may be subject to excessive error if used below this value. [2.20, 2.24]
(Added 1999)

minimum clear interval. – The shortest distance between adjacent graduations when the graduations are not parallel. (Also see "clear interval.") [3.30, 3.31, 3.32, 3.33, 3.34, 3.35, 3.36, 3.38, 5.50, 5.51, 5.56(b)]

minimum delivery. – The least amount of weight that is to be delivered as a single weighment by a belt-conveyor scale system in normal use. [2.21]

minimum load cell verification interval. – *See* v_{min}

minimum tolerance. – Minimum tolerances are the smallest tolerance values that can be applied to a scale. Minimum tolerances are determined on the basis of the value of the minimum graduated interval or the nominal or reading face capacity of the scale. (Also see definition for basic tolerances.) [2.20, 2.22, 2.24]

minimum totalized load. – The least amount of weight for which the scale is considered to be performing accurately. [2.21]

moisture content (wet basis). – The mass of water in a grain or seed sample (determined by the reference method) divided by the mass of the grain or seed sample expressed as a percentage (%). [5.56(a), 5.56(b)]

money drop. – An increment of fare indication. The "initial money drop" is the first increment of fare indication following activation of the taximeter. [5.54]

money-operated type. – A device designed to be released for service by the insertion of money, or to be actuated by the insertion of money to make deliveries of product. [1.10]

motor-fuel. – Liquid used as fuel for internal-combustion engines. [3.30]

motor-fuel device or motor-fuel dispenser or retail motor-fuel device. – A device designed for the measurement and delivery of liquids used as fuel for internal-combustion engines. The term "motor-fuel dispenser" means the same as "motor-fuel device"; the term "retail motor-fuel device" applies to a unique category of device. (Also see definition of "retail device.") [3.30, 3.32, 3.37]

multi-class. – A description of a grouping of grain classes, from the same grain type, in one calibration. A multi-class grain calibration may include (1) all the classes of a grain type (all-class calibration), or (2) some of the classes of a grain type within the calibration. [5.56(a), 5.57.]
(Added 2007)

multi-interval scale. – A scale having one weighing range which is divided into partial weighing ranges (segments), each with different scale intervals, with each partial weighing range (segment) determined automatically according to the load applied, both on increasing and decreasing loads. [2.20]
(Added 1995)

multi-jet water meter. – A water meter in which the moving element takes the form of a multiblade rotor mounted on a vertical spindle within a cylindrical measuring chamber. The liquid enters the measuring chamber through several tangential orifices around the circumference and leaves the measuring chamber through another set of tangential orifices placed at a different level in the measuring chamber. These meters register by recording the revolutions of a rotor set in motion by the force of flowing water striking the blades. [3.36]
(Added 2003)

multiple. – An integral multiple; that is, a result obtained by multiplying by a whole number. (Also see "multiple of a scale.") [1.10]

multiple cell application load cell. – A load cell intended for use in a weighing system which incorporates more than one load cell. A multiple cell application load cell is designated with the letter "M" or the term "Multiple." (Also see "single cell application load cell.") [2.20]
(Added 1999)

multiple range scale. – A scale having two or more weighing ranges with different maximum capacities and different scale intervals for the same load receptor, each range extending from zero to its maximum capacity. [2.20]
(Added 1995)

multiple of a scale. – In general, the multiplying power of the entire system of levers or other basic weighing elements. (On a beam scale, the multiple of the scale is the number of pounds on the load-receiving element that will be counterpoised by one pound applied to the tip pivot of the weighbeam.) [2.20]

multi-revolution scale. – An automatic-indicating scale having a nominal capacity that is a multiple of the reading-face capacity and that is achieved by more than one complete revolution of the indicator. [2.20]

multiple-tariff taximeter. – One that may be set to calculate fares at any one of two or more rates. [5.54]

N

NBP. – Normal Boiling Point of a cryogenic liquid at 14.696 lb/in^2 absolute. [3.34]

NTP. – Normal Temperature and Pressure of a cryogen at a temperature of 21 °C (70 °F) and a pressure of 101.325 kPa (14.696 lb/in^2 absolute). [3.34]

NTP density and volume correction factor. – A correction factor used to adjust the liquid volume of a cryogenic product at the time of measurement to the gas equivalent at NTP. [3.34]

natural gas. – A gaseous fuel, composed primarily of methane, that is suitable for compression and dispensing into a fuel storage container(s) for use as an engine fuel. [3.37]

(Added 1994)

negotiated rate. – A rate selection that, when applied, results in a fixed (non-incrementing) amount for passenger charges and is based on a value that has been agreed upon by the operator and passenger. [5.54]

(Added 2016)

n$_{max}$ (maximum number of scale divisions). – The maximum number of scale divisions for which a main element or load cell complies with the applicable requirements. The maximum number of scale divisions permitted for an installation is limited to the lowest n$_{max}$ marked on the scale indicating element, weighing element, or load cell. [2.20, 2.21, 2.24]

(Added 1997)

no-load reference value. – A positive weight value indication with no load in the load-receiving element (hopper) of the scale. (Used with automatic bulk-weighing systems and certain single-draft, manually-operated receiving hopper scales installed below grade and used to receive grain.) [2.20]

nominal. – Refers to "intended" or "named" or "stated," as opposed to "actual." For example, the "nominal" value of something is the value that it is supposed or intended to have, the value that it is claimed or stated to have, or the value by which it is commonly known. Thus, "1-pound weight," "1-gallon measure," "1-yard indication," and "500-pound scale" are statements of nominal values; corresponding actual values may be greater or lesser. (Also see nominal capacity of a scale.) [1.10]

nominal capacity. – The nominal capacity of a scale is (a) the largest weight indication that can be obtained by the use of all of the reading or recording elements in combination, including the amount represented by any removable weights furnished or ordinarily furnished with the scale, but excluding the amount represented by any extra removable weights not ordinarily furnished with the scale, and excluding also the capacity of any auxiliary weighing attachment not contemplated by the original design of the scale, and excluding any fractional bar with a capacity less than 2½ % of the sum of the capacities of the remaining reading elements, or (b) the capacity marked on the scale by the manufacturer, whichever is less. (Also see "nominal capacity, batching scale"; "nominal capacity, hopper scale.") [2.20]

nominal capacity, batching scale. – The nominal capacity of a batching scale is the capacity as marked on the scale by the scale manufacturer, or the sum of the products of the volume of each of the individual hoppers, in terms of cubic feet, times the weight per cubic foot of the heaviest material weighed in each hopper, whichever is less. [2.20]

nominal capacity, hopper scale. – The nominal capacity of a hopper scale is the capacity as marked on the scale by the scale manufacturer, or the product of the volume of the hopper in bushels or cubic feet times the maximum weight per bushel or cubic foot, as the case may be, of the commodity normally weighed, whichever is less. [2.20]

non-automatic checkweigher. – A weighing instrument that requires the intervention of an operator during the weighing process, used to subdivide items of different weights into one or more subgroups, such as identifying packages that have acceptable or unacceptable fill levels according to the value of the difference between their weight and a pre-determined set point. [2.24]

Notes: Determining the weighing result includes any intelligent action of the operator that affects the result, such as deciding and taking an action when an indication is stable or adjusting the weight of the weighed load.

Deciding the weighing result is acceptable means making a decision regarding the acceptance of each weighing result on observing the indication or releasing a print-out. The weighing process allows the operator to take an action which influences the weighing result in the case where the weighing result is not acceptable.
(Added 2004)

non-automatic weighing instrument. – A weighing instrument or system that requires the intervention of an operator during the weighing process to determine the weighing result or to decide that it is acceptable. [2.20, 2.24]

> **Notes:** Determining the weighing result includes any intelligent action of the operator that affects the result, such as deciding and taking an action when an indication is stable or adjusting the weight of the weighed load.

> Deciding the weighing result is acceptable means making a decision regarding the acceptance of each weighing result on observing the indication or releasing a print-out. The weighing process allows the operator to take an action which influences the weighing result in the case where the weighing result is not acceptable.

(Added 2004) (Amended 2005)

nonretroactive. – "Nonretroactive" requirements are enforceable after the effective date for:

1.   devices manufactured within a state after the effective date;

2.   both new and used devices brought into a state after the effective date; and

3.   devices used in noncommercial applications which are placed into commercial use after the effective date.

Nonretroactive requirements are not enforceable with respect to devices that are in commercial service in the state as of the effective date or to new equipment in the stock of a manufacturer or a dealer in the state as of the effective date. *(Nonretroactive requirements are printed in italic type.)* [1.10]
(Amended 1989)

nose-iron. – A slide-mounted, manually-adjustable pivot assembly for changing the multiple of a lever. [2.20]

notes. – A section included in each of a number of codes, containing instructions, pertinent directives, and other specific information pertaining to the testing of devices. Notes are primarily directed to weights and measures officials.

O

odometer. – A device that automatically indicates the total distance traveled by a vehicle. For the purpose of this code, this definition includes hub odometers, cable-driven odometers, and the distance-indicating or odometer portions of "speedometer" assemblies for automotive vehicles. [5.53]

official grain samples. – Grain or seed used by the official as the official transfer standard from the reference standard method to test the accuracy and precision of grain moisture meters. [5.56(a), 5.56(b)]

official with statutory authority. – The representative of the jurisdiction(s) responsible for certifying the accuracy of the device. [2.20, 2.21, 2.22]
(Added 1991)

operating tire pressure. – The pressure in a tire immediately after a vehicle has been driven for at least 5 miles or 8 kilometers. [5.53, 5.54]

over-and-under indicator. – An automatic-indicating element incorporated in or attached to a scale and comprising an indicator and a graduated scale with a central or intermediate "zero" graduation and a limited range of weight graduations on either side of the zero graduation, for indicating weights greater than and less than the predetermined values for which other elements of the scale may be set. (A scale having an over-and-under indicator is classed as an automatic-indicating scale.) [2.20]

overregistration and underregistration. – When an instrument or device is of such a character that it indicates or records values as a result of its operation, its error is said to be in the direction of overregistration or underregistration, depending upon whether the indications are, respectively, greater or less than they should be. Examples of devices having errors of "overregistration" are: a fabric-measuring device that indicates more than the true length of material passed through it; and a liquid-measuring device that indicates more than the true amount of the liquid delivered by the device. Examples of devices having errors of "underregistration" are: a meter that indicates less than the true amount of product that it delivers; and a weighing scale that indicates or records less than the true weight of the applied load. [1.10]

P

parallax. – The apparent displacement, or apparent difference in height or width, of a graduation or other object with respect to a fixed reference, as viewed from different points. [1.10]

parking meter. – A coin-operated device for measuring parking time for vehicles. [5.55]

passenger vehicles. – Vehicles such as automobiles, recreational vehicles, limousines, ambulances, and hearses. [5.53]

performance requirements. – Performance requirements include all tolerance requirements and, in the case of nonautomatic-indicating scales, sensitivity requirements (SR). (Also see definitions for "tolerance" and "sensitivity requirement.") [1.10]

point-of-sale system. – An assembly of elements including a weighing or measuring element, an indicating element, and a recording element (and may also be equipped with a "scanner") used to complete a direct sales transaction. The system components, when operated together, must be capable of the following:

1.   determining the weight or measure of a product or service offered;

2.   calculating a charge for the product or service based on the weight or measure and an established price/rate structure;

3.   determining a total cost that includes all associated charges involved with the transaction; and

4.   providing a sales receipt.

[2.20, 3.30, 3.32, 3.37]

(Added 1986) (Amended 1997 and 2015)

poise. – A movable weight mounted upon or suspended from a weighbeam bar and used in combination with graduations, and frequently with notches, on the bar to indicate weight values. (A suspended poise is commonly called a "hanging poise.") [2.20]

postal scale. – A scale (usually a computing scale) designed for use to determine shipping weight or delivery charges for letters or parcels delivered by the U.S. Postal Service or private shipping companies. A weight classifier may be used as a postal scale. [2.20]

(Added 1987)

prepackaging scale. – A computing scale specially designed for putting up packages of random weights in advance of sale. [2.20]

prescription scale. – A scale or balance adapted to weighing the ingredients of medicinal and other formulas prescribed by physicians and others and used or intended to be used in the ordinary trade of pharmacists. [2.20]

pressure type (device). – A type of device designed for operation with the liquid under artificially produced pressure. [3.30, 3.31]

primary indicating or recording elements. – The term "primary" is applied to those principal indicating (visual) elements and recording elements that are designed to, or may, be used by the operator in the normal commercial use of a device. The term "primary" is applied to any element or elements that may be the determining factor in arriving at the sale representation when the device is used commercially. (Examples of primary elements are the visual indicators for meters or scales not equipped with ticket printers or other recording elements and both the visual indicators and the ticket printers or other recording elements for meters or scales so equipped.) The term "primary" is not applied to such auxiliary elements as, for example, the totalizing register or predetermined-stop mechanism on a meter or the means for producing a running record of successive weighing operations, these elements being supplementary to those that are the determining factors in sales representations of individual deliveries or weights. (Also see "indicating element" and "recording element.") [1.10]

prover method. – A method of testing milk tanks that utilizes approved volumetric prover(s) for measuring the test liquid removed from or introduced into the tank. [4.42]

prover oil. – A light oil of low vapor pressure used as a sealing medium in bell provers, cubic-foot bottles, and portable cubic-foot standards. [3.33]

proving indicator. – The test hand or pointer of the proving or leak-test circle on the meter register or index. [3.33, 3.36.]

R

"r" factor. – A computation for determining the suitability of a vehicle scale for weighing vehicles with varying axle configurations. The factor was derived by dividing the weights in FHWA Federal Highway Bridge Gross Weight Table B by 34 000 lbs. (The resultant factors are contained in Table UR.3.2.1.) [2.20]

radio frequency interference (RFI). – Radio frequency interference is a type of electrical disturbance that, when introduced into electronic and electrical circuits, may cause deviations from the normally expected performance. [1.10]

random error(s). – The sample standard deviation of the error (indicated values) for a number of consecutive automatic weighings of a load, or loads, passed over the load receptor, shall be expressed mathematically as:

$$s = \sqrt{\frac{1}{n-1}\sum\left(x_i - \bar{x}\right)^2} \ \text{ or } \ s = \sqrt{\frac{1}{n-1}\left(\sum x_i^2 - \frac{\left(\sum x_i\right)^2}{n}\right)}$$

where: x = error of a load indication
          n = the number of loads

[2.24]

ranges, weight. – See "weight ranges." [2.20]

rated capacity. – The rate of flow in cubic meters per hour of a hydrocarbon gas vapor-measuring device as recommended by the manufacturer. This rate of flow should cause a pressure drop across the meter not exceeding ½-inch water column. [3.33]

rated scale capacity. – That value representing the weight that can be delivered by the device in one hour. [2.21]

ratio test. – A test to determine the accuracy with which the actual multiple of a scale agrees with its designed multiple. This test is used for scales employing counterpoise weights and is made with standard test weights substituted in all cases for the weights commercially used on the scale. (It is appropriate to use this test for some scales not employing counterpoise weights.) [2.20]

reading face. – That portion of an automatic-indicating weighing or measuring device that gives a visible indication of the quantity weighed or measured. A reading face may include an indicator and a series of graduations or may present values digitally, and may also provide money-value indications. [1.10, 2.20]

(Amended 2005)

reading-face capacity. – The largest value that may be indicated on the reading face, exclusive of the application or addition of any supplemental or accessory elements. [1.10]

recorded representation. – The printed, embossed, or other representation that is recorded as a quantity by a weighing or measuring device. [1.10]

recording element. – An element incorporated in a weighing or measuring device by means of which its performance relative to quantity or money value is permanently recorded on a tape, ticket, card, or the like, in the form of a printed, stamped, punched, or perforated representation. [1.10, 2.21]

recording scale. – One on which the weights of applied loads may be permanently recorded on a tape, ticket, card, or the like in the form of a printed, stamped, punched, or perforated representation. [2.20]

reference weight car. – A railcar that has been statically weighed for temporary use as a mass standard over a short period of time, typically the time required to test one scale.

   Note: A test weight car that is representative of the types of cars typically weighed on the scale under test may be used wherever reference weight cars are specified. [2.20]

(Added 1991) (Amended 2012)

remanufactured device. – A device that is disassembled, checked for wear, parts replaced or fixed, reassembled and made to operate like a new device of the same type. [1.10]

(Added 2001)

remanufactured element. – An element that is disassembled, checked for wear, parts replaced or fixed, reassembled and made to operate like a new element of the same type. [1.10]

(Added 2001)

remote configuration capability. – The ability to adjust a weighing or measuring device or change its sealable parameters from or through some other device that is not itself necessary to the operation of the weighing or measuring device or is not a permanent part of that device. [2.20, 2.21, 2.24, 3.30, 3.37, 5.56(a)]

(Added 1993)

repaired device. – A device to which work is performed that brings the device back into proper operating condition. [1.10]

(Added 2001)

repaired element. – An element to which work is performed that brings the element back into proper operating condition. [1.10]

(Added 2001)

retail device. – A measuring device primarily used to measure product for the purpose of sale to the end user. [3.30, 3.32, 3.37]

(Amended 1987 and 2004)

retroactive. – "Retroactive" requirements are enforceable with respect to all equipment. Retroactive requirements are printed herein in upright roman type. (Also see "nonretroactive.") [1.10]

road test. – A distance test, over a measured course, of a complete taximeter assembly when installed on a vehicle, the mechanism being actuated as a result of vehicle travel. [5.53, 5.54]

rolling circumference. – The rolling circumference is the straight line distance traveled per revolution of the wheel (or wheels) that actuates the taximeter or odometer. If more than one wheel actuates the taximeter or odometer, the rolling circumference is the average distance traveled per revolution of the actuating wheels. [5.53, 5.54]

S

scale. – See specific type of scale. [2.20]

scale area, belt-conveyor. – See belt-conveyor scale systems area. [2.21]
(Added 2001)

scale division, number of (n). – Quotient of the capacity divided by the value of the verification scale division. [2.20]

$$n = \frac{Capacity}{e}$$

scale division, value of (d). – The value of the scale division, expressed in units of mass, is the smallest subdivision of the scale for analog indication or the difference between two consecutively indicated or printed values for digital indication or printing. (Also see "verification scale division.") [2.20, 2.22]

scale section. – A part of a vehicle, axle-load, livestock, or railway track scale consisting of two main load supports, usually transverse to the direction in which the load is applied. [2.20]

seal. – See "approval seal," "security seal." [1.10]

section capacity. – The section capacity of a scale is the maximum live load that may be divided equally on the load pivots or load cells of a section. [2.20]
(Added 2001)

section test. – A shift test in which the test load is applied over individual sections of the scale. This test is conducted to disclose the weighing performance of individual sections, since scale capacity test loads are not always available and loads weighed are not always distributed evenly over all main load supports. [2.20]

security means. – A method used to prevent access by other than qualified personnel, or to indicate that access has been made to certain parts of a scale that affect the performance of the device. [2.21]

security seal. – A uniquely identifiable physical seal, such as a lead-and-wire seal or other type of locking seal, a pressure-sensitive seal sufficiently permanent to reveal its removal, or similar apparatus attached to a weighing or measuring device for protection against or indication of access to adjustment. (Also see "approval seal.") [1.10]
(Amended 1994)

selector-type. – A system of indication or recording in which the mechanism selects, by means of a ratchet-and-pawl combination or by other means, one or the other of any two successive values that can be indicated or recorded. [1.10]

semi-automatic zero-setting mechanism. – See "semi-automatic zero-setting mechanism" under "zero-setting mechanism." [2.20]

sensitivity (of a nonautomatic-indicating scale). – The value of the test load on the load-receiving element of the scale that will produce a specified minimum change in the position of rest of the indicating element or elements of the scale. [2.20]

D-24

sensitivity requirement (SR). – A performance requirement for a non automatic-indicating scale; specifically, the minimum change in the position of rest of the indicating element or elements of the scale in response to the increase or decrease, by a specified amount, of the test load on the load-receiving element of the scale. [2.20]

shift test. – A test intended to disclose the weighing performance of a scale under off-center loading. [2.20]

side. – That portion of a pump or dispenser which faces the consumer during the normal delivery of product. [3.30]
(Added 1987)

simulated-road test. – A distance test during which the taximeter or odometer may be actuated by some means other than road travel. The distance traveled is either measured by a properly calibrated roller device or computed from rolling circumference and wheel-turn data. [5.53, 5.54]

simulated test. – A test using artificial means of loading the scale to determine the performance of a belt-conveyor scale. [2.21]

single cell application load cell. – A load cell intended for use in a weighing system which incorporates one or more load cells. A single cell application load cell is designated with the letter "S" or the term "Single." (Also see "multiple cell application load cell.") [2.20]
(Added 1999)

single-tariff taximeter. – One that calculates fares at a single rate only. [5.54]

skirting. – Stationary side boards or sections of belt conveyor attached to the conveyor support frame or other stationary support to prevent the bulk material from falling off the side of the belt. [2.21]

slow-flow meter. – A retail device designed for the measurement, at very slow rates (less than 40 L (10 gal) per hour), of liquid fuels at individual domestic installations. [3.30]

small-delivery device. – Any device other than a large-delivery device. [3.34, 3.38]

span (structural). – The distance between adjoining sections of a scale. [2.20]
(Added 1988)

specification. – A requirement usually dealing with the design, construction, or marking of a weighing or measuring device. Specifications are directed primarily to the manufacturers of devices. [1.10]

static monorail weighing system. – A weighing system in which the load being applied is stationary during the weighing operation. [2.20]
(Added 1999)

strain-load test. – The test of a scale beginning with the scale under load and applying known test weights to determine accuracy over a portion of the weighing range. The scale errors for a strain-load test are the errors observed for the known test loads only. The tolerances to be applied are based on the known test load used for each error that is determined. [2.20, 2.22]

subordinate graduation. – Any graduation other than a main graduation. (Also see "graduation.") [1.10]

subsequent distance or time intervals. – The intervals corresponding to money drops following the initial money drop. [5.54]

substitution test. – A scale testing process used to quantify the weight of material or objects for use as a known test load. [2.20]
(Added 2003)

substitution test load. – The sum of the combination of field standard test weights and any other applied load used in the conduct of a test using substitution test methods. [2.20]

(Added 2003)

surface gauge. – A combination of (1) a stationary indicator, and (2) a movable, graduated element designed to be brought into contact with the surface of the liquid from above. [4.42]

systematic (average) error $(\overline{x})$. – The mean value of the error (of indication) for a number of consecutive automatic weighings of a load, or loads, passed over the load-receiving element (e.g., weigh-table), shall be expressed mathematically as:

$$\overline{X} = \frac{\sum x}{n}$$

where: x =   error of a load indication
          n =   the number of loads

[2.24]

T

tail pulley. – The pulley at the opposite end of the conveyor from the head pulley. [2.21]

take-up. – A device to provide sufficient tension in a conveyor belt so that the belt will be positively driven by the drive pulley. – A counter-weighted take-up consists of a pulley free to move in either the vertical or horizontal direction with dead weights applied to the pulley shaft to provide the tension required. [2.21]

tare mechanism. – A mechanism (including a tare bar) designed for determining or balancing out the weight of packaging material, containers, vehicles, or other materials that are not intended to be included in net weight determinations. [2.20]

tare-weighbeam elements. – The combination of a tare bar and its fractional bar, or a tare bar alone if no fractional bar is associated with it. [2.20]

taximeter. – A device that automatically calculates, at a predetermined rate or rates, and indicates the charge for hire of a vehicle. [5.54]

test chain. – A device used for simulated tests consisting of a series of rollers or wheels linked together in such a manner as to assure uniformity of weight and freedom of motion to reduce wear, with consequent loss of weight, to a minimum. [2.21]

test liquid. – The liquid used during the test of a device. [3.30, 3.31, 3.34, 3.35, 3.36, 3.37, 3.38]

test object. – An object whose dimensions are verified by appropriate reference standards and intended to verify compliance of the device under test with certain metrological requirements. [5.58]

test puck. – A metal, plastic, or other suitable object that remains stable for the duration of the test, used as a test load to simulate a package. Pucks can be made in a variety of dimensions and have different weights to represent a wide range of package sizes. Metal versions may be covered with rubber cushions to eliminate the possibility of damage to weighing and handling equipment. The puck mass is adjusted to an accuracy specified in N.1.2. Accuracy of Test Pucks or Packages. [2.24]

(Amended 2004)

test train. – A train consisting of or including reference weight cars and used to test coupled-in-motion railway track scales. The reference weight cars may be placed consecutively or distributed in different places within a train. [2.20] (Added 1990) (Amended 1991)

test weight car. – A railroad car designed to be a stable mass standard to test railway track scales. The test weight car may be one of the following types: a self-contained composite car, a self-propelled car, or a standard rail car. [2.20]

(Added 1991)

testing. – An operation consisting of a series of volumetric determinations made to verify the accuracy of the volume chart that was developed by gauging. [4.42]

time recorder. – A clock-operated mechanism designed to record the time of day. Examples of time recorders are those used in parking garages to record the "in" and "out" time of day for parked vehicles. [5.55]

timing device. – A device used to measure the time during which a particular paid-for service is dispensed. Examples of timing devices are laundry driers, car-wash timers, parking meters, and parking-garage clocks and recorders. [5.55]

tolerance. – A value fixing the limit of allowable error or departure from true performance or value. (Also see "basic tolerances.") [1.10]

training idlers. – Idlers of special design or mounting intended to shift the belt sideways on the conveyor to assure the belt is centered on the conveying idlers. [2.21]

transfer standard. – A measurement system designed for use in proving and testing cryogenic liquid-measuring devices. [3.38]

tripper. – A device for unloading a belt conveyor at a point between the loading point and the head pulley. [2.21]

U

uncoupled-in-motion railroad weighing system. – A device and related installation characteristics consisting of (1) the associated approach trackage, (2) the scale (i.e., the weighing element, the load-receiving element, and the indicating element with its software), and (3) the exit trackage, which permit the weighing of railroad cars uncoupled in motion. [2.20]

(Added 1993)

underregistration. – See "overregistration" and "underregistration." [1.10]

unit price. – The price at which the product is being sold and expressed in whole units of measurement. [1.10, 3.30]

(Added 1992)

unit train. – A unit train is defined as a number of contiguous cars carrying a single commodity from one consignor to one consignee. The number of cars is determined by agreement among the consignor, consignee, and the operating railroad. [2.20]

unit weight. – One contained within the housing of an automatic-indicating scale and mechanically applied to and removed from the mechanism. The application of a unit weight will increase the range of automatic indication, normally in increments equal to the reading-face capacity. [2.20]

user requirement. – A requirement dealing with the selection, installation, use, or maintenance of a weighing or measuring device. User requirements are directed primarily to the users of devices. (Also see Introduction, Section D.) [1.10]

usual and customary. – Commonly or ordinarily found in practice or in the normal course of events and in accordance with established practices. [1.10]

utility-type water meter. – 1) A device used for the measurement of water, generally applicable to meters installed in residences or business establishments. excluding batching meters. [3.36]
(Added 2011)

V

value of minimum graduated interval. – (1) The value represented by the interval from the center of one graduation to the center of the succeeding graduation. (2) The increment between successive recorded values. (Also see "graduated interval.") [1.10]

vapor equalization credit. – The quantity deducted from the metered quantity of liquid carbon dioxide when a vapor equalizing line is used to facilitate the transfer of liquid during a metered delivery. [3.38]

vapor equalization line. – A hose or pipe connected from the vapor space of the seller's tank to the vapor space of the buyer's tank that is used to equalize the pressure during a delivery. [3.38]

vehicle on-board weighing system. – A weighing system designed as an integral part of or attached to the frame, chassis, lifting mechanism, or bed of a vehicle, trailer, industrial truck, industrial tractor, or forklift truck. [2.20]
(Amended 1993)

vehicle scale. – A scale adapted to weighing highway, farm, or other large industrial vehicles (except railroad freight cars), loaded or unloaded. [2.20]

verification scale division, value of (e). – A value, expressed in units of weight (mass) and specified by the manufacturer of a device, by which the tolerance values and the accuracy class applicable to the device are determined. The verification scale division is applied to all scales, in particular to ungraduated devices since they have no graduations. The verification scale division (e) may be different from the displayed scale division (d) for certain other devices used for weight classifying or weighing in pre-determined amounts, and certain other Class I and II scales. [2.20]

visible type. – A type of device in which the measurement takes place in a see-through glass measuring chamber. [3.30]

v_{min} (minimum load cell verification interval). – The smallest load cell verification interval, *expressed in units of mass** into which the load cell measuring range can be divided. [2.20, 2.24]
[*Nonretroactive as of January 1, 2001]
(Added 1996) (Amended 1999)

W

weighbeam. – An element comprising one or more bars, equipped with movable poises or means for applying counterpoise weights or both. [2.20]

weigh-belt system. – A type of belt-conveyor scale system designed by the manufacturer as a self-contained conveyor system and that is installed as a unit. A unit is comprised of integral components and, at minimum, includes a: conveyor belt; belt drive; conveyor frame; and weighing system. A weigh-belt system may operate at single or multiple flow rates and may use variable-speed belt drives. [2.21]
(Added 2015)

weigh-labeler. – An automatic weighing system that determines the weight of a package and prints a label or other document bearing a weight declaration for each discrete item (usually a label also includes unit and total price declarations). Weigh-labelers are sometimes used to weigh and label standard and random packages (also called "Prepackaging Scales"). [2.24]

(Amended 2004)

weigh module – The portion of a load-receiving element supported by two sections. The length of a module is the distance to which load can be applied. [2.20]

(Added 2013)

weighment. – A single complete weighing operation. [2.20, 2.21]

(Added 1986)

weight, unit. – See "unit weight." [2.20]

weight classifier. – A digital scale that rounds weight values up to the next scale division. These scales usually have a verification scale division (e) that is smaller than the displayed scale division. [2.20]

(Added 1987)

weight ranges. – Electrical or electro-mechanical elements incorporated in an automatic indicating scale through the application of which the range of automatic indication of the scale is increased, normally in increments equal to the reading-face capacity. [2.20]

wet basis. – See "moisture content (wet basis)." [5.56(a), 5.56(b)]

wet hose. – A discharge hose intended to be full of product at all times. (Also see "wet-hose type.") [3.30, 3.31, 3.38]

(Amended 2002)

wet-hose type. – A type of device designed to be operated with the discharge hose full of product at all times. (Also see "wet hose.") [3.30, 3.32, 3.34, 3.37, 3.38]

(Amended 2002)

wheel-load weighers. – Compact, self-contained, portable weighing elements specially adapted to determining the wheel loads or axle loads of vehicles on highways for the enforcement of highway weight laws only. [2.20]

wholesale device. – Any device other than a retail device. (Also see "retail device.") [3.30, 3.32]

wing pulley. – A pulley made of widely spaced metal bars in order to set up a vibration to shake loose material off the underside (return side) of the belt. [2.21]

Z

zero-load balance. – A correct weight indication or representation of zero when there is no load on the load-receiving element. (Also see "zero-load balance for an automatic-indicating scale," "zero-load balance for a nonautomatic-indicating scale," "zero-load balance for a recording scale.") [2.20]

zero-load balance, automatic-indicating scale. – A condition in which the indicator is at rest at, or oscillates through approximately equal arcs on either side of, the zero graduation. [2.20]

zero-load balance, nonautomatic-indicating scale. – A condition in which (a) the weighbeam is at rest at, or oscillates through approximately equal arcs above and below, the center of a trig loop; (b) the weighbeam or lever system is at rest at, or oscillates through approximately equal arcs above and below, a horizontal position or a position midway between limiting stops; or (c) the indicator of a balance indicator is at rest at, or oscillates through approximately equal arcs on either side of, the zero graduation. [2.20]

zero-load balance for a recording scale. – A condition in which the scale will record a representation of zero load. [2.20]

zero-load reference (belt-conveyor scales). – A zero-load reference value represents no load on a moving conveyor belt. This value can be either; a number representing the electronic load cell output, a percentage of full scale capacity, or other reference value that accurately represents the no load condition of a moving conveyor belt. The no load reference value can only be updated after the completion of a zero load test.[2.21]

(Added 2002)

zero-setting mechanism. – Means provided to attain a zero balance indication with no load on the load-receiving element. The types of zero-setting mechanisms are: [2.20, 2.22, 2.24]

automatic zero-setting mechanism (AZSM). – Automatic means provided to set the zero-balance indication without the intervention of an operator. [2.22]

(Added 2010)

automatic zero-tracking (AZT) mechanism. – See "automatic zero-tracking (AZT) mechanism." (NOTE: AZT maintains zero with specified limits. "Zero-setting sets/establishes zero with limits based on scale capacity.) [2.20, 2.22, 2.24]

initial zero-setting mechanism. – Automatic means provided to set the indication to zero at the time the instrument is switched on and before it is ready for use. [2.20]

(Added 1990)

manual zero-setting mechanism. – Nonautomatic means provided to attain a zero balance indication by the direct operation of a control. [2.20]

semiautomatic zero-setting mechanism. – Automatic means provided to attain a direct zero balance indication requiring a single initiation by an operator. [2.20]

(Amended 2010)

zero-setting mechanism (belt-conveyor scale). – A mechanism enabling zero totalization to be obtained over a whole number of belt revolutions. [2.21, 2.23]

(Added 2002)

zero-tracking mechanism. – See "automatic zero-tracking mechanism" under "zero-setting mechanism." [2.20, 2.22, 2.24]

zone of uncertainty. – The zone between adjacent increments on a digital device in which the value of either of the adjacent increments may be displayed. [2.20]